# NON-NEUTRAL PLASMA PHYSICS V

# Previous Proceedings in the Series of Workshops on Non-Neutral Plasma Physics

| | Year | Publisher | ISBN |
|---|---|---|---|
| VI | 2001 | AIP Conference Proceedings 606 | 0-7354-0050-4 |
| V | 1999 | AIP Conference Proceedings 498 | 1-56396-913-0 |
| IV | 1997 | *not published* | |
| III | 1994 | AIP Conference Proceedings 331 | 1-56396-441-4 |
| II | 1992 | *not published* | |
| I | 1988 | AIP Conference Proceedings 175 | 0-88318-375-7 |

## Other Related Titles from AIP Conference Proceedings

**669** Plasma Physics: 11th International Congress on Plasma Physics: ICPP2002
Edited by Ian S. Falconer, Robert L. Dewar, and Joe Khachan, June 2003,
Print: 0-7354-0133-0; CD-ROM: 0-7354-0134-9

**649** Dusty Plasmas in the New Millennium: Third International Conference on the Physics of Dusty Plasmas
Edited by R. Bharuthram, M. A. Hellberg, P. K. Shukla, and F. Verheest,
December 2002, 0-7354-0106-3

**635** Atomic Processes in Plasmas: 13th APS Topical Conference on Atomic Processes in Plasmas
Edited by David R. Schultz, Fred W. Meyer, and Fay Ownby, October 2002,
0-7354-0090-3

**595** Radio Frequency Power in Plasmas: 14th Topical Conference
Edited by Tak Kuen Mau and John deGrassie, November 2001, 0-7354-0038-5

**563** Plasma Physics: IX Latin American Workshop
Edited by Hernán Chuaqui and Mario Favre, May 2001, 1-56396-999-8

**457** Trapped Charged Particles and Fundamental Physics
Edited by Daniel H. E. Dubin and Dieter Schneider, January 1999, 1-56396-776-6

To learn more about these titles, or the AIP Conference Proceedings Series, please visit the webpage **http://proceedings.aip.org**

# NON-NEUTRAL PLASMA PHYSICS V

Workshop on Non-Neutral Plasmas

*Santa Fe, New Mexico    7-11 July 2003*

*EDITORS*

Martin Schauer
*Los Alamos National Laboratory*
*Los Alamos, New Mexico*

Travis Mitchell
*University of Delaware*
*Newark, Delaware*

Richard Nebel
*Los Alamos National Laboratory*
*Los Alamos, New Mexico*

**SPONSORING ORGANIZATIONS**
Los Alamos National Laboratory
U.S. Department of Energy
U.S. Office of Naval Research

Melville, New York, 2003
**AIP CONFERENCE PROCEEDINGS ■ VOLUME 692**

Editors:

Martin Schauer
Los Alamos National Laboratory
MS H803
Los Alamos, NM 87545
USA

E-mail: schauer@lanl.gov


Travis Mitchell
Department of Physics and Astronomy
University of Delaware
223 Sharp Lab
Newark, DE 19716
USA

E-mail: tbmitche@udel.edu


Richard Nebel
Los Alamos National Laboratory
MS K717
Los Alamos, NM 87545
USA

E-mail: rnebel@lanl.gov

L.C. Catalog Card No. 2003113873
ISBN 0-7354-0165-9
ISSN 0094-243X
Printed in the United States of America

# CONTENTS

## SECTION 2
## ANTIMATTER PLASMAS AND RECOMBINATION

## SECTION 3
### BEAMS, STRONGLY-COUPLED PLASMAS, SPECIAL TOPICS

## SECTION 4
### TOROIDAL SYSTEMS

# Preface

The seventh Workshop on Non-Neutral Plasma Physics was held July 8-11, 2003, in Santa Fe, New Mexico. Roughly seventy scientists from the United States, Europe, and Japan participated in four sessions of invited talks and two poster sessions for contributed papers. Most of these presentations are included in this proceedings volume, the fifth in this series. No proceedings were published for two of the previous meetings.

Since the field of non-neutral plasma physics grew out of research into collective effects in particle beams, it is perhaps appropriate that the opening section comprises papers on collective modes, transport, and fluid dynamics. Included here are reports on the emerging understanding of the importance of trapped-particle-mediated effects in a host of plasma properties as well as theoretical and experimental investigations on various collective modes both as a fundamental plasma physics problem and as a diagnostic tool. Also, continuing work on vortices in pure electron plasmas is described in this section demonstrating the strong link between fluid dynamics and non-neutral plasma physics research.

Section 2 is concerned exclusively with antimatter plasmas. The most exciting development in this area since the previous workshop held in San Diego was the announcement of the formation of cold antihydrogen atoms by the ATHENA[1] and ATRAP[2] experiments. The key to this development was the ability to separately accumulate large plasmas of antiprotons and positrons and then overlap them in some manner. Interestingly, this work has lead to the development of commercial positron sources that are available for a variety of applications in wide-ranging research fields. The papers in this section deal with the ongoing efforts in this area as well as work on diagnostics applicable to antimatter plasmas.

Sections 3 and 4 contain papers on wide-ranging topics: Section 3 contains several contributions on beam physics, reports on numerical tools to study non-neutral plasmas, and some articles of relevance to fusion research. Finally, section 4 deals with the growing area of non-neutral plasmas contained in toroidal systems. Two experiments presently under construction are described, and exciting new results from established experiments on electron injection into these systems and the possible role that collective effects play in the operation of these systems are presented.

*Martin M. Schauer*

---

[1] Amoretti, M. *et al, Nature* **419**, 456 (2002).
[2] Gabrielse, G. *et al.*, *Phys. Rev. Lett.* **89**, 213401 (2002).

# Organizing Committee

Dan Barnes, Chair
Los Alamos National Laboratory

C. Fred Driscoll, Vice Chair
University of California,
San Diego

John Bollinger
National Institute of Standards and
Technology

Ron Davidson
Princeton Plasma Physics
Laboratory

Dan Dubin
University of California,
San Diego

Joel Fajans
University of California, Berkeley

John Goree
University of Iowa

Jeff Hangst
University of Aarhus

Yasuhito Kiwamoto
Kyoto University

Robert Pollock
Indiana University

Roberto Pozzoli
University of Milan

Charles Roberson
Office of Naval Research

Martin Schauer
Los Alamos National Laboratory

Dieter Schneider
Lawrence Livermore National
Laboratory

Lutz Schweikhard
Ernst-Moritz-Arndt University

Matthew Stoneking
Lawrence University

Cliff Surko
University of California,
San Diego

# SECTION 1

# COLLECTIVE MODES, TRANSPORT, AND FLUID DYNAMICS

# Trapped-Particle-Mediated Damping and Transport

C. Fred Driscoll*, Andrey A. Kabantsev*, Terance J. Hilsabeck* and
Thomas M. O'Neil*

*Dept. of Physics, University of California at San Diego, La Jolla CA USA 92093-0319

**Abstract.** Weak axial variations in $B(z)$ or $\phi(z)$ in Penning-Malmberg traps cause some particles to be trapped locally. This causes a velocity-space separatrix between trapped and passing populations, and collisional separatrix diffusion then causes mode damping and asymmetry-induced transport. This separatrix dissipation scales with collisionality as $v^{1/2}$, so it dominates in low collisionality plasmas. The confinement lifetime in the "CamV" apparatus was dominated by a weak magnetic ripple with $\delta B/B \sim 10^{-3}$, and it appears likely that the ubiquitous $(L/B)^{-2}$ lifetime scalings and other applied asymmetry scalings represent similar TPM effects. TPM transport will limit the containment of large numbers of positrons or $\bar{p}$s, since TPM loss rates generally scale as total charge $Q^2$, independent of length.

## INTRODUCTION

Two years ago, "trapped particle asymmetry" modes were reported to occur when an applied "squeeze" voltage causes some particles to be trapped axially; and a simple theory explained the observed mode frequencies [1]. Now, it appears likely that trapped-particle-mediated (TPM) effects are dominant in plasma lifetime scalings, in transport from applied asymmetries, and in diocotron mode damping. This talk will give an overview of what is known [2, 3, 4, 5], where more experiments are needed, and where the theory is lacking.

Electric or magnetic trapping probably occurs in all "long" apparatuses: unintended wall potential variations of 0.1 Volts are common, and it is sobering to note that $\delta B/B = 10^{-3}$ will trap 3% of the particles. Initial experiments (and all theory to date) considered electric trapping; but magnetic trapping is probably more common and important.

Early experiments focused on the new modes (now called "trapped particle diocotron" modes); but the important effect is particles scattering across the trapping separatrix. This breaks the $v_z$ adiabatic invariant, allowing 2D potential energy to flow to 3D kinetics, and enabling external asymmetries to generate strong transport. The effect is dominant in low-collisionality plasmas because this separatrix dissipation scales with collisionality as $v^{1/2}$, whereas most other effects scale as $v^1$. Here, the collisionality can be electron-electron, electron-neutral, or externally stimulated. The effect can be also be described as dissipation of asymmetry-induced equilibrium currents, as in the analysis of bootstrap current in Tokamaks.

CP692, *Non-Neutral Plasma Physics V*, edited by M. Schauer et al.
© 2003 American Institute of Physics 0-7354-0165-9/03/$20.00

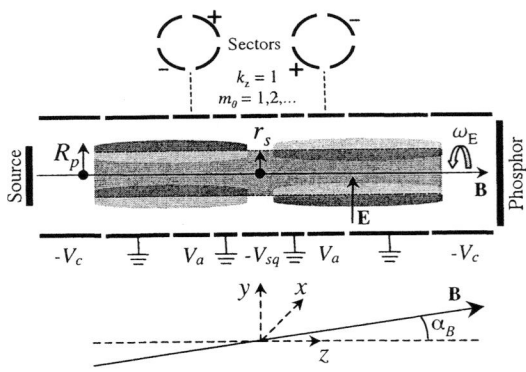

**FIGURE 1.** Schematic of electron plasma with a trapped particle mode in the cylindrical containment system.

Thus, it now appears likely that most of the $(L/B)^{-2}$ lifetime scalings from "background asymmetries" [6] can be given interpretation in terms of the (partially) known scalings for TPM transport. The measurements of transport from *applied* electric and magnetic asymmetries [7, 8, 9, 10] also should be compared to TPM predictions. "Anomalous" damping of diocotron modes [7, 11, 12] is almost certainly related to TPM effects, since TPM damping scales as $B^{-3}$. TPM transport also has important implications for containment of large numbers of positrons or pBars, since the TPM loss rate for magnetic asymmetries scales as total charge $Q^2$, *independent of length*.

Theory provides a reasonable picture of trapped-particle-mode damping with electric trapping [5], but modes in the magnetic trapping case remain enigmatic. Theory can not yet explain the observed particle transport scalings for either case, but this appears imminent for electric trapping. Diocotron mode damping has not been worked out theoretically.

## ELECTRIC TRAPPING: NEW MODE

The experiments are performed on magnetized pure electron plasmas confined in the cylindrical "CamV" apparatus, as shown in Fig. 1. The electron plasmas have density $n \sim 10^7 \mathrm{cm}^{-3}$, length $L_p \sim 40$ cm, radius $R_p \sim 1.5$ cm, and temperature $T \sim 1$ eV.

Controlled electric trapping from an applied central "squeeze" voltage $-V_{sq}$ causes electrons with axial velocity less than the separatrix velocity to be trapped in one end or the other; here, $v_s$ is defined by $v_s^2(r) \equiv \frac{2e}{m}V_{sq}(r)$. For small $V_{sq}$, a fraction $N_L^{(tr)}/N_L \sim 1.2\,(V_{sq}/\phi_p)$ of the electrons are trapped, predominantly at $r \sim R_p$; here, $N_L \equiv \int 2\pi r\, dr\, n$.

This trapping enables novel "trapped particle diocotron modes" with various $m_\theta = 1, 2, \ldots$; but we focus here on $m_\theta = 1$. The mode frequency $f_a$ ranges from the edge rotation frequency $f_E(R_p) \sim (100 \text{ kHz})B[\text{kG}]^{-1}$ at low $V_{sq}$, down to the $k_z = 0$ diocotron

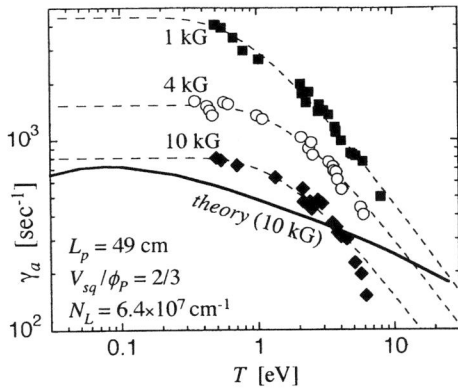

**FIGURE 2.** Measured damping rate $\gamma_a$ versus temperature for 3 magnetic fields; with theory prediction for 1 field.

mode frequency $f_d$ at $V_{sq} \gtrsim \phi_p$. The modes are anti-symmetric in $z$, with trapped particles on either end executing $\mathbf{E} \times \mathbf{B}$ drift oscillations that are $180°$ out of phase, while passing particles move along field lines in response to the potential from the trapped particles.

At large $V_{sq}$, the plasma is essentially cut in half, and on either side of the barrier the plasma supports $k_z = 0$ (diocotron) drift orbits which are $180°$ out of phase. Even for small $V_{sq}$, the trapped particle mode is essentially uniform with $z$ on either side of the barrier, changing sign at the barrier. A simple kinetic theory model with a zero-length trapping barrier [1] predicts mode frequencies agreeing with measurements to within about 10%.

This trapped particle mode is readily excited by $m_\theta = 1$ $z$-antisymmetric voltages, as shown in Fig. 1. The excited mode then rings down exponentially, and the damping rate $\gamma_a$ is unambiguously obtained. Figure 2 shows the measured $\gamma_a$ for three magnetic fields as the plasma temperature is varied; the dash lines represent the generic functional form $a[1 - \exp(-b/T)]$. The modes are strongly damped at low temperatures, but the damping decreases precipitously as $T$ increases.

Theory analysis of damping from collisional scattering across the trapping separatrix gives factor-of-two agreement with experiments; but significant discrepancies remain. Since particles on either side of the separatrix are involved in completely different types of motion, there is a discontinuity in the perturbed particle distribution function. As a result, electron-electron collisions produce a large flux of particles across the separatrix. The continual trapping and detrapping of particles results in radial transport of particles and in mode damping, and is readily observed in computer simulations [13].

These collisions at rate $v$ have been treated by a Fokker-Planck collision operator [5], in an analysis similar to that used for the dissipative trapped-ion instability by Rosenbluth, Ross, and Kostomarov [14]. Velocity space diffusion acting for one mode period smoothes out the separatrix discontinuity over a width $\delta v_t \approx \bar{v} \sqrt{v/f_E}$, and the damping includes this dependence. The predicted damping rate can be expressed as

$$\gamma_a = \frac{\frac{2\sqrt{\pi}B}{m_\theta} \int_0^{R_w} rdr |\delta\phi| \bar{v} \sqrt{\frac{v}{f_E^*}} \left[ \frac{2\pi e f_M}{T} - \frac{c}{Br} \frac{\partial f_M}{\partial r} \frac{m_\theta}{f_E^*} \right]_{v=v_s}}{\int_0^{R_w} dr \frac{|\delta\phi|^2}{f_*^2} \frac{\partial n_t}{\partial r}}, \tag{1}$$

where $\phi(r)$ is the mode potential, $f_E^*(r) \equiv m_\theta f_E(r) - f_a$, $n_t$ is the trapped density, and $f_M$ is the Maxwellian distribution.

Figure 2 shows that this predicts a somewhat less abrupt temperature dependence than is actually observed. This may be related to another significant discrepancy: experiments show non-zero ($\sim 20°$) phase shifts in the mode eigenfunction $\phi(r)$, whereas no shifts are predicted. The square root provides a significant enhancement, since $v/f_E$ is small. The damping rate is expected to have a strong and complicated temperature dependence through the density of particles at the separatrix velocity $v_s(r)$, through the collisional frequency $v$, and through the Debye shielding length $\lambda_D$.

## ELECTRIC TRAPPING AND TILT: TRANSPORT

When $\theta$-asymmetries exist in the electric or magnetic confinement fields, they create torques which change the canonical angular momentum $P_\theta$ of the plasma, causing the plasma radius to vary. These asymmetry-induced torques are stronger when the symmetric squeeze trapping is present. If the asymmetry is not static, the sign of the torque can be positive or negative. The "rotating wall" confinement technique utilizes wall voltages rotating faster than $f_E$ to obtain plasma compression [15, 16]. For the present experiments, the $\theta$-asymmetries are static in the lab frame and exert a negative torque on the electrons, resulting in bulk radial expansion.

Here, we focus on the $m_\theta = 1$, $k_z = 1$ asymmetry induced by a magnetic tilt, with $B = B(\hat{z} + \alpha_{Bx}\hat{x} + \alpha_{By}\hat{y})$; or by the electric "tilt" induced by static $m_\theta = 1$ voltages $V_a$ applied antisymmetrically in $z$ (Fig. 1). The asymmetry-induced transport rate is defined by the rate of plasma expansion

$$v_p \equiv \frac{1}{\langle r^2 \rangle} \frac{d\langle r^2 \rangle}{dt} \approx \frac{-1}{P_\theta} \frac{dP_\theta}{dt}. \tag{2}$$

The expansion rate is found to be proportional to the tilt angle $\alpha_B^2$, as shown in Fig. 3. Here, $v_p(\alpha_{Bx})$ is quadratic about $\alpha_{Bx} = 0$, but the minimum of $v_p(\alpha_{By})$ is offset by the separate electric tilt $\alpha_{Ey}$ from an applied $V_{ay}$. Indeed, electric and magnetic tilts add vectorially when the proper $z$-averaged electrostatic offsets [17] are calculated, as

$$\alpha_E \equiv (0.51) \left( \frac{4R_w}{L_p} \right) \left( \frac{V_a}{eN_L} \right) \left( \frac{L_a}{L_p} \right). \tag{3}$$

Here, $V_a$ is applied to sectors of length $L_a$, and the factor 0.51 represents the $m_\theta = 1$ Fourier coefficient for the (four) $25°$ sectors used. The deviation from this quadratic scaling at larger $v_p$ (dashed line) is due to an increase of the plasma temperature caused by fast radial expansion.

6

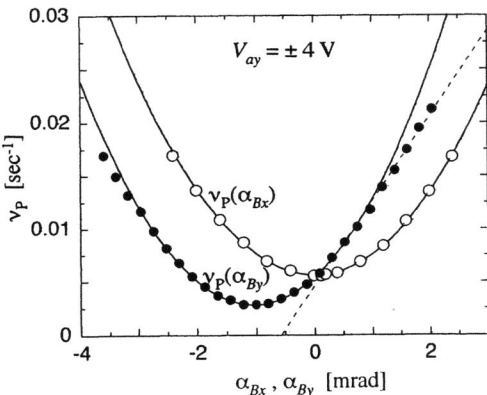

**FIGURE 3.** Measured transport rate $v_p$ from a magnetic tilt with simultaneously applied electric symmetry.

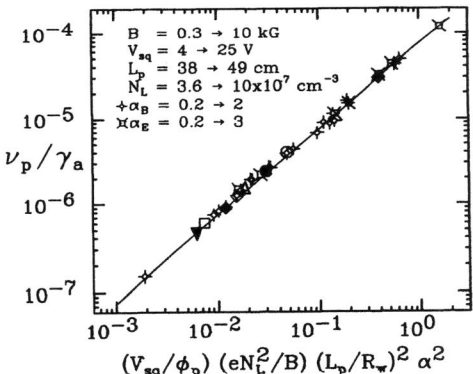

**FIGURE 4.** Measured normalized transport rate $v_p/\gamma_a$ vs scalings for all plasma parameters.

The expansion rate $v_p$ shows rather complicated dependencies on plasma parameters $n$, $T$, $R_p$, $L_p$ and $B$, *unless* the ratio $v_p/\gamma_a$ is considered. Essentially, we find that all the complicated dynamics is in the separatrix dissipation which causes $\gamma_a$. Measuring $\gamma_a$ coincident with $v_p$ (by exciting the trapped particle mode and watching it decay) allows us to accurately obtain the ratio $v_p/\gamma_a$.

Figure 4 shows that the ratio $v_p/\gamma_a$ has only simple power-law dependencies on plasma parameters, as $L_p^2$, $B^{-1}$, and $N_L^2$, where $N_L \sim \pi R_p^2 n$. We note that all temperature dependence is in $\gamma_a$, that $V_{sq}$ is normalized to the plasma potential $\phi_p$ at $r = 0$, and that $\phi_p \propto N_L$.

With the applied electric trapping and the applied tilt, a single trapped-particle-mediated damping and transport process is dominant; and this process exhibits stunningly simple and accurate scalings over 3 decades in $v_p/\gamma_a$, representing 4 decades in

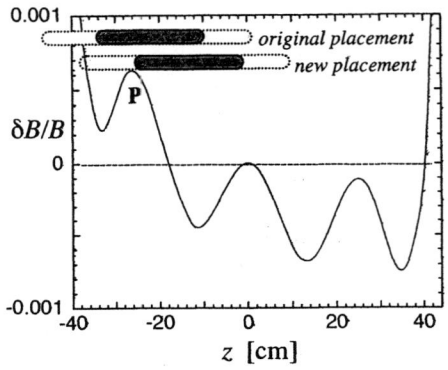

**FIGURE 5.** Modified electrode placement relative to magnetic ripples exhibits 5× less background transport.

$v_p$. However, the theory of this transport scaling is still incomplete.

## MAGNETIC TRAPPING AND TILT: TRANSPORT BUT NO MODE

Applying a central magnetic "squeeze" $B(z)$ instead of the electric squeeze also causes particle trapping in either end, and causes enhanced transport from electric or magnetic tilts. However, experiments have not yet identified a corresponding trapped particle mode. Presumably, this is because the magnetic trapped particle "mode" has $f_a^{(M)} = 0$, or $\gamma_a^{(M)}/f_a^{(M)} \sim 1$. This eliminates the conceptual advantage of relating $v_p$ to $\gamma_a$; but experiments demonstrate conclusively that scatterings across the magnetic separatrix produce transport, as in the electric trapping case.

These magnetic trapping effects have been studied using an axially centered coil which generates a magnetic mirror of strength $\beta \equiv (\delta B/B) \lesssim 4\%$ at $B = 1$ kG; and also using the ripples of strength $\beta \approx 10^{-3}$ inherent in our superconducting solenoid.

Surprisingly, these magnetic ripples with $\delta B/B \sim 10^{-3}$ are sufficient for TPM transport to dominate the "background losses." Figure 5 plots the vendor-calculated ripples in the CamV superconducting magnet, together with two axial placements of the electrode stack (shown dotted). In the original placement, the magnetic mirror $P$ occurred within the electron containment region (shown grey). Moving the electrodes by +9 cm moved the peaks to the ends of the plasma, and *reduced the background transport by 5×*. This, together with more subtle probes described below, conclusively establishes these weak mirrors as generators of asymmetry-induced transport.

One expects particles with small pitch angle to be trapped, i.e. those with $v_z < \beta^{1/2} v_\perp$. The fraction of these trapped particles is expected to scale as $\beta^{1/2}$, giving $0.03 \lesssim N_L^{(tr)}/N_L \lesssim 0.2$. Moreover, there are theoretical and experimental reasons [18] to expect that the mirror field causes the electrostatic potential to vary along a field line

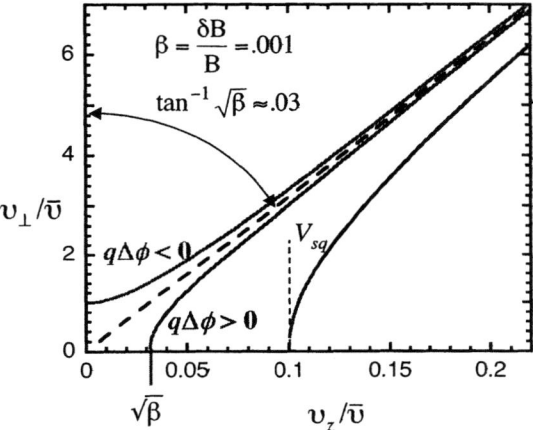

**FIGURE 6.** The magnetic separatrix with potentials.

by $\Delta\phi$; and an applied electric squeeze voltage $V_{sq}$ would add in analogously. Thus, the magnetic/electric separatrix is given by

$$v_z^2 = \left(\frac{\delta B}{B}\right) v_\perp^2 + \frac{-2e}{m}\left[\Delta\phi + V_{sq}(r)\right]. \tag{4}$$

These hyperbolic separatrices are shown in Fig. 6 for $\beta = 10^{-3}$, showing the $\beta^{1/2}$ reduction in relevant $v_z$ velocities. The lack of radial separation between trapped and untrapped particles makes $f_a^{(M)} = 0$ plausible, since the charge separation and $\mathbf{E} \times \mathbf{B}$ drifts characterizing the electric mode may not occur. These trapped particles can be directly detected by selective dumping techniques, but the relevant parallel velocities are substantially less than $\bar{v}$, so measurements to date are only qualitative.

More incisively, the separatrix can be mapped out by enhancing $v_z$ separatrix scatterings with a resonant RF field. Figure 7 shows the transport enhancement when a RF wiggle is applied near a magnetic minimum. Here, we have utilized the "sheath transport" resonance to interact with particles with small $v_z$: electrons receive a nonadiabatic kick if they have $v_z \approx L^* f_{RF}$, where $L^* \equiv 2R_w/j_{01}$ is the axial extent of the wiggle electric fields. The transport response peak in Fig. 7 is as expected for a Maxwellian distribution of particles along the naive $(\Delta\phi = 0)$ magnetic separatrix; surprisingly, recent calculations show that $\Delta\phi$ does not affect this resonance curve [18].

Alternately, adding an electric squeeze at the $z$-position of a magnetic mirror moves the separatrix so as to exclude particles with small $v_z$. This causes a *reduction* in $v_p$ for small $V_{sq}$, as $v_z \lesssim \bar{v}$ particles are excluded from the magnetic separatrix; and causes an increase in $v_p$ for large $V_{sq}$ as the radially localized electric separatrix becomes dominant. These experimental probes of the separatrix are all quantitatively consistent.

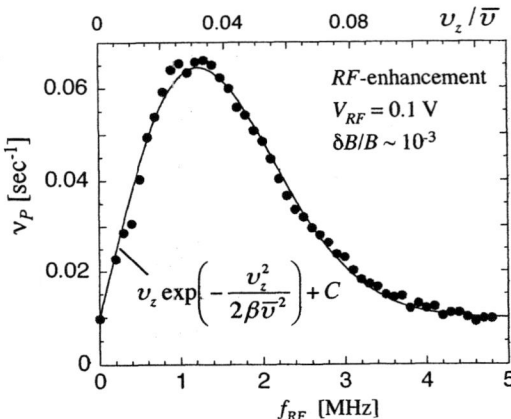

**FIGURE 7.** RF wiggle causes resonant $v_z$ separatrix-crossings and enhanced transport.

## COMMON CHARACTERISTICS: SEPARATRIX DISSIPATION

The magnetic TPM effects appear to be analogous to the electric TPM effects, except for the radial localization of the electric separatrix and the absence of a detectable magnetic mode. For electric trapping with $V_{sq}$ acting on magnetic tilt asymmetry $\alpha$, we find

$$v_p^{(E)} \propto L_p^2 B^{-(1.5 \to 2)} N_L^{1 \to 2} T^{-?} V_{sq}^? \alpha^2, \tag{5}$$

where ? represents non-power-law scalings. Most of the complication lies in the separatrix dissipation process, and measurement of $\gamma_a$ allows this to be written

$$\frac{v_p^{(E)}}{\gamma_a^{(E)}} = (6.3 \times 10^{-5}) \left(\frac{L_p}{R_w}\right)^2 \left(\frac{eN_L^2}{B}\right)^1 \left(\frac{N_L^{tr}}{N_L}\right)^1 \alpha^2; \tag{6}$$

this is valid for both magnetic and electric tilts. For magnetic trapping from $\delta B/B$ acting on tilt $\alpha$ we find

$$v_p^{(M)} \propto L_p^2 B^{-(1.5 \to 2)} eN_L^2 T^{-?} \left(\frac{\delta B}{B}\right)^0 \alpha^2. \tag{7}$$

We *hypothesize* that this represents

$$\frac{v_p^{(M)}}{\gamma_a^{(M)}} = (??) \left(\frac{L_p}{R_w}\right)^2 \left(\frac{eN_L^2}{B}\right)^1 \left(\frac{N^{tr}}{N_L}\right)^0 \alpha^2, \tag{8}$$

although $\gamma_a^{(M)}$ is only conceptual at present, since no magnetic mode has been observed. The two "damping" processes are generically similar, as seen by comparing

$$\gamma_a^{(M)}(T, B, n, L_p, R_p, \frac{\delta B}{B}) \quad \text{vs} \quad \gamma_a^{(E)}(T, B, n, R_p, L_p, V_{sq}). \tag{9}$$

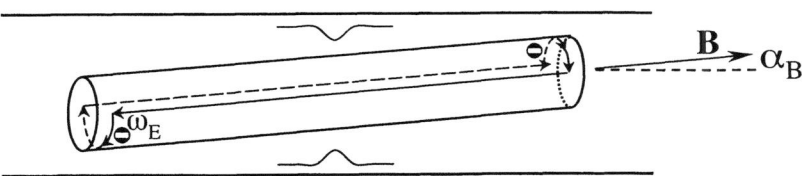

**FIGURE 8.** A tilted equilibrium has DC equilibrium currents which will be dissipated by a central squeeze.

The abrupt $T$ dependence of Fig. 2 and the $B^{-0.5}$ to $B^{-1}$ scaling of Eq. (1) probably apply to $\gamma_a^{(M)}$ also, since their nature is defined by the same process of scattering across a separatrix. Most striking as a difference is that the magnetic separatrix transport effects are *independent of $\delta B/B$,* down to our minimum of $\beta = 10^{-3}$. Experimentally, this means that adding an external magnet to increase the mirror peak in Fig. 6 does not increase the asymmetry-induced transport. This surprising result is conceptually reasonable, in that the entire Maxwellian distribution of particles participates in magnetic separatrix crossings, no matter how small the separatrix angle $\beta^{1/2}$. Presumably, this process ceases only when the collisional (or non-linear) separatrix width [5] becomes comparable to the trapping width.

An alternate view of these TMP processes emphasizes the DC currents which must exist in tilted equilibria, as sketched in Fig. 8. The electron density at top-right is high, because it is close to the wall image charges. As these electrons $\mathbf{E} \times \mathbf{B}$ rotate, they flow axially down the front of the column, to form a high density at lower-left; they then reverse their axial flow along the back of the column. This gives zero net axial current of passing particles unless there is also a diocotron mode displacement of the entire column, in which case there are "sloshing currents" at frequency $f_d$ which are readily detected [10].

Dissipation of these asymmetry-induced currents through collisional scattering across electric and/or magnetic separatrices, at a large rate characterized by $\sqrt{\nu/f_E^*}$, is the essence of the TPM mode damping and transport, as illustrated in Fig. 9.

## POSSIBLE EXAMPLES

It appears likely that TPM effects are pervasive, for two reasons: the rate is enhanced in low collisionality plasmas by $\sqrt{\nu/f_E^*}$, and magnetic trapping can be important even for $\delta B/B \sim 10^{-3}$. Thus, we suggest that TPM effects may be dominant in a variety of experimental situations.

The oft-observed (and oft-violated) $L^2/B^2$ scaling [6] for "anomalous" background transport probably results from magnetic asymmetries acting on magnetically trapped populations in moderate ridigity plasmas. The most direct demonstration of this is the $5\times$ reduction obtained on CamV by removing the $\beta = 10^{-3}$ mirror point. The EV apparatus

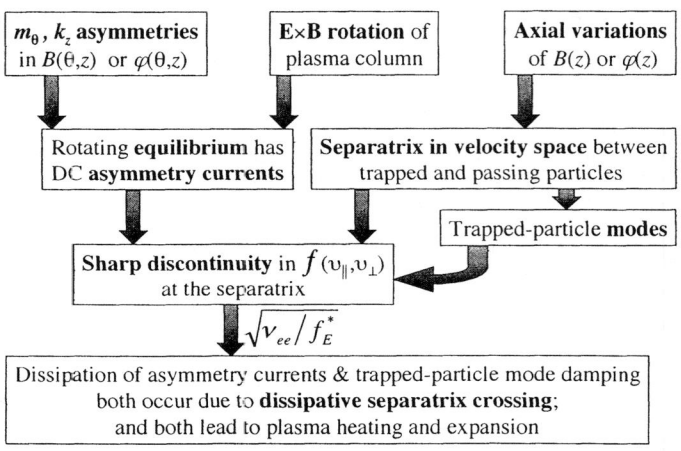

FIGURE 9.    Conceptual outline of TPM effects.

has a gentle (axially extended) magnetic peak of 0.5%.

More quantitative comparison requires knowledge of the trapping and of the background asymmetries. Multiple, or off-center, or extended trapping barriers presumably give rise to couplings to asymmetries with $(L/\pi) \, k_z = 2, 3, ...$, and this will change the scalings of Eq. (8).

Figure 10 presents an overview of electron and ion "background" expansion rates (scaled by $\sqrt{M_i/m_e}$), plotted versus the "rigidity" $\mathscr{R} \equiv f_b/f_E$. The original electron data from the V' and EV apparatuses at $n \sim 10^7 \ \mathrm{cm}^{-3}$ and $T \sim 1 \ \mathrm{eV}$ gave the dashed

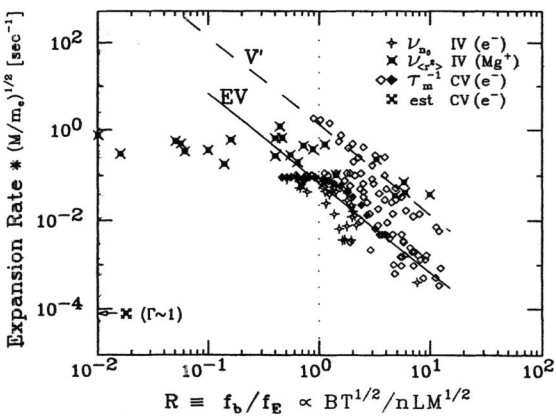

FIGURE 10.    Scaled "background" loss rates versus rigidity for electrons and ions on 4 apparatuses.

12

and solid lines, showing the puzzling $(L/B)^2$ scaling over about 4 decades. The higher density, warm electron data from IV was generically similar, but the floppy ions on IV show little correspondence with $\mathcal{R}^{-2}$. Electrons in CV at $n \sim 10^8$ look like a "swarm of killer bees," although individual temperature sequences often show an abrupt drop-off with temperature (e.g. solid diamonds). This abrupt temperature dependence is probably analogous to that of $\gamma_a^{(E)}$ in Fig. 2.

Neither "rigidity" nor "trapped particles" are valid terms for the $\mathcal{R} \ll 1$ ion points and the venerable Sendai [20] electron point ($n \sim 2 \times 10^{10}$, $T \sim 6°K$, $\mathcal{R} \sim 10^{-4}$, $\Gamma \sim 1$). For the Sendai point the collisionality is exceedingly high: we obtain $\nu = 3 \times 10^{10}$, compared to $f_b \sim 3 \times 10^5$ and $f_E \sim 5 \times 10^8$. Clearly no single process applies to all of Fig. 10.

Nevertheless, the scalings of Eqs. (5-8) show striking correspondence to many prior results on asymmetry-induced transport. For magnetic field dependence, we expect $v_p \propto \gamma_a(B) B^{-1} \propto (B^{-1.5}$ to $B^{-2})$, using the $\gamma_a(B)$ scaling of electric trapping. The observed length dependence of $v_p$ varies from $v_p \propto L_p^2 N_L^2$ for fixed magnetic tilt $\alpha_B$ to $v_p \propto L_p^{-2} N_L^0$ for electric asymmetries, since fixed $V_a$ and fixed $L_a$ give $\alpha_E \propto L_p^{-2}$ in Eq. (3).

The abrupt decrease in transport observed for $\mathcal{R} \geq 10$ in recent experiments [8] is probably due to the temperature dependence of separatrix effects, as in Fig. 2. More importantly, $\mathcal{R}$ is not a globally relevant scaling parameter for transport due to trapped-particle separatrix crossings, since unperturbed particle parameters such as $f_b$ and $f_E$ cannot describe the nonlinearities of trapped orbits and non-Maxwellian separatrix velocity distributions.

Some of the most precise measurements of transport have been reported by Eggleston [9], using applied asymmetries which vary accurately as $\sin(\theta)\sin(n\pi z/L)\sin(2\pi ft)$. Here, resonant particles are thought to be important, but the data does not match the theory of simple collisional scattering out of resonance. It appears likely that resonances are occurring in *magnetically trapped* particles with $v_z \sim \beta^{1/2}\bar{v} \sim .07\bar{v}$, and that collisions cause trapped/untrapped transitions, generating stronger transport. Here, experimental enhancement of separatrix crossings (as in Fig. 7) may help identify TPM effects.

Trapped-particle-mediated effects may also be occurring in recent experiments on transport from applied quadrupole magnetic asymmetries [10]. Here, a resonance is observed with $v_z^2$ being 5 (or more) times less than $\bar{v}^2$. This may possibly represent a "$\beta = 1/5$" reduction in $v_z^2$; but the trapping characteristics of this system would be substantially more complex than any considered here.

Neutral collisions often give puzzling effects, including recent observations [21] of $v_p \propto B^{-1.5}$. Here, we note that $e - N$ collisions also contribute to separatrix crossings, so one would expect to observe expansion at a rate

$$v_p \propto \sqrt{\frac{\nu_{ee} + \nu_{eN}}{f_E^*}} \approx \sqrt{\frac{\nu_{ee}}{f_E^*}} \left(1 + \frac{1}{2}\frac{\nu_{eN}}{\nu_{ee}}\right). \qquad (10)$$

That is, TPM transport scalings of $B^{1.5}$ may be observed, even though $v_p$ increases linearly with pressure (with an offset).

Finally, we note that TPM effects cause strong exponential damping of the nominally stable diocotron modes (frequency $f_m$, damping $\gamma_m$, with $k_z = 0$, $m_\theta = 1,2...$), when

13

$\theta$-asymmetries áre also present. This may be viewed as collisional dissipation of the asymmetry- plus diocotron-induced sloshing currents discussed above. We find that this damping [22] scales as $\gamma_m/f_m \propto v_p \alpha^2$. Combined with $v_p \propto \sqrt{v/f} \, B^{-2} N_L^2 \alpha^2$ from Eq. (5) and $f_m \propto N_L B^{-1}$, this implies $\gamma_m \propto \sqrt{v/f} \, B^{-3} N_L^3 \alpha^4$. This $B^{-3}$ damping would be expected to dominate in experiments at low magnetic fields [21]. Moreover, for the dominant electrostatic asymmetry presumed in Ref. [11], Eq. (3) gives $\gamma_m \propto \sqrt{v/f} \, B^{-3} N_L^{-1} V_a^4$; and this $N_L^{-1}$ scaling was indeed observed.

Targeted experiments incorporating separatrix manipulation and diagnostic techniques will be required to clarify the role of TPM transport and damping over the wide range of plasma parameters, trapping geometries, and asymmetry types in current experiments. Hopefully, this will combine with emerging theory to give a broader picture of TPM effects.

## ACKNOWLEDGMENTS

This work is supported by National Science Foundation grant NSF-PHY9876999 and Office of Naval Research Grant N00014-96-1-0239.

## REFERENCES

1. A.A. Kabantsev, C.F. Driscoll, T.J. Hilsabeck. T.M. O'Neil, and J.H. Yu, Phys. Rev. Lett. **87**, 225002 (2001).
2. A.A. Kabantsev, J.H. Yu, R.B. Lynch, and C.F. Driscoll, Phys. Plasmas **10**, 1628 (2003).
3. A.A. Kabantsev and C.F. Driscoll, Phys. Rev. Lett. **89**, 245001 (2002).
4. A.A. Kabantsev and C.F. Driscoll, Rev. Sci. Instrum. **74**, 1925 (2003).
5. T.J. Hilsabeck, A.A. Kabantsev, C.F. Driscoll, and T.M. O'Neil, Phys. Rev. Lett. **90**, 245002 (2003).
6. C.F. Driscoll, K.S. Fine, and J.H. Malmberg, Phys. Fluids **29**, 2015 (1986).
7. K.S. Fine, UCSD Ph.D. dissertation (1988).
8. J.M. Kriesel and C.F. Driscoll, Phys. Rev. Lett. **85**, 2510 (2000).
9. D.L. Eggleston and B. Carillo, Phys. Plasmas **10**, 1308 (2003).
10. E. Gilson and J. Fajans, Phys. Rev. Lett. **90**, 015001 (2003).
11. E. Sarid, E. Gilson, and J. Fajans, Phys. Rev.Lett. **89**, 105002 (2002).
12. S.F. Paul, *et al.*, "$m = 1$ Diocotron Mode Damping in the Electron Diffusion Gauge (EDG) Experiment," in *Non-Neutral Plasma Physics IV*, AIP Conf. Proc. **606**, (F. Anderegg *et al.*, eds.), 305 (2002).
13. G.W. Mason, Phys. Plasmas **10**, 1231 (2003).
14. M.N. Rosenbluth, D.W. Ross, and D.P. Kostomarov, Nucl. Fusion **12**, 3 (1972).
15. X.-P. Huang, F. Anderegg, E.M. Hollmann, T.M. O'Neil, and C.F. Driscoll, Phys. Rev. Lett. **78**, 875 (1997).
16. F. Anderegg, E.M. Hollmann, and C.F. Driscoll, Phys. Rev. Lett. **81**, 4875 (1998).
17. J.M. Kriesel, UCSD Ph.D. dissertation (1999).
18. J. Fajans, Phys. Plasmas **10**, 1209 (2003).
19. G.W. Hart, Phys. Fluids B **3**, 2987 (1991).
20. J.H. Malmberg *et al.*, "The Cryogenic Pure Electron Plasma," in *Proc. of 1984 Sendai Symposium on Plasma Nonlinear Phenomena* (N. Sato, ed.), 31 (1984).
21. E.H. Chao *et al.*, Phys. Plasmas **7**, 831 (2000).
22. A.A. Kabantsev, to be published.

# Thermal Fluctuations:
# Modes versus the Continuum

## Roy W. Gould

*California Institute of Technology, Pasadena California 91125, USA*

**Abstract.** The thermal fluctuation spectrum of the signal received on a patch electrode is examined and it is shown that the spectrum shows both the modes of the plasma and a continuous spectrum related to the independent-particle motions of plasma electrons. Modes whose axial phase velocity are more than 3-4 times the electron thermal speed are lightly Landau-damped and are clearly separated from the continuum. Long wavelength modes are "acoustic" in nature. If the axial phase velocity of a mode becomes less than 1-2 times the electron thermal speed, then the mode becomes strongly Landau-damped and it merges into the continuum. The mode velocities are of the order of $\omega_p a$ , where $a$ is the plasma radius, so that the plasma radius must be at least several deBye lengths in order to have lightly damped modes. In general, the spectrum is a mixture of a continuous spectrum together with a finite number of modes which are Landau-damped by varying amounts, depending on their phase velocity relative to the electron thermal speed. Only in the extreme limit, $\omega_p a \ll v_{th}$ does the continuous spectrum tend to a Gaussian of width $k\,v_{th}$, characteristic of independent particles. The effect of the "load impedance" on the measurements is also discussed.

## INTRODUCTION

Two procedures have been described recently for determining the temperature of pure electron plasmas at or near thermal equilibrium by measuring the spectrum of the fluctuating charge on a patch electrode. The first [1] method employs the narrow resonant peaks associated with the modes of the plasma, and the second [2] makes use of the broad continuous spectrum associated with the independent particle motion. A simple model of the plasma column, using the warm plasma dielectric function, is used to calculate the input admittance $Y_p(\omega)$ of a patch electrode. $Y_p(\omega)$ gives the patch current when a voltage is applied, and it reflects the dynamical processes within the plasma. For example, when the frequency of the voltage applied to the electrode is close to one of the mode frequencies of the plasma, the patch current can be very large if the modes are lightly damped.

Since many non-neutral plasmas are at, or close to, thermal equilibrium, a thermodynamic argument can be used to obtain the fluctuation spectrum from $Y_p(\omega)$. The plasma can be treated as an electrical circuit and Nyquist's theorem [3] for electric circuits can be used to obtain the fluctuation spectrum. Alternatively, one could use the fluctuation-dissipation theorem [4] to obtain the same result. According to Nyquist's theorem, the fluctuation spectrum of the patch current is related to the dissipative part

CP692, *Non-Neutral Plasma Physics V*, edited by M. Schauer et al.
© 2003 American Institute of Physics 0-7354-0165-9/03/$20.00

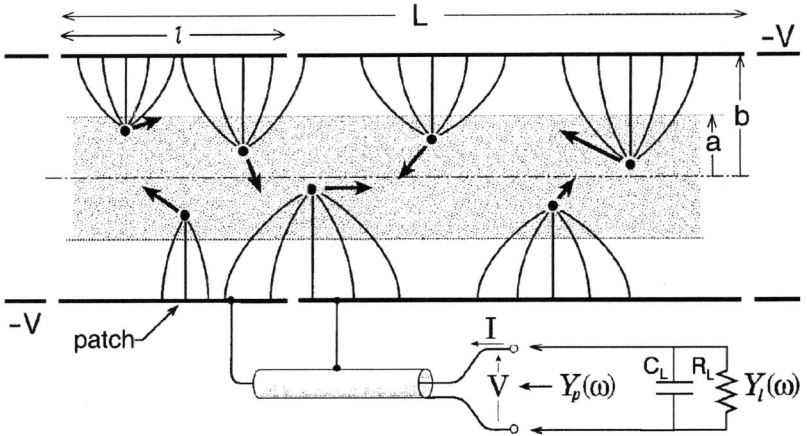

FIGURE 1. Schematic of nonneutral plasma with cylindrical patch electrode and coaxial cable connection to the measurement instrumentation, represented by $Y_l(\omega)$.

of the plasma admittance, $Re\{Y_p(\omega)\}$. Fig. 1. shows schematically a typical experimental geometry. The patch electrode is connected to the measuring apparatus with a coaxial cable. The input to the measuring apparatus represents a "load" $Y_l(\omega)$ on the plasma and is assumed to consist of a resistance $R_l$ and a capacitance $C_l$ in parallel. $C_l$ also includes the capacitance of the coaxial cable. The fluctuation spectrum can be obtained by measuring either the voltage $V(\omega)$ at the coax terminals, or the current $I(\omega)$ which flows to the load. Nyquist's theorem gives either the fluctuating open-circuit voltage which appears across the terminals, or the fluctuating short-circuit current which flows. The latter is

$$I_{sc}^2(\omega) = 4\kappa T_p Re\{Y_p(\omega)\}, \tag{1}$$

where $\kappa$ is Boltzmann's constant and $T_p$ is the plasma temperature. In general, the load is neither a short-circuit, nor a open-circuit, so the effect of the load admittance must be taken into account. At the end of the paper various load effects are discussed. However, to simplify the present discussion, it is assumed that the load is simply the capacitance $C_L$ and that $|Y_p| \ll |Y_l|$. Then the spectral density of the potential which appears across $C_L$ is $V^2(\omega) = I_{sc}^2(\omega)/\omega^2 C_L^2 = q_{sc}^2/C_L^2$, with $q_{sc}$ being the fluctuating charge on the sector electrode. First the calculation of $Y_p(\omega)$ is outlined, and then some illustrative results are given.

## ADMITTANCE CALCULATION

In this section the calculation of the input admittance $Y_p(\omega)$ is outlined. To simplify the calculation, a number of reasonable assumptions and approximations are made: a) the patch electrode extends over a length, $l$, at one end of the plasma, b) the plasma has constant density up to radius $a$, and the density is zero outside this radius, c) the plasma length is $L$, and periodic boundary conditions are used which are appropriate to specular reflection of electrons at each end, d) the plasma is described by its dielectric tensor, $\boldsymbol{K}$. $\nabla \cdot \boldsymbol{K} \cdot \nabla \phi = 0$ is solved for the potential, $\phi$. $K_1$, $K_2$, and $K_3$ are the the perpendicular, Hall, and parallel components of $\boldsymbol{K}$, respectively. Assumption (c) allows us to write the potential and radial electric field, with $k_n = n\pi/L$, as

$$\phi = \sum_{mn} \phi_{mn} \cos(k_n z) e^{im\theta - i\omega t} \tag{2a}$$

$$E_r = \sum_{mn} E_{rmn} \cos(k_n z) e^{im\theta - i\omega t} \tag{2b}$$

Solving the potential equation leads to $\phi_{mn}(r) \sim AI_m(k_n r) + BK_m(k_n r)$ for $a \leq r \leq b$, and $\phi_{mn}(r) \sim J_m(Tr)$ for $r < a$ with $T^2 = -k_n^2 K_3/K_1$. $T$ is the radial wave number. Assuming that the potential $\phi_{mn}(b)$, the potential of one Fourier component at the wall, is known these solutions can be fitted together so as to give $E_{rmn}(b)$, the radial electric field at the wall. We define the quantity $\chi_{mn}(\omega) = -bE_{rmn}(b)/\phi_{mn}$, which is the logarithmic derivative of the potential at the wall. $\chi_{mn}$ gives the response of the plasma to an applied potential on the wall. $\chi_{mn}$ is a function of the frequency, $\omega$, and is useful in calculating the current flowing to the sector probe, $I(\omega)$. This current is a displacement current at the surface of the patch electrode, whose area is $S_p$,

$$I(\omega) = -i\omega\epsilon_o \int_{S_p} (-E_r) \, dS, \tag{3}$$

which can be evaluated in terms of $E_{rmn}(b)$ using Eq. (2b). If the patch electrode is at potential $V(\omega)e^{-i\omega t}$, the coefficients $\phi_{mn}(b)$ can be evaluated and the input admittance of the patch electrode can be shown to be

$$Y_p(\omega) = \frac{I(\omega)}{V(\omega)} = \frac{2i\omega\epsilon_o S_p^2}{S_{tot} b} \sum_{mn} |M_{mn}|^2 \chi_{mn}(\omega) \tag{4}$$

where $M_{mn} = \frac{1}{S_p} \int_{S_p} \cos(k_n z) e^{im\theta} dS \leq 1$, is a sector factor, $S_{tot} = 2\pi bL$,

$$\chi_{mn}(\omega) = G_3 \frac{F(\omega) + G_1}{F(\omega) + G_2}, \tag{5}$$

$$F(\omega) = K_1 Ta\left[\frac{J'_m(Ta)}{J_m(Ta)}\right] + mK_2, \quad G_1 = ka \frac{I'_m(kb)K'_m(ka) - K'_m(kb)I'_m(ka)}{K'_m(kb)I_m(ka) - I'_m(kb)K_m(ka)},$$

$$G_2 = ka \frac{I_m(kb)K'_m(ka) - K_m(kb)I'_m(ka)}{K_m(kb)I_m(ka) - I_m(kb)K_m(ka)}, \quad G_3 = kb \frac{K'_m(kb)I_m(ka) - I'_m(kb)K_m(ka)}{K_m(kb)I_m(ka) - I_m(kb)K_m(ka)},$$

and $T^2 = -k^2(K_3/K_1)$. Thus $Y_p(\omega)$ is a weighted sum over the various $\chi_{mn}$. Note that $G_1$, $G_2$, and $G_3$ are just functions of the geometry and the axial wave number $k = k_n$. For a frequency corresponding to a mode $\chi_{mn}$, the plasma response becomes very large. The mode frequencies are thus determined by the vanishing of the denominator of Eq. 5.

# RESULTS

In the remainder of this paper axisymmetry $(m = 0)$, and a strong magnetic field so that $K_1 \approx 1$, and $K_2 \approx 0$, will be assumed. In addition, it will be assumed that electron parallel motion obeys the kinetic equation so that $K_3 = 1 - (\omega_p^2/k^2 v_{th}^2)Z'(\omega/k v_{th})$, where $Z'$ is the derivative of the plasma dispersion function and $v_{th}^2 = 2\kappa T_p/m$. In the cold plasma limit $K_3 = 1 - \omega_p^2/\omega^2$ and for $b/a = 2$, this leads to the undamped cold plasma modes shown in Fig. 2. For low frequencies and long wavelengths, the dispersion is acoustic in nature, $\omega \sim k$, with the lowest radial mode having the highest axial phase velocity, $v_{ph} = \omega/k$, shown by the dashed line. Higher order radial modes have lower axial phase velocities and in Fig. 2. lie below the lowest mode. If the phase velocity of a mode is less than 2-3 times $v_{th}$ that mode will be strongly Landau-damped.

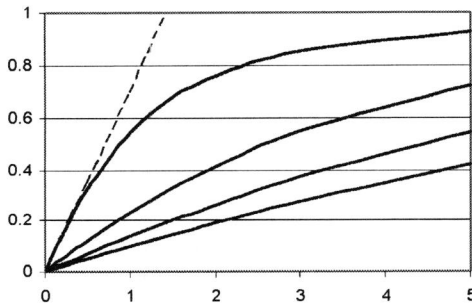

FIGURE 2. Radial mode frequencies, $\omega/\omega_p$ versus $ka$, for $m = 0$, $b/a = 2$, and $T_p = 0$. The dashed line indicates low frequency asymptote for the lowest $(l = 1)$ radial mode.

Furthermore, for $ka, kb \ll 1$ (long wavelengths), $G_1 \to 0$, $G_2 \approx G_3 \approx 1/ln(b/a)$. In this limit,

$$\chi_{0n} \approx \frac{1}{-\frac{J_0(Ta)}{Ta\,J_1(Ta)} + ln(b/a)} \tag{6}$$

with $T^2 = -K_3 k_n^2$. The plasma response, $\chi_{mn}$, becomes very large when the excitation frequency corresponds to a mode so the denominator of Eq. 6 must vanish. Thus, for a given mode, $b/a$ determines $Ta$, irrespective of $\omega$ and $k$ for long wavelengths. For a cold plasma, and for frequencies well below the plasma frequency, $K_3 \approx = -\omega_p^2/\omega^2$, and the low frequency phase velocity is $v_{ph} = \omega/k = \omega_p a/Ta$. $Ta$ is less than 2.405 for the lowest mode, with $Ta \approx 1.45$ for $b/a = 2$. Thus the ratio of velocity of the lowest mode to thermal velocity is $\sim 0.7\,\omega_p a/v_{th}$, i.e of the order of the plasma radius divided by the deBye length.

If the plasma radius is less than a deBye length, all modes will be strongly Landau-damped. As the deBye length becomes smaller, for example if the plasma temperature is reduced, the modes become less damped. In Fig. 3 we illustrate this by focusing on the contribution to $q_{sc}^2$, the charge fluctuation spectrum, from a single wave number, $k_n$.

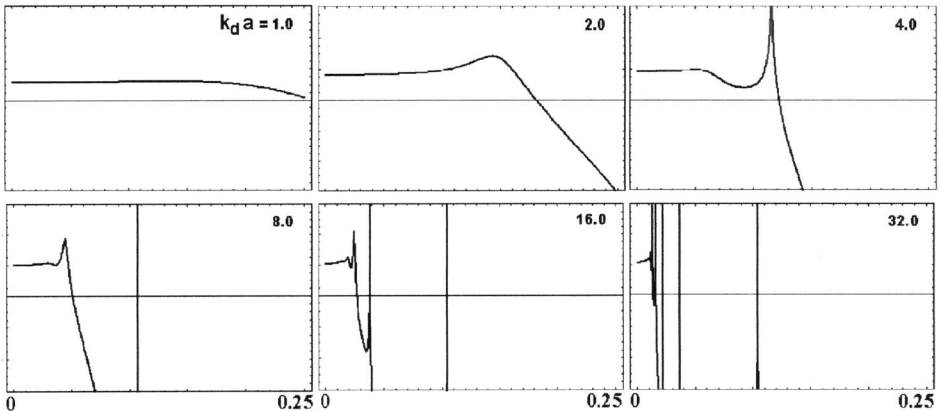

FIGURE 3. Semilog plot of the charge fluctuation spectrum on a sector probe for values of $k_d a =$ 1, 2, 4, 8, 16, and 32. The ordinate is the normalized frequency $f = \omega/\omega_p$. Parameters for this example are $b/a = 2.5$, $L/b = 11.4$. When a mode becomes too narrow to resolve, it is shown as a vertical line.

We plot, on a semilogarithmic scale, the charge fluctuation spectrum on a patch electrode for values of $k_d a = \omega_p a/v_{th} = 1, 2, 4, 8, 16$, and 32. The ordinate is the normalized frequency $f = \omega/\omega_p$. The dimensional factors in front of Eq. 4. have been omitted for this illustration and we have plotted $Re\{M_{mn}^2 \chi_{mn}(f)/f\}$, for $b/a = 2.5$ and $L/b = 11.4$. As $k_d a$ increases due to decreasing plasma temperature, modes become less damped and emerge, one at a time, from the continuous spectrum. The continuous spectrum due to non-resonant particles has a width $\sim k\, v_{th}$, which decreases with decreasing temperature.

When $k_d a < 1$, the single wave-number spectrum is nearly Gaussian (a parabola on the semi-log plot) with width $k_n\, v_{th}$ (upper left panel of Fig 3.). This is the single wave number charge fluctuation spectrum for an uncorrelated Maxwellian velocity distribution of non-resonant particles. As $k_d a$ increases due to decreasing temperature, the lowest radial mode begins to emerge from the Gaussian continuum, at first strongly damped, because the mode velocity is the same order as the thermal velocity. With further increase of $k_d a$, the continuous spectrum narrows and the lowest mode becomes less damped (peak is narrower and higher). Once can also notice a slight downward shift of the mode frequency, due to a smaller "thermal shift". Still further increase in $k_d a$ causes a further narrowing of the continuum, and higher radial modes (of lower phase velocity) emerge from the continuum. Fig. 3 is for a single wave number, $k_1$, but the results for the higher order axial modes, $k = k_2, k_3, k_4...$ are very similar except for a change in the frequency scale.

In Fig. 4. the various wave-number contributions $(k_1 k_2 k_3 k_4)$ have been added together as required in Eq. 4, for $k_d a = 2.3$. Also shown in Fig. 4. by the dashed curve is an approximate fit to the mode peaks using 4 simple poles. At frequencies below the lowest mode, one can still see the non-resonant particle contribution (labeled NRP in Fig. 4).

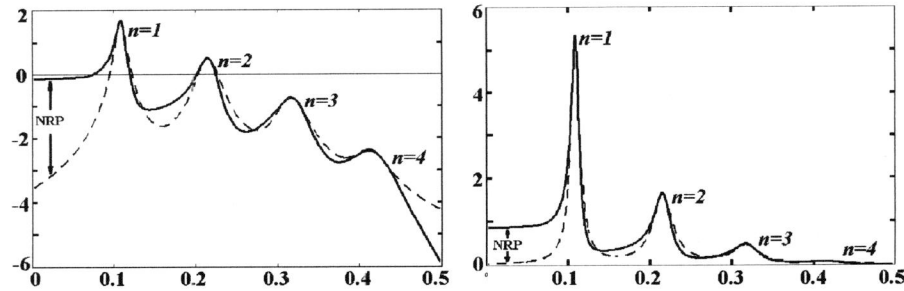

FIGURE 4. Semilog Plot (left) and linear plot (right) of the charge fluctuation spectrum versus $f = \omega/\omega_p$ for $k_d a = 2.3$. Solid curve: calculation using plasma dispersion function, Dashed curve: 4 pole approximation to calculated result.

# DISCUSSION

It has been shown that the thermal fluctuation spectrum of a nonneutral plasma has both a broad continuous spectrum at low frequencies, due to non-resonant particles, together with peaks at the resonant modes of the plasma. A paper at this workshop [6] presents experimental results which shows this.

Modes whose axial phase velocity are more than 3-4 times the electron thermal speed are lightly damped and are clearly separated from the continuum. If the axial phase velocity of a mode becomes less than 1-2 times the electron thermal speed, then the mode becomes strongly Landau-damped and it merges into the continuum. Since mode velocities are of the order of $\omega_p a$ , where $a$ is the plasma radius, the plasma radius must be at least several deBye lengths in order to have lightly damped modes. In general, the spectrum is a mixture of a continuous spectrum together with a finite number of modes which are Landau-damped by varying amounts, depending on their phase velocity relative to the electron thermal speed. Only in the extreme limit, $\omega_p a \ll v_{th}$ does the continuous spectrum tend to a Gaussian of width $k v_{th}$ , characteristic of independent particles.

In order for the resonant modes to emerge from the non-resonant continuous spectrum, the modes must not be too strongly damped by Landau damping and this generally requires that the plasma be at least a few deBye lengths across. Only if $k_d a \ll 1$, can one neglect collective effects (screening).

In order to simplify the presentation, it has been assumed that $|Y_p| \ll |Y_l|$ in the results above. This is often the case, except when the mode damping becomes very small, and $|Y_p|$ can become relatively large, and comparable with $|Y_l|$ near a mode resonance. Then the following "load effects" must be accounted for when interpreting experimental data [1]: a) the load affects the magnitude of the signal observed, b) the capacitance of the load can shift the mode frequency slightly, c) the resistance of the load can increase the damping of the mode, d) the load resistance can be an additional source of thermal noise.

While the analysis here has been for cylindrical plasmas, similar results apply for spheroidal plasmas. Although the analysis of mixed cylindrical electrode geometry and spheroidal plasma geometry can complicate an analysis, nevertheless the patch fluctuation spectra is still related to the input admittance of the plasma.

In the **RESULTS** section of this paper examples were given only for the axisymmetric $m = 0$ modes. Some non-axisymmetric modes are negative energy modes and can be resistively destabilized. However, when they are not destabilized, the fluctuation spectrum is still related to the plasma admittance, but in a more subtle way [5].

## ACKNOWLEDGEMENT

The author is grateful to N. Shiga, F. Anderegg, and other members of the U.C.S.D. group for stimulating discussions of their results.

## REFERENCES

1. Francois Anderegg, et al, in Non-Neutral Plasma Physics IV, p253, AIP 606 (2002), Francois Anderegg et al, Phys. Rev. Lett. 90 115001 (2003).
2. M. Takahashi Nakata, Grant W. Hart, Bryan G. Peterson, in Non-Neutral Plasma Physics IV p271, AIP 606 (2002), Grant Hart, FP126 Bull. A.P.S 47, No. 9, p126 (2002).
3. H. Nyquist, Phys. Rev. **32** 110, (1928).
4. H. B. Callen and T.A. Welton, Phys. Rev. **83** 34 (1951).
5. Roy W. Gould, in Non-Neutral Plasma Physics IV, p 263, AIP 606 (2002).
6. N. Shiga, et al, in Non-Neutral Plasma Physics V.

# Direct Excitation of High-Amplitude Chirped Bucket-BGK Modes

W. Bertsche*, J. Fajans* and L. Friedland*†

*Dept. of Physics, U.C. Berkeley, Berkeley CA 94720
†Permanent Address:Hebrew University, Jerusalem, Israel

**Abstract.** For the first time, high amplitude ($\Delta n/n \approx 40\%$), high Q (up to 100,000) BGK modes have been controllably excited in a plasma. The modes are created by sweeping an excitation voltage downwards in frequency, thereby dragging a phase space "bucket" of low density into the bulk of the plasma velocity distribution. The modes have no linear limit, and differ markedly from plasma waves and Trivelpiece-Gould modes.

Plasma waves [Trivelpiece-Gould (TG) waves in finite geometry [1]] are easy to generate and ubiquitous in nature. However, plasma waves are Landau damped, often quickly. At first glance this damping seems unavoidable, so it was quite surprising when Bernstein, Greene and Kruskal (BGK) predicted that there exists a broad class of waves that do not damp. These BGK modes [2] are undamped because the distribution of the particles in the wave is already in the Landau relaxed form. In principle, BGK modes can have very large amplitude.

BGK modes underpin much of kinetic wave theory, but experimental verification of the existence of undamped BGK modes has proved difficult. It is easy to create transient large-amplitude waves or structures, but the waves are typically short-lived or unstable. For instance, waves created by Wharton, Malmberg and O'Neil [3] were unstable due to a sideband instability. More recent work has not been much more successful [4, 5]. Long-lasting structures can be created by continuous drives; driven double layers, which are closely related to BGK modes, have been observed in the earth's auroral zone [6]. Danielson [7] recently reported that plasma waves eventually decay into low amplitude, but long-lasting, BGK modes.

Here we report that we can excite very high amplitude BGK modes that differ markedly from the more common TG modes. The modes are excited by an oscillating voltage applied to one end of a pure-electron plasma column confined in a standard Penning-Malmberg trap [8] [see Fig. 1(a).] The resulting density fluctuations are detected by monitoring the image charge on another cylinder, typically at the opposite end of the trap. When the plasma is cold, TG waves are observed as expected. A typical spectrum is shown in Fig. 2. But as the plasma temperature $T$ is increased, the TG waves become so heavily damped that they essentially disappear. Nonetheless, undamped waves can be excited in hot plasmas by low amplitude drives that sweep downward in frequency, from some frequency $f_s$ to a lower frequency $f_e$. Typical response curves are shown in Fig. 3. Large amplitude waves are excited for a very broad range of $f_s$ and $f_e$.

We believe that these waves are BGK modes. Table 1 highlights the differences

CP692, *Non-Neutral Plasma Physics V*, edited by M. Schauer et al.
© 2003 American Institute of Physics 0-7354-0165-9/03/$20.00

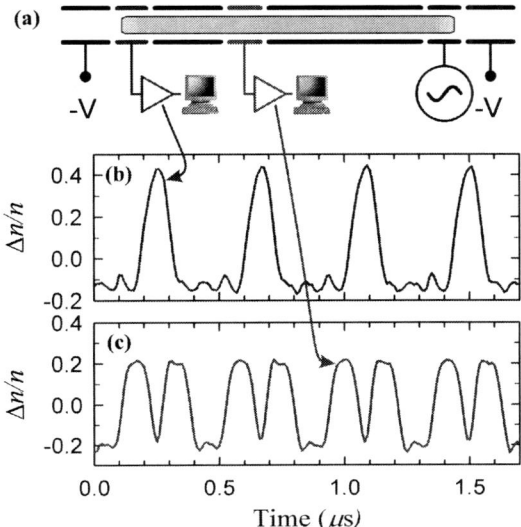

**FIGURE 1.** (a) Penning-Malmberg Trap geometry. The pure-electron plasma column is confined axially by the large negative end potentials, and radially by a 1500G magnetic field. Typical plasma densities are $\sim 10^7\,\mathrm{cm}^{-3}$, lengths $\sim 27\,\mathrm{cm}$, and radii $\sim 1\,\mathrm{cm}$, and the plasma is confined in cylinders with radius 1.905 cm. (b) Density fluctuations due to the BGK mode at the column end, and (c) near the column center. The positive fluctuation corresponds to an electron hole. Because the BGK mode must pass through the center twice (approaching and leaving) each time the pulse reaches the end, there are twice as many pulses in (c) as in (b). As expected for an open-ended reflection, the mode is twice as large at the end (b) as in the center (a).

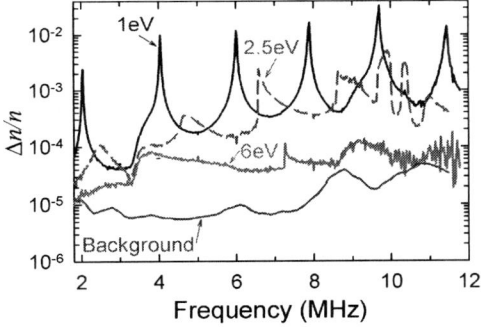

**FIGURE 2.** Observed response from a pure 0.05 V drive, for three different plasma temperatures. Also shown is the background noise from external sources. As predicted by the TG dispersion relation, $\omega^2 = a^2k^2\omega_p^2/(1+a^2k^2) + 3v_{\mathrm{th}}^2k^2$, (where $\omega$ is the wave frequency, $a$ is the radial geometric constant, $k = m\pi/L$ is the wavenumber, $m$ is a positive integer, $L$ is the plasma length, $\omega_p$ is the plasma frequency and $v_{\mathrm{th}}$ is the electron thermal velocity,) the wave dispersion relation is slightly concave down, and the frequencies increase with plasma temperature.

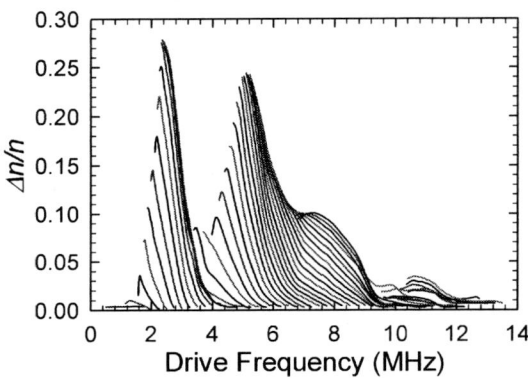

**FIGURE 3.** Response to 50 mV, 1 GHz/s swept frequency drives. Each curve plots the response as a function of time, and, hence, frequency, as the sweep progresses downward from the varying start frequencies $f_s$. Along each curve, the response is phase-locked to the drive. As the drive is swept past the peaks at 2.4 MHz and 5 MHz, phase-locking is lost and the mode amplitude collapses.

between these modes and TG modes in our experiments. First, TG modes only exist when the plasmas are sufficiently cold, while our BGK modes exist only when the plasmas are sufficiently hot; there is only a small overlap region. The TG modes occur at distinct frequencies; even when thermally broadened, the TG modes possess a well-defined linear limit. In contrast, the BGK modes have *no* linear limit. As can be seen in Fig. 3, they can be excited over a broad range of frequencies. The TG modes are typically excited by single frequency drive. The BGK modes can only be excited by a swept frequency drive. Using a swept frequency drive, Yamazawa and Michishita [9] determined that TG modes possess a hard nonlinearity: as their amplitude increases, their frequency increases slightly ($< 5\%$). BGK modes have a soft nonlinearity: as their amplitude increases, their frequency decreases by nearly 50%. Even for cold plasmas, the Q's ($\omega/\gamma$, where $\gamma$ is the damping rate) for undriven, low amplitude TG modes are never much higher than 600, and the Q's get very low for hotter plasmas. Once excited by a swept frequency drive, the Q's for the undriven BGK modes are remarkably high: typically around 6000, but occasionally as high as 100,000. Indeed, Q is often a poor measure of the lifetime of these modes as their amplitude does not decay exponentially. For example, the very-high-Q mode amplitudes are often nearly flat, except for small fluctuations, followed by a sudden collapse. High amplitude TG modes are unstable; Hart and Peterson [5], and Yamazawa and Michishita [9] demonstrated that they mode convert quickly. As one might expect from their high Q's, the BGK modes are very stable even when their density fluctuations exceed 40%. Finally, low amplitude TG modes have little harmonic content, but the harmonic content of the BGK modes is rich even at low amplitude.

We identify these modes as BGK modes because of the method by which they are excited. When the drive is first applied, electrons whose end-to-end bounce frequencies are close to the drive frequency will be captured into a trapping bucket if they have the right phase relative to the drive; approximately half the resonant electrons are captured

**TABLE 1.**

|  |  | TG Modes | BGK Modes |
|---|---|---|---|
| Mode Frequencies: | Cold | Distinct | Nonexistent |
|  | Hot | Nonexistent | Continuous |
| Excitation |  | Pure or Swept | Swept Only |
| Nonlinearity: | Type | Hard | Soft |
|  | Freq. Shifts | +5% | -50% |
| Q (typical): | Cold | 600 |  |
|  | Hot | 3 | 6000–100,000 |
| High Amplitude Stability |  | Poor | Excellent |
| Harmonic Content |  | Single Harmonic (low amp.) | Many Harmonics (all amp.) |

(see Fig. 4). The untrapped electrons are perturbed, and may form a TG mode, but this TG mode damps out quickly because of the high plasma temperature.

As the drive frequency is swept downward to frequency $f_e$, the resonant electron velocity $\bar{v} = 2Lf$ decreases. To the first approximation, few electrons can cross the separatrix between trapped and untrapped electrons if the sweep is slow; few new electrons enter the trapping bucket, and few initially-trapped electrons leave it. Thus, the trapped electrons are dragged to lower velocities, and the density of electrons in the bucket remains fixed. But for typical distribution functions, the initial density of electrons is lower at velocity $\bar{v}_s$ than at $\bar{v}_e$, and dragging the bucket downward creates a hole in phase space. Since the spatially-localized bucket bounces from end to end in phase with the drive, the phase space hole oscillates as well. The hole creates an electrostatic perturbation, which constitutes the postulated BGK modes. Because the excitation process involves dragging a bucket through velocity space, we call the generated waves "Chirped Bucket-BGK" modes.

Strong evidence for this model comes from experiments which perturb the distribution function. For example, a strong, fixed frequency drive at $f_f$ will phase-mix the velocities in the region centered around the velocity $\bar{v}_f$ resonant with the drive, effectively

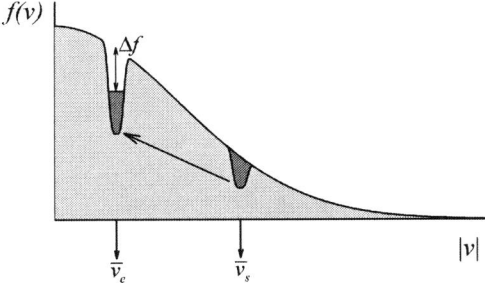

**FIGURE 4.** Spatially-averaged distribution function $f(v)$ showing the evolution of the trapping bucket as the frequency is swept from $f_s$ to $f_e$. The dark and light regions indicate the trapped and untrapped electrons, respectively.

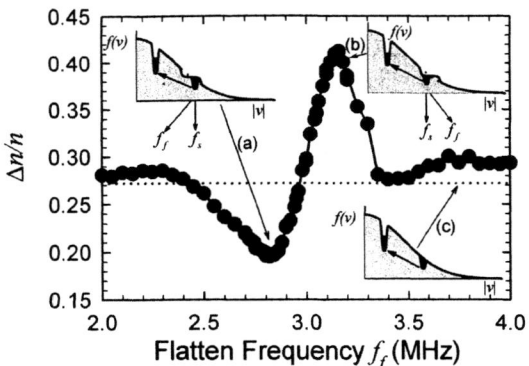

**FIGURE 5.** Response from a 1 GHz/s, 100 mV sweep from $f_s = 3$ MHz to $f_e = 2.2$ MHz that is preceded by a 100 $\mu$s long, 500 mV, fixed frequency $f_f$ drive. The fixed frequency drive flattens the distribution function around the resonant velocity linked to its frequency. Near (a), the the fixed frequency drive is below 3 MHz, near (b) it is above 3 MHz, and the dotted line (c) is the response without the fixed frequency drive.

flattening the distribution function there. The flat in the distribution function increases the number of particles with velocity slightly greater than $\bar{v}_f$, and decreases the number of particles with velocity slightly less than $\bar{v}_f$. If a sweeping drive is then initiated, the difference between the flattening frequency $f_f$ and the swept drive start frequency $f_s$ will affect the final amplitude of the BGK mode (see Fig. 5.) If $f_s < f_f$, the density will be diminished at $\bar{v}_s$, so fewer particles will be trapped into the sweeping bucket, the phase space hole will be proportionally larger, and the final amplitude will increase. In contrast, if $f_s > f_f$, the density of particles near $\bar{v}_s$ will be increased, and the final amplitude will decrease.

These flattening experiments clearly demonstrate that the system response depends on the shape of the initial distribution function. Other experiments, not shown, show that we can transport particles in velocity space; for instance, an upwardly sweeping drive will carry particles to a higher velocity, leaving a hole behind in the distribution function at the drive's initial resonant velocity $\bar{v}_s$. If a downward drive is then initiated starting from the same initial frequency, the resulting BGK mode will be larger. Repeating the upward-going drive several times before initiating the downward drive will dig an ever deeper hole, resulting in ever larger BGK modes. We can also observe trapping oscillations, proving that particles are trapped in the potential of the wave.

We have analyzed the excitation process using the multibeam model presented in Stix [10] and elsewhere, modified to include the swept excitation process. The details of the analysis will be presented in a later paper; the analysis predicts that the mode amplitude $a$ grows as

$$\sqrt{a} \propto \frac{\omega}{\omega - \omega_0} \Delta f(\omega/k), \qquad (1)$$

where $\omega$ is the mode frequency in radians/sec, $\omega_0$ is the nearest lower linear TG mode frequency, and $\Delta f$ is the depth of the hole in the distribution function: $\Delta f(\bar{v}) \approx$

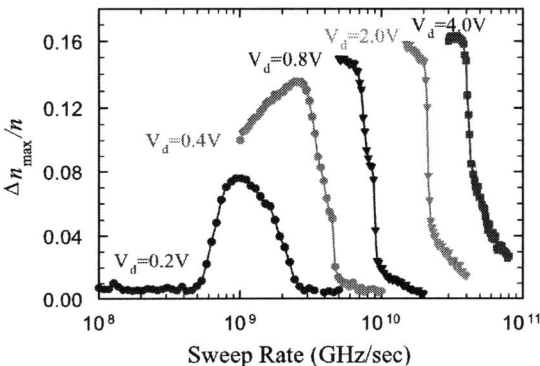

**FIGURE 6.** Peak signal for a $3 \to 2.2\,\text{GHz}$ sweep as a function of sweep rate, for several drive amplitudes.

$0.5[f(\bar{v}) - f(\bar{v}_s)]$. This expression explains many aspects of the observed excitations: (1) The multiple peaks found in Fig. 3; as $\omega$ is swept downwards towards any of the $\omega_0$, the mode amplitude increases. (2) The mode amplitude collapses near each $\omega_0$; as $\omega$ sweeps past the relevant $\omega_0$, the RHS of Eq. 1 becomes negative. (3) The ability to excite the modes almost anywhere; while there are peaks near the $\omega_0$, the mode grows for almost any downward sweep. (4) The lack of modes below the last peak; for $\omega$ less than the lowest TG mode frequency $\omega_{00}$, the RHS of Eq. 1 is necessarily negative. (5) The necessity to sweep downwards; sweeping downwards increases both terms on the RHS of Eq. 1, so the amplitude grows. Sweeping upwards makes both terms smaller, so the amplitude decreases.

Equation 1 also explains why Bucket-BGK modes and TG modes are generally not found simultaneously. For TG modes to exist, the plasma temperature has to be cool enough that the phase velocity $\omega_{00}/k$ is much greater than the thermal velocity $v_{\text{th}}$. But Bucket-BGK modes exist only when $f(\omega/k)$ is not trivially small, and when $\omega$ is greater than $\omega_{00}$. Taken together, these rules requires that $\omega_{00}/k \simeq v_{\text{th}}$. For our plasmas, the smallest $k$ is about $0.1\,\text{cm}^{-1}$, and the smallest phase velocity is about $15 \times 10^7\,\text{cm/s}$. At $1\,\text{eV}$, $v_{\text{th}} \approx 4 \times 10^7\,\text{cm/s}$, and, as expected, only TG modes exist. At $6\,\text{eV}$, $v_{\text{th}} \approx 10 \times 10^7\,\text{cm/s}$ and only Bucket-BGK modes exist. In a narrow range of intermediate temperatures, we do observe both TG and Bucket-BGK modes, but neither wave is well behaved in this regime.

A key assumption in the theory behind Eq. 1 is that particles do not escape the bucket as the driving frequency is swept. We find (Fig. 6) that this assumption is correct so long as the sweep rate is neither too slow nor too fast. If the drive is swept too slowly, particles escape out of the bucket because of collisions and/or loading in the external circuitry [7]. Preliminary evidence suggests that the time scales of these effects are related to the time scale for the decay of the undriven modes.

If the sweep rate is too high, the Bucket-BGK modes disappear. Using action-angle variables, we find that the buckets disappear entirely when the sweep rate is too high; the drive strength must be increased linearly with the sweep rate for the buckets to exists.

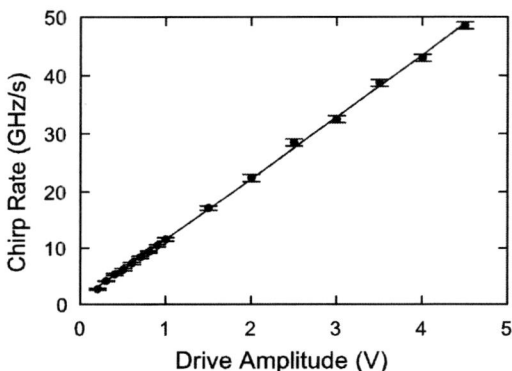

**FIGURE 7.** Maximum Sweep Rate that generates Bucket-BGK modes as a function of drive strength. The sweep rates extends from $3 \to 2.2\,\text{GHz}$

This prediction is proved by experimental measurements (Fig. 7.) The analysis is similar to that for autoresonance for Diocotron modes [11, 12].

In conclusion, we have generated very high amplitude Bucket-BGK modes. The modes have no linear limit and can be excited over a very broad range of frequencies. They are very stable, can be produced controllably, and respond appropriately to changes in the distribution function. The longevity of the modes is remarkable. Not only do the mode Q's reach 100,000, but the modes survive despite up to 60,000 reflections from the trap ends. We have developed a simple theory that explains many of the results. Our Chirped Bucket-BGK modes live much longer than the BGK modes found in earlier experiments. The reason may be related to the fact that most of the previous experiments used strong, or even impulsive, drives. Our gentle excitation process allows us to create seemingly perfect BGK modes.

This works was supported by the NSF. We thank B. Afeyan, T. O'Neil, and J. Wurtele for their comments.

## REFERENCES

1.  Trivelpiece, A., and Gould, R., *J. Appl. Phys.*, **30**, 1784 (1959).
2.  Bernstein, I. B., Greene, J. M., and Kruskal, M. D., *Phys. Rev.*, **108**, 546 (1957).
3.  Wharton, C. B., Malmberg, J., and O'Neil, T., *Phys. Fluids*, **11**, 1761 (1968).
4.  Moody, J. D., and Driscoll, C. F., *Phys. Plasmas*, **2**, 4482 (1995).
5.  Hart, G., and Peterson, B. G., "Interacting Solitons in a Nonneutral Plasma," in *Non-Neutral Plasma Physics IV: Workshop on Non-Neutral Plasmas*, edited by F. Anderegg, C. F. Driscoll, and L. Schweikhard, AIP, New York, 2002, p. 341.
6.  Ergun, R. E., Carlson, C. W., McFadden, J. P., Mozer, F. S., Muschietti, L., Roth, I., and Strangeway, R., *Phys. Rev. Lett.*, **81**, 826 (1998).
7.  Danielson, J., *Measurement of Landau Damping of Electron Plasma Waves in the Linear and Trapping Regimes*, Ph.D. thesis, UCSD (2002).

8.  Anderegg, F., Driscoll, C. F., and Schweikhard, L., editors, *Non-Neutral Plasma Physics IV: Workshop on Non-Neutral Plasmas*, AIP, New York, 2002.
9.  Yamazawa, Y., and Michishita, T., *Jpn. J. Appl. Phys.*, **40**, 5431 (2001).
10. Stix, T. H., *Waves in Plasmas*, Springer Verlag, Heidelberg, 1992.
11. Fajans, J., Gilson, E., and Friedland, L., *Phys. Rev. Lett.*, **82**, 4444 (1999).
12. Fajans, J., Gilson, E., and Friedland, L., *Phys. Plasmas*, **6**, 4497 (1999).

# Injection into Electron Plasma Traps

Vladimir Gorgadze, Thomas A. Pasquini, Joel Fajans, Jonathan S. Wurtele

*Department of Physics, UC Berkeley, Berkeley, CA 94720*

**Abstract.**
   Computational studies and experimental measurements of plasma injection into a Malmberg-Penning trap reveal that the number of trapped particles can be an order of magnitude higher than predicted by a simple estimates based on a ballistic trapping model. Enhanced trapping is associated with a rich nonlinear dynamics generated by the space-charge forces of the evolving trapped electron density. A particle-in-cell simulation is used to identify the physical mechanisms that lead to the increase in trapped electrons. The simulations initially show strong two-stream interactions between the electrons emitted from the cathode and those reflected off the end plug of the trap. This is followed by virtual cathode oscillations near the injection region. As electrons are trapped, the initially hollow longitudinal phase-space is filled, and the transverse radial density profile evolves so that the plasma potential matches that of the cathode. Simple theoretical arguments are given that describe the different dynamical regimes. Good agreement is found between simulation and theory.

## INTRODUCTION

Pure electron plasmas confined in Malmberg-Penning traps have been the subject of extensive theoretical and experimental investigations over the last two decades [1]. These traps have an axial magnetic field which confines electrons radially and an electrostatic well which confine them axially (see Fig. 1). Electrons are injected into the trap by lowering the one wall of the electrostatic well, thereby allowing electrons to flow from the cathode trap to the confinement region. An electron column subsequently forms between the negatively-biased cathode and the far wall of the electrostatic well. Next, the column is pinched off from the cathode by restoring the near wall of the electrostatic well. The electrons in confinement region are thus trapped, ending the injection phase of the experimental cycle. If the injection time is sufficiently long, it has been observed that the number of electrons increases, and their phase-space distribution evolves, until the electron column's electrostatic potential matches that of the cathode bias potential. After injection, experiments on a wide-range of phenomena, from vortex dynamics [2] to collisional relaxation [3] are conducted. The plasmas created in these traps are highly reproducible. Surprisingly, the injection process itself has not been carefully studied in previous experimental and theoretical investigations.

   Our experiments and particle-in-cell (PIC) simulations show that the number of trapped electrons is much higher than expected from a straightforward picture of the injection process. In this straightforward picture, the trapping process is purely ballistic. Electrons leave the cathode, stream across the confinement region, bounce from the far electrostatic wall, and return to be reabsorbed by the cathode. The restoration of a negative bias to the confinement wall near the cathode isolates the cathode from the main

CP692, *Non-Neutral Plasma Physics V*, edited by M. Schauer et al.
© 2003 American Institute of Physics 0-7354-0165-9/03/$20.00

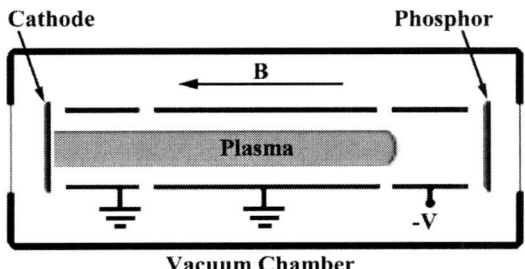

Cathode                Phosphor

B

Plasma

-V

**Vacuum Chamber**

**FIGURE 1.** Malmberg-Penning traps consist of a series of collimated conducting cylinders, or gates, aligned along a strong magnetic field. Electron plasmas are confined in these traps by appropriately biasing the trap cylinders to form an axial electrostatic well. Radial confinement is provided by a strong magnetic field. Electrons are injected into the trap by grounding an 'inject' gate near the cathode, and allowing electrons from the negatively biased cathode to enter the trap as shown. The plasma is imaged by briefly grounding the 'dump' gate, accelerating the electrons toward the phosphor screen, and recording the image with a CCD camera.

plasma column. The electrons thereby confined are not in any sense trapped before the column is pinched; these are just the electrons that happen to be transiting the trapping region the instant the column is pinched. The density of these transiting electrons will be $\sim 2J/qv$, where $q$ is the electron charge, and $v$ is the average speed of the electrons. We find that the actual number of trapped electrons is an order of magnitude higher than predicted by estimates based on this ballistic trapping. Some other mechanism must be trapping electrons in the confinement region, and these electrons must be genuinely trapped in the sense that they are confined to the central column region, rather than re-absorbed at the cathode, before the column is pinched. Thus, the physics of the injection process goes beyond a model consisting of laminar, time independent electron motion, for, in such a model, electrons could not be trapped. This paper elucidates the critical role of self-fields, with their rich nonlinear physics, in the process by which the trapped charge continues to increase over hundreds of ballistic transit times.

Several mechanisms exist that could explain the increase in density beyond the ballistic limit. One such mechanism is trapping through collision with other electrons or with background neutral gas. If an electron loses kinetic energy in a collision it will not be able to return to the cathode and will be on a trapped orbit. Calculated collision rates and experimental measurements indicate, however, that this mechanism is far too slow. Another mechanism could be a small inward displacement of the electrons. Most Malmberg-Penning traps use a spiral-wound tungsten filament that is increasingly negative in the center. Thus, an electron that moved inward might not have enough kinetic energy to return to the cathode, and would be trapped. However, most of our experiments were done with an unusual equipotential cathode that could not trap electrons by this mechanism.

Simulations of this system are straightforward, and suggest that multiple dynamic processes lead to the trapping of electrons during the injection. Very roughly, the processes tend to occur at different temporal stages of the trapping:

In the first stage, electrons accelerate into the initially empty trapping region, reflect off of the potential barrier and return to the cathode. No trapping occurs during this ballistic filling of the trap.

In the second stage, the stream of electrons leaving the cathode interacts with the counter-propagating electrons that have reflected off the end of the trap. A two-stream interaction develops which distorts the equilibrium fields in the trap. Localized time-dependent perturbations in the electrostatic potential grow to the point where they extract energy from localized groups of electrons. These electrons are unable to return to the cathode and are trapped.

In the third stage, electrons which have lost energy interact with the transiting charges, thereby causing the development of a novel, soliton-like travelling cavitation, or bubble, so-named for the appearance of a large bubble in phase space. This bubble produces strongly inhomogeneous plasma self-fields, which traverse the length of the trap many times, decelerating and trapping electrons.

In the fourth stage, which overlaps significantly with the prior stage, the effective current in the trapping region exceeds a critical value and a virtual cathode develops near the physical cathode. The size and position of the virtual cathode are affected by the phase space dynamics. As the cavitation dies off, the oscillations of the virtual cathode continue trapping electrons, albeit at a slower rate than before. In this final stage, the plasma column develops a profile determined not by the current emission from the cathode, but by the matching condition between the plasma potential and the cathode potential. Consequently, a plasma created by a filament cathode (which has a parabolic potential) maintains an approximately constant density. A plasma produced by an equipotential cathode emitting uniformly will hollow significantly as its potential begins to match that of the cathode. The hollow column is then subject to the diocotron instability, which takes a somewhat different form when it occurs in contact with an emitting cathode. In the following sections we give an overview of the experiment and simulation and describe the trapping regimes.

## EXPERIMENT AND SIMULATION

Experimental studies of trapping during injection have been performed on two Malmberg-Penning traps, as shown in Fig. 1 The bulk of the data was collected on the Berkeley Photocathode trap [4], in which the cathode is an illuminated photoemissive material held at a constant potential [5]. The experiments were repeated on a more typical trap where the cathode is a heated tungsten filament and has a parabolic potential profile, and similar results were found. The only significant difference observed occurs in the late time development of the plasma column; the column hollowing and subsequent instabilities on the photocathode machine are absent on the machine with the parabolic cathode. This discrepancy is understood to be a result of the potential matching condition between the plasma column and the cathode.

In the photocathode machine, a 1.0 cm diameter circular section of the cathode is illuminated with light of a constant intensity. For a typical injection time of $30 \mu s$ with a $-5$ V photocathode bias, the plasma column has a density of $1 \times 10^7 \, cm^{-3}$. In the thermal emission machine, the filament emits over a 1.1 cm diameter. For a typical injection time of 1ms with a $-10\,\text{V}$ bias, the plasma column has a density of $1 \times 10^7\,\text{cm}^{-3}$. In both machines, the plasma temperature is approximately $1\,\text{eV}$, corresponding to a Debye length of $0.23\,\text{cm}$.

Measurements of the electron population within the trapping region are made as follows. The electron column is injected for the desired time by grounding the trap segment closest to the cathode, the inject gate. The electron population is then isolated from the cathode by lowering the potential on the inject gate to a large negative value. This process traps approximately 80% of the electrons present in the trapping region. The plasma column is then allowed to escape by grounding the trap segment closest to the phosphor screen. Electrons are accelerated onto the phosphor, and the resulting light is recorded by a CCD camera. Analysis of the pattern gives the number of charges in the trapping region. This procedure cannot distinguish between electrons which were electrostatically trapped during the injection process and those which were circulating through the trapping region when the inject gate was lowered.

Simulations were performed using the PIC simulation XOOPIC [6]. Owing to computational expense, we limit our simulations to the short time evolution of the system ($< 1\,\text{ms}$). The electrostatic code is two-dimensional with an assumed $(r, z)$ dependence for the electrostatic field. The system is assumed to possess azimuthal symmetry. The gyroradius in the experiment is of order $10^{-3}\,\text{cm}$, and typical plasmas have a Debye length of $10^{-1}\,\text{cm}$. We thus make the further approximation that the gyromotion can be ignored. Electrons are therefore limited to motion in the axial direction, along the magnetic field lines. A uniform mesh is applied in both the axial and radial coordinates. The radial distribution still evolves, since the trapping dynamics depends on both self-fields and external fields, and, hence, on the radius.

Simulations have been made for both the equipotential cathode and the parabolic cathode. It is assumed in both cases that electron emission is constant over the area of the cathode. The experimental value of the current was used in all simulations. The temperature of the emitted electrons (assumed Maxwellian distribution) was taken as $T = 0.1\,\text{eV}$ to fit the experimental data.

The experimental and simulation results are plotted in Fig. 2, where the lower curves correspond to cathode with flat potential and the upper curves are for parabolic cathode potential. The simulations reproduce the experimental results and, importantly, provide detailed information on the phase space density. The insights gained from simulations lead to an understanding of the nonlinear processes in the trapping.

## Ballistic Filling

As previously noted, no electron trapping occurs in the first stage of the injection process. Electrons emitted from the cathode traverse the trapping region, reflect from the negatively biased end gate and return to the cathode to be reabsorbed. The number of electrons in the trapping region increases linearly with time until the first electrons to enter the trapping region are reabsorbed at the cathode. At this point, an equilibrium number of trapped charges is reached, depending only on the cathode current, the thermal distribution of emitted electrons, and the trap length. Calculations predict an elec-

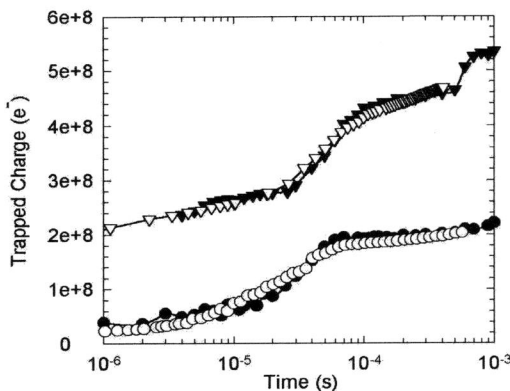

**FIGURE 2.** The total number of trapped electrons as a function of time for an equipotential cathode (circles) and a tungsten filament (triangles). Experimental data are shown in black, simulation data in white. Very short injection times ($< 1\,\mu$s) are explained by a non-trapping model. Beyond $1\,\mu$s, however, instabilities in the injection beam lead to charge trapping.

tron population of $2 \times 10^7$ for the photocathode trap and of $2.1 \times 10^7$ for the parabolic cathode trap.

Simulation results shown in Fig. 3, corresponding to the first bounce period of $0.5\mu$s, confirm that the total charge in the trapping region increases linearly to the value predicted by ballistic trapping. Electrons then continue to cycle in and out of the trap region. In phase space, also shown in Fig. 3A - 3D, the electron orbits correspond to a split ring trajectory. The phase space plot shows that none of the charges in the trapping region are actually trapped.

The equilibrium number of electrons in the trap depends only on the cathode current, the thermal distribution of emitted electrons, and the trap length. The velocity of the particles remains almost constant throughout the trap (Fig. 3D) and can be found from conservation of energy (neglecting self-fields) to be $v_e = 1.3 \times 10^6\,\text{m/s}$. Thus, the transit time for a single particle is $\tau_{bounce} = 2 \times L_{trap}/v_e \approx 3.8 \times 10^{-7}\,$s. Taking the injected current $J_N = 5 \times 10^{13}\,e/$s, calculations and simulations are both within 10% experimental value of $2.4 \times 10^7 e$ for the number of trapped electrons at $t = \sim 1\mu$s (the earliest time for which data could be taken). This "equilibrium" value is, in fact, only maintained for a few bounce times. The dynamics quickly becomes more complicated, with the streams of electrons emitted from and returning to the cathode interacting with each other, as is next discussed.

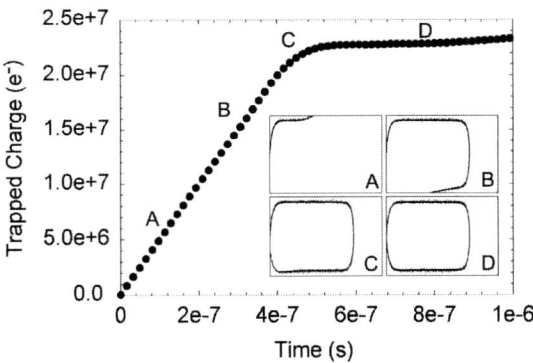

**FIGURE 3.** The total number of trapped electrons as a function of time during the first stage of injection with inset plots of the $z$-$v_z$ phase space. The labels A-D match the times of corresponding phase-space plots. During this first injection stage, no electrons are trapped.

## Two-Stream Trapping

Directly after the initial ballistic trapping, the number of particles in the trap doubles, and an oscillation develops with a period of roughly one bounce time, within $\sim 10$ bounce times. This is shown in Fig. 4 along with the phase space distribution at three different times. The simulation shows that as phase space structure appears, the potential becomes time dependent, and particles are trapped. Energy loss is needed for trapping, since an electron that circuits that trap and returns with the same or more energy will be reabsorbed at the cathode. Collisional losses are, for the plasmas in question, too slow to account for the trapping, and thus collisionless mechanisms must be examined. An obvious candidate is the development of a two-stream interaction between the electrons leaving and returning to the cathode. Indeed, the simulations reproduce the experimental results and indicate that this is the mechanism for initial trapping.

Simulations confirm that a two-stream interaction develops shortly after the initial filling stage. From the phase space plots Fig. 4B,C, it is evident that some electrons lose energy as they circuit the trap. As these electrons are unable to be reabsorbed at the cathode, there results a net increase in the number of trapped electrons. The plasma evolves by continuing to trap fresh electrons and subjecting the previously trapped electrons to further energy loss. Thus, the originally hollow longitudinal phase space begins to fill-in. The dynamic builds on itself, as initially small time-dependent fields grow ever larger.

The importance of two-stream interactions in this stage is demonstrated by numerical studies. We can eliminate two stream interactions by performing a simulation in which twice the charge is emitted from the cathode, but all the electrons are absorbed at the reflecting end of the trap. In this way the charge density is the same as in the the reflecting case, but there is only one stream of electrons. In this case, the electron stream remains quiescent, showing that energy loss and trapping are connected to the existence of two counter propagating streams of electrons. Simulations also indicate growth of Langmuir

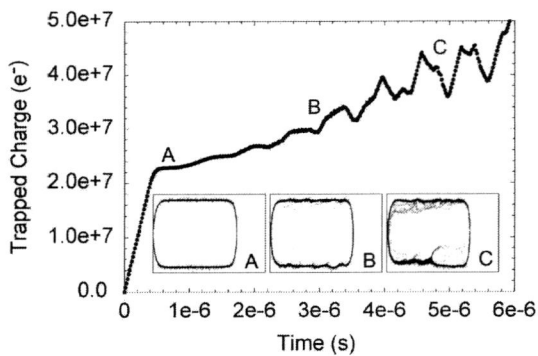

**FIGURE 4.** The total number of trapped electrons as a function of time during the second stage of injection with inset plots of the $z$-$v_z$ phase space. In this second injection stage, electrons are trapped due to a streaming instability.

waves for short traps.

A conventional theory of two-stream instability, including transverse geometry and temperature, yields the result that the two-stream interactions for beams with parameters appropriate for the center region of the trap is stable. Simulations indicate that a possible explanation for developing the instability is that the perturbations in the vicinity of the cathode are unstable. In this region the velocity is rapidly decreasing and the local plasma density is much higher than in the trap, resulting in an enhanced two-stream interaction. Another remarkable simulation result, under further study, is a strong dependence of dynamics on the energy of electrons as they are emitted from the cathode.

## Bubble Stage

After about ten bounce times, the phase space distribution develops a strongly nonlinear perturbation, which has the visual appearance of a 'bubble'. That is, electrons move around the outside of the disturbance, both gaining and losing energy. as shown in Fig. 5. The bubble's velocity is close to the mean electron velocity in the trap. The bubble persists over many round-trips in the trap, reflecting at the cathode and at the far wall. Numerically, we note that more electrons approaching the cathode along with the bubble have gained energy than have lost energy, leading to momentary depletion of the total charge in the trap. Experimentally, this effect is observable by monitoring the image charge on the inject gate cylinder (Fig 6). Each time the bubble reaches the cathode, the image charge in the inject gate dips slightly. Thus, on top of the increase in the trapped charge due to time-dependent potentials, the rotating bubble induces an oscillation in the trapped charge at, roughly, the bounce frequency.

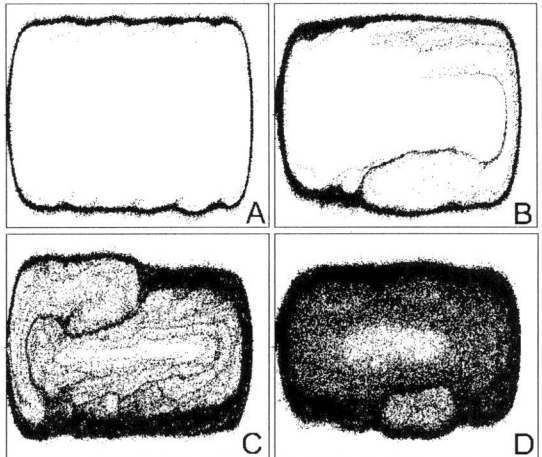

**FIGURE 5.** PIC code simulation of the $z$-$v_z$ phase space at $3\,\mu$s, $6\,\mu$s, $13\,\mu$s, and $26\,\mu$s depicting the development of the bubble instability. Horizontal axis is axial direction with cathode on the left wall and reflecting potential on the right wall (total length of the trap is 36 cm. Axial velocity $v_z$ is on vertical axe, extending from $-1.5 \cdot 10^6 m/s$ to $1.5 \cdot 10^6 m/s$. The bubble decreases in size over the next $200\,\mu$s, but continues to cause significant perturbations to phase space even beyond this time period. A movie of the phase-space evolution is available at
http://bc1.lbl.gov/CBP_pages/wurtele/UncompressedBubbleMovie1.avi.

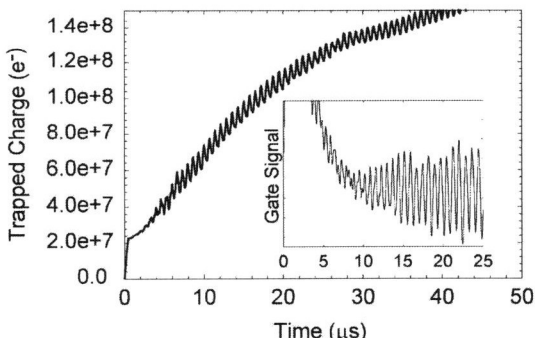

**FIGURE 6.** Plots of the simulation and experimental data (inset) showing the oscillation in charge associated with the bubble instability. The oscillation in the total charge in the trapping region is manifested by a high frequency signal measured on the inject gate.

Experimental observations and simulation show great similarity in the observable properties of the bubble. In both cases, the bubble begins to form around $10\,\mu$s after the start of injection. The bubble grows in size, evidenced by the increased amplitude of charge oscillations. The bubble disappears around $70\,\mu$s after the start of injection, and does not reappear.

# Virtual Cathode Formation and Radial Profile Evolution

Careful study of the dynamics of trapping reveals further complexity, beyond that of large nonlinear perturbations to phase space. As ever more electrons are trapped by energy loss to plasma disturbances, the plasma potential itself becomes a barrier to further trapping. Charges entering the trap are slowed by the space charge potential of the trapped electrons. Simulations show that $\sim 1$cm in front of the photocathode the potential in the center of the plasma becomes negative. When the potential becomes stongly negative, trapping ceases–there is a self-consistent barrier between the trapped region and the untrapped region. This phenomena is familiar in beam physics as a virtual cathode. In the injection proccess, it appears along with the nonlinear phase-space bubble and oscillates in intensity.

From $10\,\mu$s to $70\,\mu$s, the potential of the virtual cathode becomes increasingly negative. Initially, it fully disappears each cycle, and during the interval that it is not present, new charges enter the system and are trapped. When the virtual cathode reappears it is then more negative. At later times, it no longer vanishes but only weakens during an oscillation. The virtual cathode then limits the injection of additional electrons into the trapping region to those with ever greater initial velocity. Eventually, the system reaches a quasi-equilibrium with few electrons entering or exiting the trapping region.

Beyond $\sim 70\,\mu$s, there is a distinct difference between the evolution of the trapped charge for the equipotential cathode and the filament cathode. The filament cathode potential falls radially as $r^2$, due to the spiral design. The same is true of a plasma of constant density. Because of this, the potential profile of the cathode matches the potential profile of the trapped charge at all points. As the virtual cathode grows, it reaches its equilibrium level uniformly over the radius of the electron column.

No match exists for the equipotential photocathode. Here, the virtual cathode begins to form at different times for different radii. The center of the plasma column is more negative than the edge and, thus, the virtual cathode first limits the entrance of new electrons at $r = 0$. Charges continue to enter at larger radii, and the virtual cathode then develops outward across the plasma column. The column develops a hollow profile in an attempt to match the constant cathode potential, as shown in figure 7. Initially, the column is of nearly constant density, owing to the uniformly illumination of the cathode. Over the first 30 $\mu$s of injection, the column density increases from $4 \times 10^6\,\mathrm{cm}^{-3}$ to $1 \times 10^7\,\mathrm{cm}^{-3}$ while maintaining a uniform density cross-section to within 10%. and the column then undergoes hollowing. There is good agreement seen in Fig. 7 between the simulation (right) and experiment (left).

Further work needs to be performed to model the physics of injection at longer timescales, where, for example, the diocotron instability is seen in the hollow plasma described above.

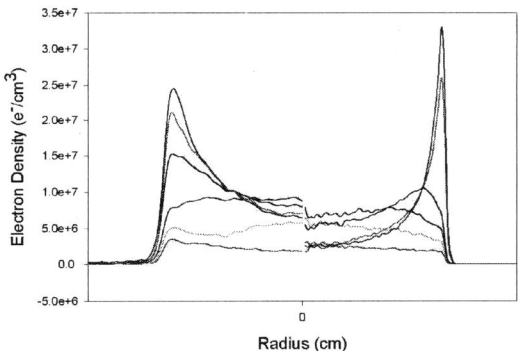

Radial Plasma Density (Experiment and Simulation)

**FIGURE 7.** Plots of the column cross-section at $1\,\mu s$, $10\,\mu s$, $30\,\mu s$, $50\,\mu s$, $500\,\mu s$, $1\,ms$, $2\,ms$ (in order of increasing maximum density) with a $-5\,V$ photocathode bias. Simulation profiles are shown on the left, experimental profiles on the right. Note that the central density is a maximum at $30\,\mu s$. Beyond $30\,\mu s$ the column begins to hollow due to the mismatched cathode and trapped charge potentials.

# ACKNOWLEDGMENTS

We gratefully acknowledge the assistance of Dr. Peter Mardahl on the XOOPIC code and simulations. This work is supported by the National Science Foundation, the Department of Energy, Division of High-Energy Physics, and the Office of Naval Research.

# REFERENCES

1. See the series of NNP Workshop Proceedings published by AIP.
2. Fine, K. S., Driscoll, C. F., Malmberg, J. H., and Mitchell, T. B., *Phys. Rev. Lett.*, **67**, 588 (1991).
3. Hyatt, A. W., Driscoll, C. F., and Malmberg, J. H., *Phys. Rev. Lett.*, **59**, 2975 (1987).
4. Durkin, D., and Fajans, J., *Rev. Sci. Instrum.*, **70**, 4539 (1999).
5. Driscoll, C. F., and Malmberg, J. H., *Phys. Fluids*, **19**, 760 (1976).
6. Verboncoeur, J. P., Langdon, A. B., and Gladd, N. T., *Comp. Phys. Comm.*, **87**, 199 (1995).

# Using Variable Frequency Asymmetries to Understand Radial Transport in a Malmberg-Penning Trap

## D.L. Eggleston

*Occidental College, Physics Department, Los Angeles, California 90041*

**Abstract.** It has long been known that asymmetric electric and magnetic fields produce radial transport in Malmberg-Penning traps, and much work has been done to understand this transport. Our approach is to apply a variable frequency electric asymmetry to a low density population of electrons and to measure the resulting radial particle flux $\Gamma$ as a function of radius $r$. The low particle density eliminates many plasma modes (which have their own frequency dependence) and allows us to focus on the transport physics. The usual azimuthal $E \times B$ drift is maintained by a biased central wire, and this arrangement also allows us to independently vary the drift frequency $\omega_R$ by adjusting either the axial magnetic field $B_z$ or the bias of the central wire $\phi_{cw}$. Up to forty wall sectors are used in order to apply an asymmetry consisting of a single fourier mode $(n, l, \omega)$, where $n$ is the axial wavenumber, $l$ is the azimuthal wavenumber, and $\omega$ is the asymmetry frequency. In the current experiments, we vary $\omega$, $n$, $\phi_{cw}$, and $B_z$. As $\omega$ is varied, the particle flux shows a resonance similar to that predicted by resonant particle theory. The peak frequency of this resonance $f_{peak}$ increases with $\omega_R$ and varies with $n$, in qualitative agreement with theory, but when quantitative comparisons are made the experimental values for $f_{peak}$ do not match those predicted by theory. Instead, the dependence of $f_{peak}$ on $\phi_{cw}$, $B_z$, and $r$ follows simple empirical scaling laws: for inward directed flux, $f_{peak}(\text{MHz}) \approx [-R\phi_{cw}(\text{V})/rB_z(\text{G})]^{1/2}$, where $R$ is the wall radius, and for outward directed flux, $f_{peak}(\text{MHz}) \approx 0.8[-\phi_{cw}(\text{V})/B_z(\text{G})]^{1/2}$. These results may provide guidance for the construction of the correct theory of asymmetry-induced transport.

## INTRODUCTION

It has been known for some time that the confinement in Malmberg-Penning traps is limited by the presence of electric or magnetic fields that break the cylindrical symmetry of the trap. Such asymmetries produce a radial component to the $E \times B$ or grad-B drift that leads to particle loss. This basic understanding is supported by confinement studies [1, 2] as well as experiments with applied electric [3, 4, 5, 6, 7, 8] and magnetic [9, 10] asymmetries.

Many of these early papers also suggested that this asymmetry-induced transport might be described by a theoretical model developed for studies of radial transport in the early tandem mirrors (see, e.g., [11, 12]) where static, asymmetric end cells produced radial grad-B drifts that largely determined the radial particle flux. A key prediction of the theory is that the resulting transport will be dominated by particles whose axial bounce motion and azimuthal drift motion causes them to move in resonance with the asymmetry. As these resonant particles repeatedly encounter the asymmetry they take radial steps in the same direction, thus allowing them to diffuse more quickly than non-

CP692, *Non-Neutral Plasma Physics V*, edited by M. Schauer et al.

resonant particles.

Here we present experiments that test this key prediction of the theory. We do this by applying a variable frequency electrostatic asymmetry to a Malmberg-Penning trap and measuring the resulting radial particle transport as a function of the asymmetry frequency. Our modified trap design avoids previously encountered complications produced by collective effects and allows for a clean test of the transport physics. While the experimental results are qualitatively consistent with theory and seem to confirm the dominant role played by resonant particles, the frequency dependence of the transport does not quantitatively match the predictions of theory.

## ASYMMETRY-INDUCED TRANSPORT THEORY AND EXPERIMENTAL APPROACH

Our experiments are performed in a modified Malmberg-Penning trap in which the plasma has been replaced by a biased wire and the transport of low density test particles is studied. This experimental approach is best understood in the context of asymmetry-induced transport theory, so we begin with a summary of the theory as recently adapted to Malmberg-Penning traps and allowing for electric field asymmetries at a non-zero angular frequency $\omega$ [13]. The theory assumes a cylindrical geometry with an axial magnetic field $B_z$. Asymmetric electric fields are applied by placing voltages on wall sectors. Under these conditions, the resulting radial particle flux for the plateau regime (suitable for small asymmetry amplitudes) is given by (see reference [13] for details)

$$\Gamma_{plateau} = -\sum_{n,l,\omega} \frac{n_0}{\sqrt{2\pi \bar{v}^2}} \frac{L}{|n|} \left| \frac{cl\phi_{nl\omega}(r)}{rB_z} \right|^2 \left[ \frac{1}{n_0} \frac{dn_0}{dr} + \sqrt{2} \frac{n\pi}{L} \frac{r\omega_c}{l\bar{v}} x \right] e^{-x^2}. \quad (1)$$

Here $\phi_{nl\omega}(r)$ is the Fourier amplitude of the asymmetry mode characterized by axial mode $n$, azimuthal mode $l$, and angular frequency $\omega$. For simplicity, we have assumed here that the temperature $T$ is constant with radius. The variable $x$ is equal to $v_{res}/\sqrt{2}\bar{v}$, where

$$v_{res} = \frac{L}{n\pi}(\omega - l\omega_R) \quad (2)$$

is the resonant velocity for the asymmetry mode $n, l, \omega$ and $\omega_R$ is the azimuthal $E \times B$ rotation frequency of the plasma column. The symbols $n_0$, $L$, $\bar{v}$, and $\omega_c$ denote the electron density, plasma length, thermal velocity, and the cyclotron frequency, respectively.

It is worth noting several features of Eq. (1). The radial flux involves a sum over all the asymmetry modes produced by the wall voltages. The square brackets contain a diffusive term $\frac{1}{n_0} \frac{dn_0}{dr}$ and a generalized mobility $\sqrt{2} \frac{n\pi}{L} \frac{r\omega_c}{l\bar{v}} x$ (note that this latter term reduces to $eE/kT$ for $\omega = 0$). The plateau regime flux is proportional to the square of the asymmetry amplitude $\phi_{nl\omega}^2$. Our previous studies of the amplitude scaling [8] of this transport suggest that we are in the plateau regime, but the results of this paper are not dependent upon that identification because both the plateau regime and the banana regime suitable for higher asymmetry amplitudes have the same frequency dependence. The previously mentioned domination of the transport by resonant particles is reflected in the $e^{-x^2}$ factor, which stems from evaluating the Maxwellian distribution function at the resonant velocity. Note that $x$ can be positive or negative since $\omega$ may be greater than or less than $\omega_R$. Here, we use the convention that $\omega > 0$ corresponds to an asymmetry that rotates with the plasma column and $\omega < 0$ corresponds to one that rotates against the column. When the second term in brackets dominates over the first, a static field asymmetry ($\omega = 0$, $x < 0$) will move electrons radially outward ($\Gamma > 0$), but an appropriately chosen asymmetry ($\omega > \omega_R$, $x > 0$) can move particles radially inward. Such inward transport has been observed in "rotating wall" experiments [3, 5, 14, 15, 16].

The presence of $\omega$ in the variable $x$ provides the experimentalist with an ideal way of testing the notion that resonant particles dominate the transport. By varying $\omega$, one can obtain any value of the resonant velocity $v_{res}$ (see Eq. (2)) while keeping other experimental parameters fixed. The resulting radial flux should then exhibit a resonance as $v_{res}$ sweeps through the distribution function. The ability to place $v_{res}$ in the bulk of the distribution function also makes it possible to obtain a measurable amount of radial transport while keeping the asymmetry amplitude low. This approach, however, is complicated by the strong $\omega$-dependence of the asymmetry potential $\phi_{nl\omega}(r)$ in the plasma. Numerical studies for typical plasma parameters [13] show that the transport-producing electric field in the plasma (i.e. $E_\theta = l\phi_{nl\omega}/r$) can vary by many orders of magnitude as adjustments of $\omega$ produce plasma phenomena ranging from standing waves to Debye shielding. These variations in $E_\theta$ (and thus in the radial flux $\Gamma$) tend to dominate or mask those produced by resonant particle effects. This produces, for example, enhanced transport when the asymmetry drives a helical standing wave of the plasma column as was observed in previous experiments [3, 14, 15, 16]. While these collective processes are interesting, they are not essential to the transport physics. In this context we note that these numerical studies also show that the variations of $E_\theta$ with $\omega$ can be reduced as the temperature is increased or the density is reduced.

These considerations led us to employ the modified trap design shown in Fig. 1. The axial magnetic field and negatively biased end cylinders of the standard trap design are retained, but the plasma is replaced by a thin biased wire (0.356 mm diam.) suspended along the axis of the trap. This wire provides a radial electric field to replace the field normally produced by the plasma column and allows low density electrons injected into this device to have the same zeroth-order dynamical motions as those in a typical non-neutral plasma (axial bounce and azimuthal $E \times B$ drift motions). The collective variations of $\phi_{nl\omega}(r)$, however, are minimized since the lower density ($10^5$ cm$^{-3}$) and higher temperature (4 eV) of the electrons give a Debye length ($\lambda_D = 4.7$ cm) larger than the trap radius ($R = 3.82$ cm). Under these conditions, the applied asymmetry potentials are essentially the vacuum potentials and their $\omega$-dependence is eliminated. In short, we have constructed a trap where the electrons will act as test particles moving in the prescribed fields. Despite these changes, the confinement time scaling with no applied asymmetries [17] shows the same $(L/B_z)^2$ dependence found in higher density experiments [1, 2], supporting the notion that the radial transport is primarily a single particle effect.

Experiments studying asymmetry-induced transport typically use azimuthally sectored cylinders in the confinement region to apply asymmetric electric fields. In our device, we have sectored the entire confinement region (five cylinders, labeled S1 through

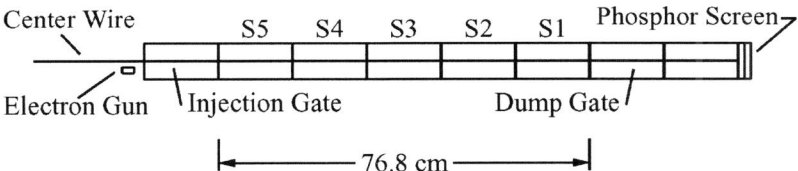

**FIGURE 1.** Schematic of the Occidental Trap. The usual plasma column is replaced by a biased wire that maintains the basic dynamical motions of low density electrons injected from an off-axis gun. The low density and high temperature of the injected electrons largely eliminates collective modifications of the vacuum asymmetry potential. The five cylinders labeled S1 through S5 are divided azimuthally into eight sectors. These forty wall sectors allow for the application of asymmetries consisting of essentially one Fourier mode.

S5 in Fig. 1, with eight azimuthal divisions each), for two reasons. The first reason is to ensure that the applied potentials stay small enough for the theory to be valid. When the potential $\phi_W$ is applied on a single sector of length $L_s$, the amplitude of the Fourier modes will be proportional to $(L_s/L)\phi_W$, where $L$ is the length of the plasma. The smaller $L_s/L$ is, the larger the wall potential necessary to produce a mode $\phi_{nl\omega}$ of given amplitude and thus a given amount of transport. The amplitude of the wall potential, however, is not unrestricted: linear theory assumes the trajectories of the electrons are not radically different from the unperturbed case, and thus requires $e\phi_W \ll kT_e$. In order to satisfy the theoretical assumptions while producing an observable amount of transport, it is thus advantageous to sector the entire confinement region.

The second reason for our modifications to the confinement region is to allow a clean, unambiguous test of theory. As previously noted, the theoretical expressions for the radial flux involve a sum over Fourier modes $n$, $l$, and $\omega$, with each mode contributing to the total transport. Experimental measurement of the flux, however, produces a single number $\Gamma$. Comparison to a theoretical expression that involves a sum over terms will therefore always be somewhat ambiguous since the combination of terms producing a given flux value is not unique. The least ambiguous case would involve a single Fourier mode, but this requires many wall sectors. Forty sectors is a number that can be reasonably handled and represents a great improvement over previous experiments. By judiciously selecting the amplitude and phase of the signals applied to each sector, we can produce an asymmetry that is essentially a single Fourier mode, with higher harmonics typically having amplitudes smaller than 10% of this mode's. For these experiments, we used a helical standing wave of the approximate form

$$\phi(r,\theta,z,t) = \phi_W \frac{r}{R} \cos\left(\frac{n\pi z}{L}\right) \cos(l\theta - \omega t) \tag{3}$$

where $\phi_W$ is the asymmetry potential at the wall (typically 0.2 V), $R$ is the wall radius (3.82 cm), $L$ is the length of the confinement region (76.8 cm), and $z$ is measured from one end of the confinement region. This asymmetry, which decomposes into oppositely propagating helical modes, will allow particles to maintain resonance with the asymmetry when they bounce off the ends of the trap and change direction. For most of this work, the wall sectors are configured to give an asymmetry where the $n = 1, l = 1$ mode is dominant. By adjusting the relative phase of the applied signals, the asymmetry can be made to rotate either with the zeroth-order azimuthal drift ($\omega > 0$) or against it ($\omega < 0$).

With the frequency dependence of $\phi_{nl\omega}(r)$ eliminated by increasing the Debye length and the sum over modes removed by the strategic use of multiple wall sectors, the expression for the flux can be simplified to

$$\Gamma = -C\,[A + Bx]\,e^{-x^2} \tag{4}$$

where $A = dn_0/dr$, $B = \sqrt{2}\,n_0 r w_c n\pi/(l\bar{v}L)$, and $C$ contains the remaining factors in Eq. (1). The remaining frequency dependence is contained in the normalized resonance velocity $x$. As the asymmetry frequency is varied, the flux will have extrema when $d\Gamma/dx = 0$. Applying this to Eq.(4) and solving for $x$ gives

$$x_{peak} = \frac{1}{2}\left[-\frac{A}{B} \pm \sqrt{\left(\frac{A}{B}\right)^2 + 2}\,\right] \tag{5}$$

or

$$\omega_{peak} = l\omega_R + \frac{\sqrt{2}}{2}\frac{n\pi}{L}\bar{v}\left[-\frac{A}{B} \pm \sqrt{\left(\frac{A}{B}\right)^2 + 2}\,\right] \tag{6}$$

Note that the solutions for $x_{peak}$ depend only on the ratio $A/B$. The two solutions correspond to two flux peaks of opposite signs, with the plus sign corresponding to flux minima and the minus sign corresponding to flux maxima. Because of the Gaussian dependence on $x$ in Eq. (4), often only one of these peaks will be sizable.

The remaining features of the trap have been discussed in detail elsewhere [17, 18]. Electrons injected into the trap from an off-axis gun are quickly dispersed into an annular distribution. At a chosen time (here, 1600 ms after injection), the asymmetries are switched on for a period of time $\delta t$ (here, 100 ms) and then switched off. At the end of the experiment cycle, the electrons are dumped axially onto a phosphor screen and the resulting image is digitized using a $512 \times 512$ pixel charge-coupled device camera. A radial cut through this image gives the density profile $n_0(r)$ of the electrons, where calibration is provided by a measurement of the total charge being dumped. Profiles are taken both with the asymmetry on and off, and the resulting change in density $\delta n_0(r)$ is obtained. If the asymmetry amplitude is small enough and the asymmetry pulse length $\delta t$ short enough, then $\delta n_0(r)$ will increase linearly in time. We may then approximate $dn_0/dt \simeq \delta n_0(r)/\delta t$ and calculate the radial particle flux $\Gamma(r)$ (assuming $\Gamma(r = a) = 0$):

$$\Gamma(r) = -\frac{1}{r}\int_a^r r'dr' \cdot \frac{dn_0}{dt}(r') \tag{7}$$

Here $a$ is the radius of the central wire. The entire experiment is then repeated for a series of asymmetry frequencies and the resulting flux vs. radius and frequency data saved for analysis.

# EXPERIMENTAL RESULTS

A typical result is shown in Fig. 2 where we plot the radial flux $\Gamma$ vs. asymmetry frequency $f$ for four selected radii. The radial density profile is shown in the inset. There are several things to note in this figure. First, note that the predominant flux peaks occur only for positive frequencies. This is in marked contrast to other experiments [3, 14, 15, 16] where driven plasma modes (which occur for both positive and negative frequencies) dominated the transport and produced flux peaks for both positive and negative frequencies. The data thus supports the conclusion that we have effectively limited the role of collective modes in the experiment. Second, the fact that the flux peaks occur only for positive frequencies is in qualitative agreement with the theory. As Eq. (4) shows, the frequency dependence of the theoretical flux is constrained by the factor $e^{-x^2}$, a Gaussian curve centered where $\omega = l\omega_R$. Since $\omega_R$ is a positive quantity and the Gaussian width $\Delta f = \sqrt{2}n\bar{v}/L$ here is about 1.5 MHz, the flux produced by negative frequencies is expected to be small. Finally, note that both positive and negative fluxes are observed and that the flux peaks occur at different frequencies. This also seems to be qualitatively consistent with the theory. When the density gradient is large, the first term in the square brackets of Eq. (4) will dominate and we expect a bell-shaped curve, opposite in sign to $dn_0/dr$, centered around $\omega = l\omega_R$. This behavior is shown by the curves for $r/R$ equal to 0.19, 0.30 and 0.52 . Note that in our experiment $\omega_R$ is set by the center wire bias $\phi_{cw}$ and decreases with radius

$$\omega_R = \frac{-c\phi_{cw}}{r^2 B_z \ln(R/a)} \tag{8}$$

and thus it is expected that the flux resonances will shift to lower frequencies with radius, as observed. Near the top of the density profile, the gradient is near zero, so the second term in the square brackets of Eq. (4) will dominate. We then expect an $xe^{-x^2}$ behavior, consistent with the shape of the $r/R = 0.39$ curve.

To further check that the resonances are associated with $\omega_R$, we have varied the center wire bias $\phi_{cw}$ and the axial magnetic field $B_z$. As expected, the flux resonances maintain their general shapes but shift to higher frequencies as the magnitude of $\phi_{cw}$ is increased and to lower frequencies as $B_z$ is increased (see reference [19] for details).

As seen in Eq. (6), the theory also predicts a variation of $f_{peak}$ with axial mode number $n$. To check this, we applied $n = 2$ and $3$ asymmetries to the wall sectors. The results are shown in Fig. 3 along with data from the $n = 1$ configuration. Here we use an alternate method of displaying the frequency dependence of the flux: we plot the frequency at which the flux has an extremum, $f_{peak}$, versus radius. Data corresponding to both positive and negative flux peaks are shown. The points in the upper left portion of the graph give the frequencies of the negative flux peaks (flux minima) while the points in the lower right portion of the graph correspond to the positive flux peaks (flux maxima). The fall off of $f_{peak}$ with radius is clear, as expected from the dependence of $\omega_R$ on $r$. The upward shift of $f_{peak}$ with $n$ is also clear, and the inset of the figure shows that $f_{peak}$ increases linearly with $n$ for three representative radii. The $n$-dependence of Eq. (6) is not simple (note that $B \propto n$), but deviates only slightly from linearity for experimental values of $A/B$ and $n$ ($-1 < nA/B < 2$). The observed $n$-dependence is thus consistent

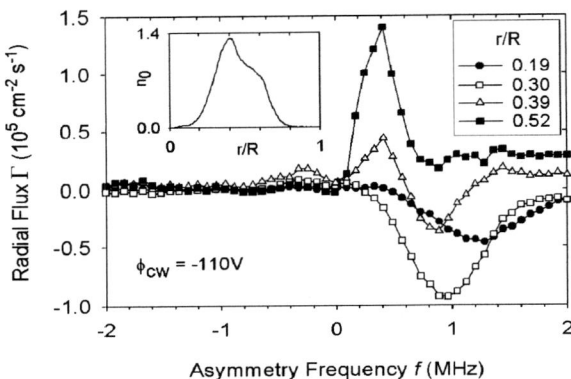

**FIGURE 2.** Radial particle flux at four selected radii as a function of asymmetry frequency for center wire bias $\phi_{cw} = -110$ V, magnetic field $B = 364$ G, and Fourier mode numbers $n = 1, l = 1$. The shapes of the flux curves are qualitatively consistent with that expected from theory. Positive frequencies correspond to experiments where the asymmetry is rotating in the direction of the electron's azimuthal $E \times B$ drift and negative frequencies correspond to counter-drift rotations. The electron density $n_0$ ($10^5$ cm$^{-3}$) versus scaled radius $r/R$ is shown in the inset.

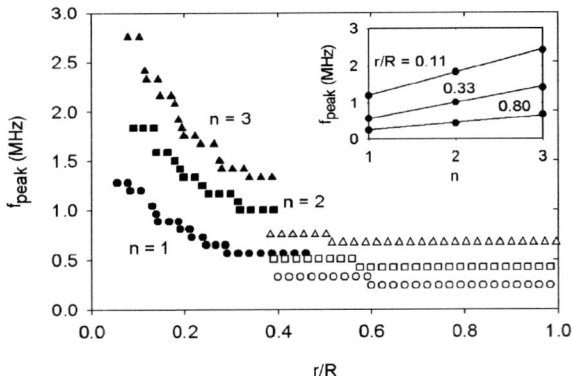

**FIGURE 3.** $f_{peak}$ versus $r$ for three values of the axial mode number $n$. The filled symbols in the upper left portion of the graph give the frequencies at which the flux is a minimum while the open symbols in the lower right portion of the graph correspond to frequencies of the flux maxima. A linear increase of $f_{peak}$ with $n$ is shown in the inset for three representative values of scaled radius $r/R$. Here, $\phi_{cw} = -146$ V and $B_z = 364$ G.

46

with Eq. (6) for the flux minima data. For the flux maxima data, however, Eq. (6) predicts that $f_{peak}$ will decrease with $n$, in contrast to the observed increase with $n$. Nevertheless, the dependence of $f_{peak}$ on $n$ shows that the transport we are studying depends on axial variation in the asymmetry, in contrast to earlier work on radial transport induced by $n = 0$ diocotron waves [20]. We have also verified that $f_{peak}$ shifts upward appropriately for $l = 2$ asymmetries.

The dependence of $f_{peak}$ on $\phi_{cw}$, $B_z$, and $r$ obeys an empirical scaling law, as shown in Fig. 4. Data with values of $\phi_{cw}$ from -20 V to -140 V and $B_z$ values from 243 G to 607 G were scaled according to $f_{scaled} = f_{peak}\sqrt{\frac{B_z}{-\phi_{cw}}}$ and plotted versus $r/R$. The number of radial points plotted for each case was reduced for clarity. Although there is some scatter in the data, a fairly good universal curve is formed. The frequencies for the flux minima also scale inversely with the square root of the radius, as shown by the solid line. The scaling law for these points is thus

$$f_{peak}(\text{MHz}) = \sqrt{\frac{-\phi_{cw}(\text{V})}{B_z(\text{G})}\frac{R}{r}}.$$

The frequencies for the flux maxima decrease slightly with radius, but the scatter in the data is too large to validate a particular radial dependence. If we ignore the radial variation, we obtain the rough scaling law

$$f_{peak}(\text{MHz}) = 0.8\sqrt{\frac{-\phi_{cw}(\text{V})}{B_z(\text{G})}}.$$

As we have seen, the parametric dependence of $f_{peak}$ on $\phi_{cw}$, $B_z$, $r$, and, for the flux minima, $n$ is qualitatively consistent with theory (i.e. $f_{peak}$ increases and decrease appropriately). A quantitative comparison of experiment and theory, however, reveals serious discrepancies. To illustrate this, we have used experimental values to evaluate Eq. (6) and found $f_{peak}$ as a function of radius $r$ for the case where $n = 1, l = 1, B_z = 364$ G and $\phi_{cw}$ = -146 V. The results are shown in Fig. 5 along with the experimental data and the calculated $f_R$. While the experimental frequencies for the flux minima (solid circles) match the theory for $r/R \simeq 0.4$, they clearly diverge from theory at smaller radii. More seriously, the theory for these negative flux peaks constrains $f_{peak}$ to be greater than the rotation frequency $f_R$, but the experimental data clearly crosses the $f_R$ line, shown dotted in the figure. Flux minima data for $n = 2$ and 3 show similar behavior and a similar level of agreement. The data for the frequencies of the flux maxima (open circles) do not match the theory at any point. The theory for the positive flux peak shows $f_{peak}$ going smoothly through zero to include negative values, while the experimental values always remain positive. Finally, as noted above, the theory has $f_{peak}$ for the flux maxima decreasing with $n$, whereas the experiment shows an increase with $n$.

There are several simplifying assumptions made in the theory which might be invoked to account for the discrepancy between theory and experiment. The theory used here assumes that the plasma particles specularly reflect at the ends of the trap and that this reflection point is the same at all radii (i.e. $L$ is not a function of radius). The theory also assumes that the rotation frequency $\omega_R$ is not a function of axial position $z$. While these assumptions are clearly violated in our experiment, our estimates of these effects give corrections that are too small to account for the observed discrepancies.

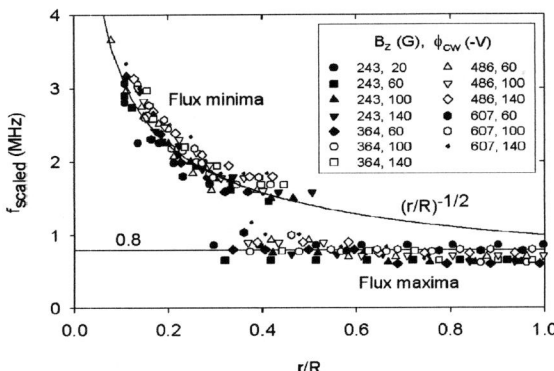

**FIGURE 4.** Empirical scaling of $f_{peak}$. Data having various values of $B_z$ and $\phi_{cw}$ were scaled according to $f_{scaled} = f_{peak}\sqrt{\dfrac{B_z}{-\phi_{cw}}}$ and plotted versus the scaled radius $r/R$. Solid lines show simple "universal" curve fits to the resulting data points.

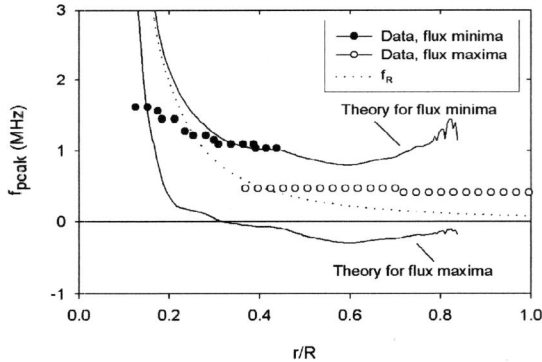

**FIGURE 5.** Comparison of experimental and theoretical values for $f_{peak}$. Experimental density profiles are used in Eq. (6) to produce the theory curves shown by the solid lines. Experimental data is also shown for this case, where $B_z = 364$ G and $\phi_{cw} = -146$ V. The rotation frequency $f_R$ is shown by the dotted line for comparison.

The theory used here also assumes that the plasma temperature $T$ is constant with radius. A radial temperature variation would add a term $\frac{n_0}{T}\frac{dT}{dr}\left(x^2 - \frac{1}{2}\right)$ to the square brackets of Eq. (4) and thus change the theoretical predictions for $f_{peak}$. Measurements of $T(r)$ in our experiment, however, show that this correction would also be too small to account for the discrepancies.

# CONCLUSION

We have measured the frequency dependence of asymmetry-induced transport under very simple experimental conditions and compared the results to resonant particle theory. Our results are qualitatively consistent with the theory and support the idea that resonant particles dominate the transport, but the quantitative predictions of the simple theory employed here do not match the experiments. Apparently, the current theory does not give a complete description of this transport.

# ACKNOWLEDGMENTS

This work was supported by U.S. Department of Energy grant No. DE-FG03-98ER54457. The contributions of Brenda Carrillo and Brian Fowler are gratefully acknowledged.

# REFERENCES

1. C. F. Driscoll and J. H. Malmberg, Phys. Rev. Lett. **50**, 167 (1983).
2. C. F. Driscoll, K. S. Fine, and J. H. Malmberg, Phys. Fluids **29**, 2015 (1986).
3. D. L. Eggleston, T. M. O'Neil, and J. H. Malmberg, Phys. Rev. Lett. **53**, 982 (1984).
4. J. Notte and J. Fajans, Phys. Plasmas **1**, 1123 (1994).
5. X.-P. Huang, F. Anderegg, E. M. Hollman, C. F. Driscoll, and T. M. O'Neil, Phys. Rev. Lett. **78**, 875 (1997).
6. D. L. Eggleston, in *Non-Neutral Plasma Physics III*, edited by John J. Bollinger, Ross L. Spencer, and Ronald C. Davidson, (American Institute of Physics, Melville, NY, 1999), p. 241.
7. J. M. Kriesel and C. F. Driscoll, Phys. Rev. Lett. **85**, 2510 (2000).
8. D. L. Eggleston and B. Carrillo, Phys. Plasmas **9**, 786 (2002).
9. D. L. Eggleston, J. H. Malmberg, and T. M. O'Neil, Bull. Am. Phys. Soc. **30**, 1379 (1985).
10. E. Gilson and J. Fajans, in *Non-Neutral Plasma Physics IV*, edited by Francois Anderegg, C. Fred Driscoll, and Lutz Schweikhard, (American Institute of Physics, Melville, NY, 2002), p. 378.
11. D. Ryutov and G. Stupakov, Sov. J. Plasma Phys. **4**, 278 (1978).
12. R. Cohen, Comments Plasma Phys. Cont. Fusion **4**, 157 (1979).
13. D. L. Eggleston and T. M. O'Neil, Phys. Plasmas **6**, 2699 (1999).
14. J. H. Malmberg, C. F. Driscoll, B. Beck, D. L. Eggleston, J. Fajans, K. Fine, X.-P. Huang, and A. W. Hyatt in *Non-Neutral Plasma Physics*, edited by C. W. Roberson and C. F. Driscoll, (American Institute of Physics, Melville, NY, 1988), p. 28.
15. F. Anderegg, E.M. Hollmann, and C.F. Driscoll, Phys. Rev. Lett. **81**, 4875 (1998).
16. E.M. Hollmann, F. Anderegg, and C.F. Driscoll, Phys. Plasmas **7**, 2776 (2000).
17. D. L. Eggleston, Phys. Plasmas **4**, 1196 (1997).
18. D. L. Eggleston, Phys. Plasmas **1**, 3850 (1994).
19. D. L. Eggleston and B. Carrillo, Phys. Plasmas **10**, 1308 (2003).
20. J. S. DeGrassie and J. H. Malmberg, Phys. Rev. Lett. **39**, 1077 (1977).

# Diocotron Instability in ELTRAP

G. Bettega*, F. Cavaliere*, M. Cavenago†, F. De Luca**, I. Kotelnikov‡, R. Pozzoli**, M. Romé** and Yu. Tsidulko‡

*I.N.F.M., Dipartimento di Fisica, Università degli Studi di Milano, Milano, Italy
†I.N.F.N., Laboratori Nazionali di Legnaro, Legnaro, Italy
**I.N.F.M., I.N.F.N., Dipartimento di Fisica, Università degli Studi di Milano, Milano, Italy
‡Budker Institute of Nuclear Physics, Novosibirsk, Russian Federation

**Abstract.** Electrostatic modes in an electron plasma confined in the Penning-Malmberg trap EL-TRAP (Physics Department of Milan University) are investigated, analyzing the electric signals on the electrodes, and using the CCD diagnostics. In different experimental conditions, a well-defined mode (possibly the $m = 1$ diocotron mode) is found, with a frequency proportional to $(Q/B)$ ($Q$ is the total charge, $B$ the magnetic field strength). The amplitude starts to increase at the beginning of the hold phase of the cycle, reaches a maximum, and then decreases to the background noise level. The decrease of the amplitude corresponds to a decrease of the frequency, indicating plasma loss at the wall. A fast rise of the potential barrier easily excites the mode. Measurements performed with increasing ramp times show that the growth rate decreases, and at a long enough ramp time the instability does not arise.

## ELECTROSTATIC DIAGNOSTICS

The operation of the trap [1] consists in the standard inject-hold-dump cycle: the electron plasma, generated by a thermionic source, is injected in the confinement volume by a small accelerating voltage (a few Volt) and trapped on the bottom of a deep (tens of Volt) potential well. During the hold phase the oscillations of the plasma are monitored detecting the induced charge signals at the walls. These plasma modes are analyzed using smooth and sectored cylinders, as shown in the Fig. 1. At the end of the hold phase the plasma is dumped onto a phosphor screen and the produced light is collected by a CCD camera which gives a two-dimensional $z$-integrated picture of the plasma. The design of the electrostatic probes strictly relates to the kind of charge oscillation under investigation: the $\pi$-sectored (S2) cylinder is sensitive to the plasma off-axis rotation, the $\pi/2$-sectored one (S4) to the quadrupole $m = 2$ perturbation. The S2 and S4 sectored electrodes show less sensitivity towards higher odd and even wave numbers modes respectively; the axial oscillations at the plasma frequency are studied by means of the induced charge signals on the the smooth cylinders.

## MEASUREMENTS ON THE DIOCOTRON MODE

At the end of the injection phase the plasma column is detached from the electron source by the negative voltage applied to the confining cylinder. Plasma oscillations are

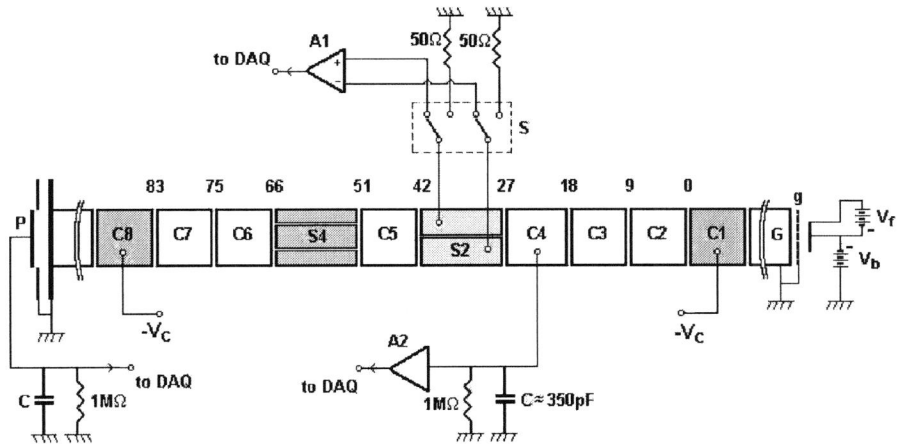

**FIGURE 1.** Overall schematic of the electrostatic diagnostics: the sectored cylinders S2 and S4 provide sensibility to the azimuthal modes of oscillation. The source connections (power supplies for heating and biasing) and a couple of amplification circuits are shown. The software driven relé S connects the sectors for a very short time during the measurements in order to avoid resistive effects. An aluminum layer on the phosphor screen is used as a charge collector (the CCD camera and the phosphor high voltage polarization circuits are not shown in the figure).

observed when the potential on the plug-in electrode (the cylinder on the source side) is raised from zero to the end value $V_c$ with a given slew rate. In our experiments we detect axial charge oscillations connecting a low noise and high gain amplification circuit to a smooth confining cylinder: using software filtering techniques and FFT transforming the signal, we get a high amplitude peak in the frequency domain, well above the noise spectrum. The time evolution of frequency and amplitude of this signal are followed in real time. The frequency of the signal is of the order of tens of kHz and depends on the ratio $(n/B)$ as shown in Fig. 2a. The observed $(n/B)$ dependence is typical of the plasma diocotron perturbations. From the measurements of the plasma density performed with the charge collector the diocotron frequency is obtained usign the expression [2]

$$\nu_{m=1} = \frac{\lambda_p}{4\pi^2\varepsilon_0 R_W^2 B} = \frac{e}{4\pi\varepsilon_0} \frac{R_P^2}{R_W^2} \frac{n}{B} \tag{1}$$

The $(n/B)$ dependence of the frequency of the signal and the agreement of its values with Eq. (1) (see Fig. 2b) demonstrate that the detected axial charge oscillations are caused by the rotation of the column: this is due to the three-dimensional nature of the perturbation arising in a finite length column.

The frequency shows an initial plateau and then monotonically decreases until disappears in the background noise. Starting from the beginning of the hold phase the amplitude increases up to end of the frequency plateau, then it decreases together with the frequency. These time dependent behaviors are shown in Fig. 3a and 3b. The saturation

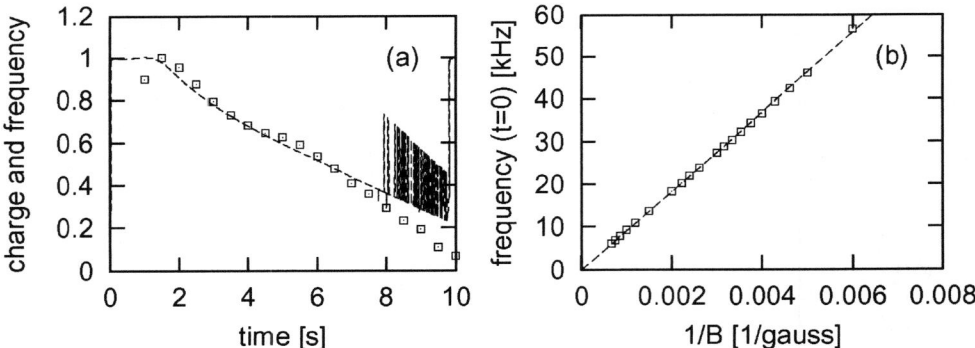

**FIGURE 2.** (a) Comparison between the frequency of the charge oscillation and the total confined charge. The time evolution of the frequency is obtained with a single plasma. The total charge data are obtained dumping many plasmas held for increasing trapping times. (b) Frequency of the mode at the beginning of the hold phase. The data show that the frequency is inversely proportional to the magnetic field.

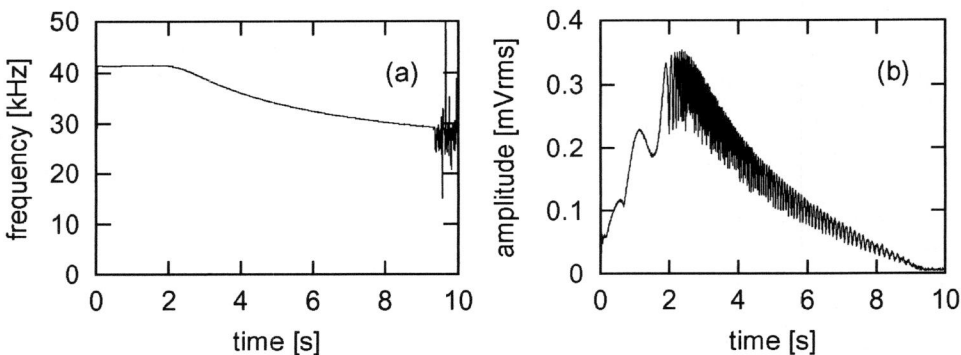

**FIGURE 3.** (a) Time evolution of the frequency of the plasma off-axis rotation. (b) Time evolution of the amplitude of the mode. The data refer to a 30 cm long plasma. The magnetic field strength is 330 gauss.

of the mode and its decreasing frequency indicate particle loss at the walls. This occurrence is confirmed by the optical diagnostics: at the end of the frequency plateau the plasma touches the walls. As shown in Fig. 3b the instability occurs on the time scale of seconds and shows a trend characterized by linear growth rate with superimposed very low frequency oscillations.

Studying the effects of the impedance of the amplification circuit on the plasma life time, we have experimentally tested that the observed instability is not a *resistive wall* effect [3]. While an external impedance connected to a sectored electrode easily excites the diocotron mode and drives the plasma towards the walls in a very short time, in our case the input impedance of the measurement circuit that grounds the smooth cylinder during the experiment, does not change the life time of the plasma.

The experiments with variable magnetic fields show that the growth rate of the

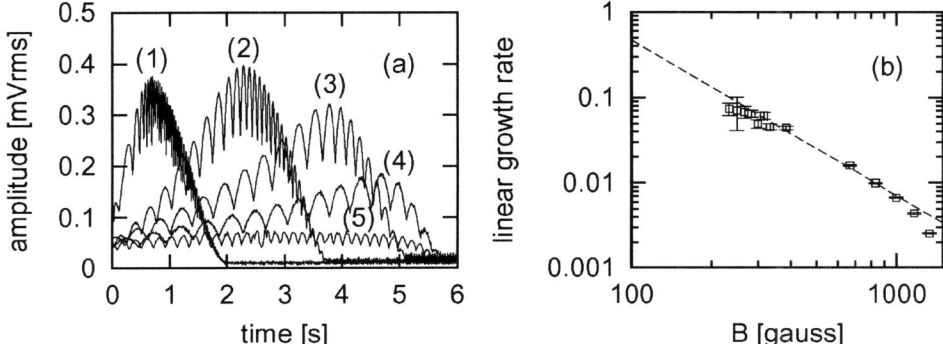

**FIGURE 4.** (a) Time behavior of the amplitude of the detected signal versus the magnetic field strength. For high B (>1.5 kgauss) the oscillation has a nearly constant amplitude. Data are taken with a 50 cm plasma. (b) The data on the linear growth rate are well fitted by a $B^{-2}$ law (the reported data refer to a plasma length of 75 cm).

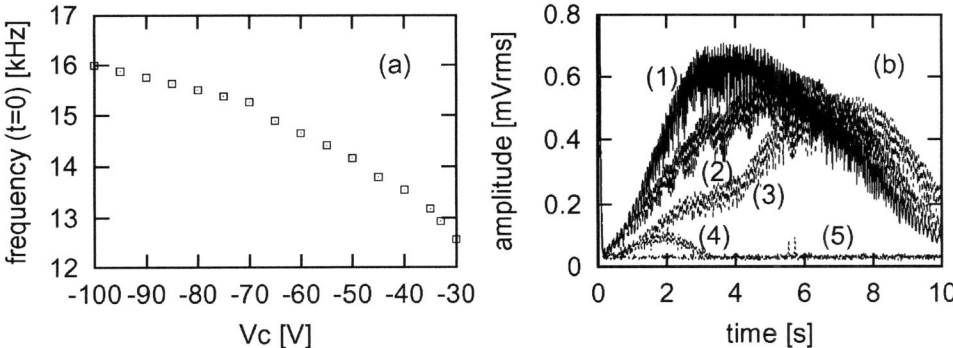

**FIGURE 5.** (a) Dependency of the rotation frequency taken at t=0 on the confining voltages. (b) The effect of the rise time of the confining voltage on the instability. (1) 4 ms ramp; (2) 10 ms ramp; (3) 15 ms ramp; (4) 20 ms ramp; (5) 25 ms ramp.

instability decreases with the magnetic field: for high values of the magnetic field strength (>1.5 kgauss) the amplitude remains nearly constant for the whole plasma life (see Fig. 4a). We have found that the linear growth rate of the instability follows a $B^{-2}$ law for all the available plasma lengths (see Fig. 4b).

We have investigated the effects of time and amplitude variations of the confining voltages on the charge oscillation. The simplest effect is the variation of the frequency with the confining potential $V_c$: as shown in Fig. 5a, a higher value of $V_c$ produces a shorter and denser plasma with higher rotation frequency. The most important effect comes from the application of a ramp potential to the confining cylinder at the end of the injection phase. We found that the maximum amplitude of the instability is related to the ramp time as shown in Fig. 5b. Increasing the ramp time the maximum amplitude of the mode is reduced and using a very slow ramp on the plug the instability does not develop.

# ACKNOWLEDGMENTS

The authors are grateful to G. Malvezzi for his valuable help in the construction of various mechanical parts of the ELTRAP device.

# REFERENCES

1. M. Amoretti, G. Bettega, F. Cavaliere, M. Cavenago, F. De Luca, R. Pozzoli, and M. Romé, Rev. Sci. Inst. (2003), to be published.
2. R. C. Davidson, *"An Introduction to the Physics of Nonneutral Plasmas"* Addison-Wesley, Redwood City, USA (1990).
3. W. D. White, J. H. Malmberg and C. F. Driscoll, Phys. Rev. Lett. **49**, 1822 (1983).

# Evaluation of the axial loss rate of a plasma confined by a Malmberg-Penning trap

Yongbin Chang and C. A. Ordonez

*Department of Physics, University of North Texas, Denton, Texas 76203*

**Abstract.** A theoretical approach is presented for predicting the collision-based axial loss rate of a plasma confined by a Malmberg-Penning trap, just after the axial well depth is made shallower. The assumptions used in developing the approach include a square well, uniform plasma properties, a particle mean-free-path that is much larger than the plasma dimensions, and a velocity distribution that is Maxwellian except within the regions of direct loss in velocity space. Example predictions are provided, and comparisons with existing theories are made.

A Malmberg-Penning trap produces an electric potential well, which can serve to provide electric confinement of a single plasma species in one dimension, and a uniform magnetic field, which serves to provide plasma confinement in the other two dimensions [1, 2]. Suppose that a Malmberg-Penning trap confines a single plasma species that is thermally relaxed, and the electric potential well is suddenly made shallow enough for collision-based axial transport to be the dominant plasma loss mechanism temporarily. Such a condition may exist when obtaining an overlap of oppositely signed plasma species in traps having a nested configuration [3], which have been used to produce antihydrogen [4, 5]. Most of the existing theories that might be used for predicting the associated loss rate are reviewed in Ref. [6]. The theories, which were developed primarily for describing hot fusion plasmas in magnetic mirrors, typically apply in the limit that the plasma temperature (in energy units) is small compared to the depth of the one-dimensional potential energy well. Here, a theory is presented that applies for any ratio of the plasma temperature to the depth of the potential energy well following a sudden reduction in the magnitude of the depth of the axial electric potential well (just after non-collisional direct axial losses become negligible).

Two interacting species of plasma particles are considered, a test particle species and a field particle species. All particles are considered to be point particles that experience successive, independent, binary, elastic, central-force collisions. If $N$ test particles are trapped in a well, the test-particle loss rate is given by

$$\frac{dN}{dt} = \int n\, n_{\mathrm{F}}\, \langle \sigma g \rangle_{(v, v_{\mathrm{F}})}\, dV_w, \tag{1}$$

where the integral is a volume integral over the volume contained by the well, $n$ and $n_{\mathrm{F}}$ are the test and field particle densities, respectively, $\langle\ \rangle_{(v, v_{\mathrm{F}})}$ denotes an average over the velocity distributions of the test particles and field particles, respectively, $\sigma = \sigma(v, v_{\mathrm{F}})$ is the cross section for a collision between a test particle and a field particle to cause the test particle to be lost from the well, $v$ is the test particle velocity before the collision,

CP692, *Non-Neutral Plasma Physics V*, edited by M. Schauer et al.

$v_F$ is the field particle velocity before the collision, and $g = |v - v_F|$ is the relative speed of the two particles before the collision. Except where noted, the test particle species and the field particle species are treated as separate plasma species so that the theory can be applied to two interacting plasma species. To apply the theory to a single plasma species interacting with itself, $n_F$ is set equal to $n$, and all other properties of the two species are considered to be the same. For simplicity, the well in which the test particles are confined is approximated as a square well, and uniform plasma properties are considered. Also, it is assumed that the test particle mean free path is much larger than the plasma dimensions, and $\langle \sigma g \rangle_{(v,v_F)}$ has approximately the same value at any location in the well. Then, the loss rate is characterized by

$$\dot{n} = \frac{dn}{dt} = \frac{1}{V_w}\frac{dN}{dt} = n n_F \langle \sigma g \rangle_{(v,v_F)}, \tag{2}$$

where $V_w$ is the volume contained by the well.

To derive an expression for $\sigma$, it is helpful to consider a differential density $dn$ of test particles, which have the same velocity, $v$, and a differential density $dn_F$ of field particles, which have the same velocity, $v_F$. The differential loss rate is $d\dot{n} = dn\,dn_F\,\sigma(v,v_F)\,g(v,v_F)$. In an inertial frame of reference moving at velocity $v_F$ with respect to the laboratory frame, the field particles are at rest, and the differential loss rate becomes $d\dot{n} = dn\,dn_F\,\sigma(g,0)\,g(g,0)$, where $g = v - v_F$. Define a cylindrical coordinate system that is at rest with respect to the field particles and that is associated with a set of unit vectors denoted $\hat{b}'$, $\hat{\phi}'$, and $\hat{k}'$. Take the direction of $\hat{k}'$ to be parallel to $g$, and consider a single collision between a test particle and a field particle. Choose the origin of the coordinate system such that if the test particle is located at the origin just before the test particle begins to interact with the field particle, then the test particle and the field particle would experience a head-on collision. Suppose that just before the test particle and the field particle begin to interact, the test particle is located at cylindrical coordinates $(b, \phi, 0)$. Then, $b$ is the impact parameter for the collision. It should also be noted that $\phi$ is equally likely to have any value between 0 and $2\pi$. The cross section for the collision to cause the test particle to be lost from the well is expressed as

$$\sigma = \int_0^{2\pi} \int_0^{b_{max}} \Theta_{esc}\, b\, db\, d\phi = \int_0^{2\pi} \int_0^{b_{max}} \Theta_{esc}\, \sigma_{max}\, f_b(b)\, f_\phi(\phi)\, db\, d\phi = \langle \Theta_{esc}\, \sigma_{max} \rangle_{(b,\phi)}. \tag{3}$$

Here, $\Theta_{esc}$ is defined to equal unity if the collision would cause the test particle to be lost from the well and zero otherwise, $b_{max}$ is defined to have a value that can be arbitrarily chosen but which must be larger than the largest impact parameter that can cause the test particle to be lost as a result of a collision, $\sigma_{max} = \pi b_{max}^2$, $f_b(b) = 2b/b_{max}^2$ is the impact parameter distribution function normalized to unity, and $f_\phi(\phi) = 1/(2\pi)$ is the azimuthal angle distribution function normalized to unity. With Eq. (3), the differential loss rate is written as $d\dot{n} = dn\,dn_F\,\langle \Theta_{esc}\,\sigma_{max}\rangle_{(b,\phi)}\,g$. The differential density of the test particles is written as $dn = nf(v)dv$, where the test particle velocity distribution function, $f(v)$, is normalized to unity. The differential density of the field particles is written in a similar form as $dn_F = n_F f_F(v_F)dv_F$, where the field particle velocity distribution function, $f_F(v_F)$, is normalized to unity. Integrating the differential loss rate gives

$$\dot{n} = n \, n_{\mathrm{F}} \, \langle \sigma g \rangle , \qquad (4)$$

where

$$\langle \sigma g \rangle = \langle \Theta_{\mathrm{esc}} \sigma_{\mathrm{max}} g \rangle_{(b,\phi,\mathbf{v},\mathbf{v}_{\mathrm{F}})} = \int_{-\infty}^{\infty} \int_{-\infty}^{\infty} \int_{-\infty}^{\infty} \int_{-\infty}^{\infty} \int_{-\infty}^{\infty} \int_{-\infty}^{\infty} \int_{0}^{2\pi} \int_{0}^{b_{\mathrm{max}}} \Theta_{\mathrm{esc}} \sigma_{\mathrm{max}} g$$

$$\times f_b(b) f_\phi(\phi) f(v_x, v_y, v_z) f_{\mathrm{F}}(v_{\mathrm{Fx}}, v_{\mathrm{Fy}}, v_{\mathrm{Fz}}) db \, d\phi \, dv_x dv_y dv_z dv_{\mathrm{Fx}} dv_{\mathrm{Fy}} dv_{\mathrm{Fz}}. \qquad (5)$$

To come up with an expression for the test particle velocity distribution function, it is assumed that at times $t < 0$, the test particle velocity distribution is Maxwellian. If the magnitude of the depth of the one-dimensional electrostatic well that confines the test particle plasma is suddenly reduced at time $t = 0$, the evolution of the velocity distribution is describable relative to a time period, $t_p$, defined as the time required for plasma particles in the tails of the distribution to be lost from the well after no longer being confined by the well. It is assumed that during times $0 < t < t_p$, the effect of collisions on the evolution of the velocity distribution is negligible, and all the particles that exist in loss regions in velocity space leave the well. Then, at time $t = t_p$, the velocity distribution of the plasma is a cutoff Maxwellian:

$$f_{t=t_p}(\mathbf{v}) = \left( \frac{m}{2\pi k_{\mathrm{B}} T} \right)^{3/2} \exp\left( -\frac{mv^2}{2k_{\mathrm{B}} T} \right) \Theta(v_{z,\mathrm{esc}} - |v_z|) \qquad (6)$$

Here, $T$ is the test particle temperature, $m$ is the test particle mass, $k_{\mathrm{B}}$ is Boltzmann's constant, $v_{z,\mathrm{esc}} = \sqrt{2U_{\mathrm{w}}/m}$, where $U_{\mathrm{w}}$ is the depth of the potential energy well that the plasma particles are trapped in for times $t \geq 0$, and $\Theta$ is the Heaviside step function. A cutoff Maxwellian velocity distribution, Eq. (6), is used in Eq. (5) for the test particle plasma, and the present theory applies specifically at time $t = t_p$. For times $t > t_p$, the plasma loss rate can be expected to decrease with time. Consequently, a loss rate prediction made with the present theory may be considered to provide an upper limit for the collision-based axial plasma loss rate at times $t > t_p$. To consider a single plasma species interacting with itself, the velocity distribution for the field particles is taken to be the same as for the test particles.

An expression for $\Theta_{\mathrm{esc}}$ is arrived at by considering the condition for a test particle to be lost from the well. The condition is $|v_z'| > v_{z,\mathrm{esc}}$, where $v_z'$ is the axial component of the velocity of the test particle after it experiences a collision with a field particle. The expression used for $\Theta_{\mathrm{esc}}$ is

$$\Theta_{\mathrm{esc}} = \Theta(|v_z'| - v_{z,\mathrm{esc}}). \qquad (7)$$

For $b_{\mathrm{max}}$, any arbitrary value can be chosen so long as the value is larger than the largest impact parameter that can cause the test particle to be lost as a result of a collision. Thus, the limit $b_{\mathrm{max}} \to \infty$ can be taken. However, it is desirable to use as small a value for $b_{\mathrm{max}}$ as possible if Eq. (5) is evaluated numerically to minimize the computation time. Also, if a Coulomb interaction potential is considered, then a finite value for $b_{\mathrm{max}}$ is necessary if interactions at unrealistically long distances affect the calculation too much. For the present work, the value for $b_{\mathrm{max}}$ is set equal to the smaller of five quantities:

$$b_{\mathrm{max}} = \min\left( \lambda_{\mathrm{DT}}, \lambda_{\mathrm{DF}}, r_{\mathrm{T}}, r_{\mathrm{F}}, b_{\mathrm{upp}} \right). \qquad (8)$$

57

Here $\lambda_{DT}$ and $\lambda_{DF}$ are the Debye lengths associated with the test particles and field particles, respectively, $r_T$ and $r_F$ are the Larmor radii associated with a test particle and a field particle, respectively, and $b_{upp}$ is defined below. The expression used for the Debye length of the test particle species is $\lambda_{DT} = [k_B T/(4\pi k Z^2 e^2 n)]^{1/2}$, and that for the field particles is $\lambda_{DF} = [k_B T_F/(4\pi k Z_F^2 e^2 n_F)]^{1/2}$, where $T_F$ is the temperature of the field particles, $Z$ and $Z_F$ are the test particle and field particle charge states, respectively, e is the unit charge, and $k$ is Coulomb's constant [$k = 1/(4\pi\varepsilon_0)$ in SI units and $k = 1$ in cgs units]. For a single plasma species interacting with itself, $\lambda_{DT} = \lambda_{DF}$. The expression used for the cyclotron radius of the test particle is $r_T = m(v_x^2 + v_y^2)^{1/2}/(\ell Z e B)$ and that for the field particle is $r_F = m_F(v_{Fx}^2 + v_{Fy}^2)^{1/2}/(\ell Z_F e B)$, where $m_F$ is the mass of a field particle, $\ell$ is a constant associated with the units used ($\ell = 1$ in SI units, $\ell = 1/c$ in cgs units, and $c$ is the speed of light), and $B$ is the magnetic field strength.

A definition for $b_{upp}$ is arrived at by considering a specific type of binary interaction. The interaction potential energy for a Coulomb interaction is $U = \kappa/r$, where $\kappa = Z Z_F e^2 k$ and $r$ is the separation between the pair of interacting charged particles. For a Coulomb interaction,

$$b = \frac{\kappa \cot(\theta/2)}{\mu} \frac{2\kappa\mu}{g^2} = \frac{2\kappa\mu}{m^2(\Delta v)^2}\sin\theta \leq \frac{2\kappa\mu}{m^2(\Delta v)^2} \leq \frac{2\kappa\mu}{m^2(\Delta v_z)^2} \leq \frac{2\kappa\mu}{m^2(|v_z'| - |v_z|)^2}. \quad (9)$$

Here, $\theta$ is the test particle scattering angle in the center-of-mass frame of reference and $\mu = m m_F/(m + m_F)$ is the reduced mass. For the test particle to be lost from the well, the condition $|v_z'| > v_{z,esc}$ must be satisfied. Consequently, the particle can only be lost from the well if the impact parameter is less than

$$b_{upp} = \frac{2\kappa\mu}{m^2(v_{z,esc} - |v_z|)^2}. \quad (10)$$

Assume the velocity of a test particle is $v = v_x\hat{x} + v_y\hat{y} + v_z\hat{z}$, and that of a field particle is $v_F = v_{Fx}\hat{x} + v_{Fy}\hat{y} + v_{Fz}\hat{z}$ before the two particles collide. Here, $\hat{x}$, $\hat{y}$ and $\hat{z}$ are unit vectors in the laboratory frame of reference. The relative velocity before the collision is

$$g = v - v_F = (v_x - v_{Fx})\hat{x} + (v_y - v_{Fy})\hat{y} + (v_z - v_{Fz})\hat{z}, \quad (11)$$

The post-collision velocity of the test particle is

$$v' = v - (2\mu/m)g\sin^2(\theta/2) + (\mu/m)g\sin\theta\left(\left(\frac{v}{v_\perp} - \frac{v_\| g}{v_\perp g}\right)\cos\phi + \frac{g \times v}{g v_\perp}\sin\phi\right). \quad (12)$$

where $v_\| = v \cdot g/g$ and $v_\perp = \sqrt{v^2 - v_\|^2}$. The scattering angle $\theta$ is expressed in term of the impact parameter and particle velocities before the collision as

$$\theta = 2\arctan\left(\frac{\kappa}{\mu b g^2}\right). \quad (13)$$

Various theoretical expressions have been developed that might be used for calculating collisional losses of plasma particles from a one-dimensional electrostatic well [6].

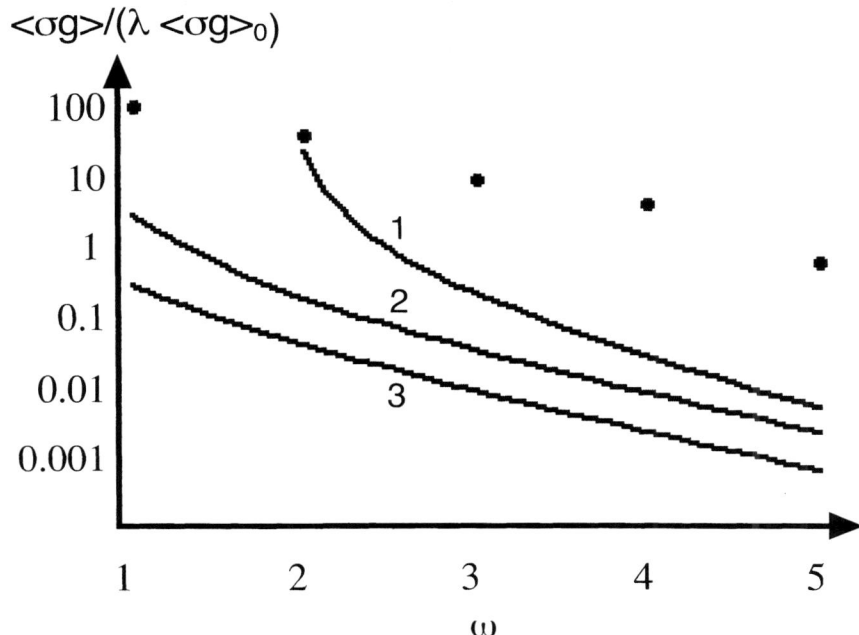

$\langle\sigma g\rangle/(\lambda \langle\sigma g\rangle_0)$

**FIGURE 1.** Comparison of Eqs. (5) (dots), (14) (curve 3), (15) (curve 1), and (16) (curve 2) for an electron plasma having a temperature of $T = 0.5$ eV in a magnetic field of $B = 0.2$ T. The plasma density is sufficiently small for the cyclotron radius to be small compared to the Debye length.

Using an approximate solution of the Fokker-Planck equation, Pastukhov obtained an expression for the loss rate of electrons confined by a magnetic mirror device within which an electrostatic well exists [7]. In the limit that the magnetic field is uniform, Pastukhov's expression can be used to obtain the following expression for $\langle\sigma g\rangle$:

$$\langle\sigma g\rangle_P = 0.42\lambda \langle\sigma g\rangle_0 \frac{e^{-\omega}}{\omega}\left(1 + \frac{e^\omega}{2}\sqrt{\frac{\pi}{\omega}}\mathrm{erfc}(\sqrt{\omega})\right). \tag{14}$$

Here, $\lambda$ is the Coulomb logarithm, $\langle\sigma g\rangle_0 = \sqrt{2}\pi\kappa^2/[\sqrt{m}(k_B T)^{3/2}]$, $\omega = U_w/(k_B T)$, and erfc is the complimentary error function. From a theory by Chernin and Rosenbluth [8], the corresponding $\langle\sigma g\rangle$ is obtained as

$$\langle\sigma g\rangle_C = 1.13\lambda \langle\sigma g\rangle_0 \frac{e^{-\omega}}{\omega}\left(\frac{1}{\ln(4\omega) - 2} + \frac{\ln(1.7\omega)}{[\ln(4\omega) - 2]^2}\right). \tag{15}$$

59

From a theory by Khudik [9], the corresponding $\langle \sigma g \rangle$ is obtained as

$$\langle \sigma g \rangle_{\mathrm{K}} = 1.89 \lambda \langle \sigma g \rangle_0 \frac{e^{-\omega}}{\omega} \left( 1 - \frac{1.43}{\omega} + \frac{3.46}{\omega^2} \right). \tag{16}$$

For calculating the Coulomb logarithm, $\lambda = \ln \Lambda$ is used. For a binary collision between a test particle and a field particle, $\Lambda = 2 r_{\max} E_c / \kappa$, where $E_c$ is the center-of-mass energy. A single species interacting with itself is considered, and $r_{\max}$ is calculated according to $r_{\max} = \min(\lambda_{\mathrm{D}}, r_c)$, where $\lambda_{\mathrm{D}} = \lambda_{\mathrm{DT}} = \lambda_{\mathrm{DF}}$ is the Debye length, and $r_c$ is the cyclotron radius. The cyclotron radius is calculated using $r_c = \sqrt{m k_{\mathrm{B}} T} / (\ell Z e B)$, and the center-of-mass energy is calculated using $E_c = k_{\mathrm{B}} T$.

As an example, an electron plasma of temperature $T = 0.5$ eV in a magnetic field $B = 0.2$ T is considered. The density is taken to be sufficiently small for the cyclotron radius $r_c$ to be much smaller than the Debye length. In such a limit, the value of $\langle \sigma g \rangle$ becomes independent of the plasma density. Figure 1 shows plots of $\langle \sigma g \rangle / (\lambda \langle \sigma g \rangle_0)$ obtained using Eqs. (5), (14), (15), and (16). The computation time to obtain numerical convergence for a calculation of Eq. (5) using a Monte Carlo integration technique increases with $\omega$, and only calculations with $\omega \leq 5$ were manageable. The comparison in Fig. 1 appears to indicate that the log plot of the normalized loss rate calculated using Eq. (5) has the same slope relative to the prior theories in the limit, $\omega \gg 1$. However, the loss rate calculated using Eq. (5) is on the order of 100 times larger than calculated with the prior theories. A possible explanation for the difference is that Eq. (5) may only be considered to provide an upper limit for $t > t_p$ as a result of using a cutoff Maxwellian velocity distribution.

This material is based upon work supported by the National Science Foundation under Grant No. PHY-0099617 and the Texas Advanced Research Program under Grant No. 3594-0003-2001.

# REFERENCES

1. J. H. Malmberg and J. S. deGrassie, Phys. Rev. Lett. **35**, 577 (1975).
2. D. H. E. Dubin and T. M. O'Neil, Rev. Mod. Phys. **71**, 87 (1999).
3. C. A. Ordonez, D. D. Dolliver, Y. Chang and J. R. Correa, Phys. Plasmas **9**, 3289 (2002), and references therein.
4. M. Amoretti, C. Amsler, G. Bonomi, A. Bouchta, P. Bowe, C. Carraro, C. L. Cesar, M. Charlton, M. J. T. Collier, M. Doser, V. Filippini, K. S. Fine, A. Fontana, M. C. Fujiwara, R. Funakoshi, P. Genova, J. S. Hangst, R. S. Hayano, M. H. Holzscheiter, L. V. Jorgensen, V. Lagomarsino, R. Landua, D. Lindelof, E. Lodi Rizzini, M. Macri, N. Madsen, G. Manuzio, M. Marchesotti, P. Montagna, H. Pruys, C. Regenfus, P. Riedler, J. Rochet, A. Rotondi, G. Rouleau, G. Testera, A. Variola, T. L. Watson, and D. P. van der Werf, Nature **419**, 456 (2002).
5. G. Gabrielse, N. S. Bowden, P. Oxley, A. Speck, C. H. Storry, J. N. Tan, M. Wessels, D. Grzonka, W. Oelert, G. Schepers, T. Sefzick, J. Walz, H. Pittner, T. W. Hansch, and E. A. Hessels, Phys. Rev. Lett. **89**, 213401 (2002).
6. R. F. Post, Nucl. Fusion **27**, 1579 (1987).
7. V. P. Pastukhov, Nucl. Fusion **14**, 3 (1974).
8. D. P. Chernin and M. N. Rosenbluth, Nucl. Fusion **18**, 47 (1978).
9. V. N. Khudik, Nucl. Fusion **37**, 189(1997)

# Diocotron Instabilities in an Electron Column Induced by a Small Fraction of Transient Positive Ions

Andrey A. Kabantsev and C. Fred Driscoll

*Physics Department, University of California at San Diego, La Jolla CA 92093-0319 USA*

**Abstract.** It is well known that a small fraction of positive ions can destabilize diocotron modes on electron plasmas. However, the historical (and recent) interpretation of experimental results in terms of 2D (or modified 2D) theories of ion-induced instabilities is apparently erroneous. Here, we experimentally characterize a strong exponential instability with no threshold, obtaining growth rates orders of magnitude larger than predicted. The positive ion population is maintained either by continuous external injection of ions or by ionization of the background gas within hot electron plasmas. In both cases, the observed exponential growth rate $\gamma_m$ is directly proportional to the ion creation rate $\nu_+$, i.e., $\gamma_m = \kappa_m \nu_+$, with $\kappa_m \approx (10^1\text{-}10^3)/N_e$ for $m_\theta = 1,2,3$. Experimental results also suggest that non-2D effects, including end confinement fields, are important. This strong instability may have important implications for the anti-hydrogen creation technique of propelling anti-protons through trapped $e^+$ clouds.

## INTRODUCTION

Theory demonstrates that two-component nonneutral plasmas consisting of trapped electrons and a small fraction of ions can exhibit unstable $k_z = 0$ diocotron modes if there is a resonance between the electron diocotron modes and the radial motion of the ion [1]. This *exponential* ion-resonance instability [2] was suspected to be the dominant loss mechanism in toroidal experiments [3,4]. This original analysis [2] treated the case of trapped ions; the case of untrapped (transient) ions was investigated later in a linear electron trap configuration both experimentally [5] and theoretically [6]. Several significant differences were found between these two cases: transient ions appeared to cause *linear* rather than *exponential* growth; and this linear growth occurs over a somewhat broader region around the resonance.

The 2D-analysis of non-resonant motion of transient ions in an electron column by Fajans [6] suggests that the average ion motion is well represented by the orbit of an ion which is placed initially at the electron column center. This "average ion" orbits around a fixed point in the $m_\theta = 1$ diocotron frame, and this point is displaced *outward* from the center of electron column by a small distance $\Delta$, which is a small fraction of the diocotron mode amplitude $D$, i.e., $\Delta \sim (r_p^4/\lambda)D$. Here, $r_p \equiv R_p/R_w$ is the electron column radius, and the ion magnetization parameter [2] is $\lambda \equiv B^2/2\pi n_e m_i c^2 \gg 1$. An ion wobbling around this fixed point causes an exponential

CP692, *Non-Neutral Plasma Physics V*, edited by M. Schauer et al.

growth of the diocotron mode, with a rate which can be derived from Eq. 24 in Ref. 6 as $\gamma_1{}^{th} \approx v_+ r_p{}^4/(1 - r_p{}^2)\lambda$. Here, $v_+$ is the ion production (internal ionization or external injection) rate per trapped electron. Thus, one expects growth rates much less than the ion production (injection) rate, i.e., $\gamma_1{}^{th}/v_+ \ll 1$.

However, our measurements of $\gamma_1$ and $v_+$ in the range of $10^{-5} \leq r_p{}^4/\lambda \leq 10^{-2}$ show that $\gamma_1/v_+ \gg 1$, which suggests a process not treated in prior analysis. In particular, this result implies that the lifetime-averaged displacement of the ion exceeds the diocotron radius $D$ by factor of $10^1$-$10^3$ even far away from the resonance.

Since many of the current experiments on a cold antihydrogen production require the propulsion of antiprotons through the trapped positron plasma [7,8], there is a free energy to drive similar instabilities. Due to slow antihydrogen formation rates, even weak instabilities could be troublesome. Here, we find that instabilities essentially always occur, with growth rates proportional to the number of transiting ions.

# EXPERIMENTAL SETUP

Our experiments are performed in a cylindrical Penning-Malmberg trap, as shown schematically in Fig. 1. The electron column of length $L_p \leq 50$ cm is contained inside a stack of hollow conducting cylinders of radius $R_w = 3.5$ cm, which reside in an ultrahigh vacuum with residual pressure $P \approx 0.1$ nTorr. Molecular hydrogen from the warm walls is a majority (95%) of the background neutral gas. The end cylinders $G_1$ and $G_{10}$ are negatively biased, with $V_c \geq$ -100 V, thereby providing axial confinement for the electrons. A strong axial magnetic field $B \leq 14$ kG ensures radial confinement both for the electrons and ions.

The electron column is generated by thermoionic emission from a hot tungsten filament located axially outside of the trapping region. Temporarily grounding the confinement gate $G_1$ allows the electrons to fill the trap. The resulting trapped electron column has typical density $1 \leq n_e \leq 2 \times 10^7$ cm$^{-3}$ over a bell-shaped radial profile with a characteristic radius $R_p \approx 1.2$cm, and with temperature $0.6 \leq T_e \leq 0.8$ eV before auxiliary heating. The electrons can be heated by applying short ($\leq 5$ ms) rf-burst of variable amplitude and tuned frequency to one of the end confining cylinders in resonance with the particles bounce motion [9].

In our experiments we generate ions by ionization of the background gas directly inside the electron column, or by continuous injection of ions created by filament electrons in the Grid-$G_1$ region. In both cases, the resulting fluxes of ions are directly proportional to the background neutral pressure. The beam method has the advantage of independence of the electron plasma temperature, which is especially important at high pressures due to ionization cooling. The rf-heating method has the advantage of allowing direct measurement of the ionization rate simultaneously with the mode growth rate: since every ionization adds one electron to the plasma column, the ionization rate follows directly from the charge accumulation rate (if ions are not confined in the trap).

The growth rates $\gamma_m(t)$ for the $k_z = 0$, $m_\theta = 1,2,3\ldots$ diocotron modes are measured by digitizing the amplitudes $A_m(t)$ of corresponding wall signals induced by the diocotron oscillations at the sectored electrodes $S_4$ and $S_7$. These amplitudes are verified (and calibrated) by calculating the center-of-mass displacement $D$ (or the quadrupole moment $Q$ for the $m_\theta = 2$ case) of the dumped plasma column from a CCD camera diagnostic [10]. The ionization rate $v_+(t)$ is obtained *simultaneously* with $\gamma_m(t)$, by measuring the frequency change of the $m_\theta = 1$ diocotron mode; that is, $v_+ = (1/N_e)dN_e/dt = (1/f_1)df_1/dt$, with verification by the relative change in the total number of electrons dumped onto the phosphor screen. In the experiments, we keep the scaled mode amplitudes, $d \equiv D/R_w$, small enough ($d \leq 0.03$) to keep nonlinear effects [11] in the mode frequency ($\delta f_1/f_1 \leq d^2$) well below the resolution level of our frequency measurements (0.01%).

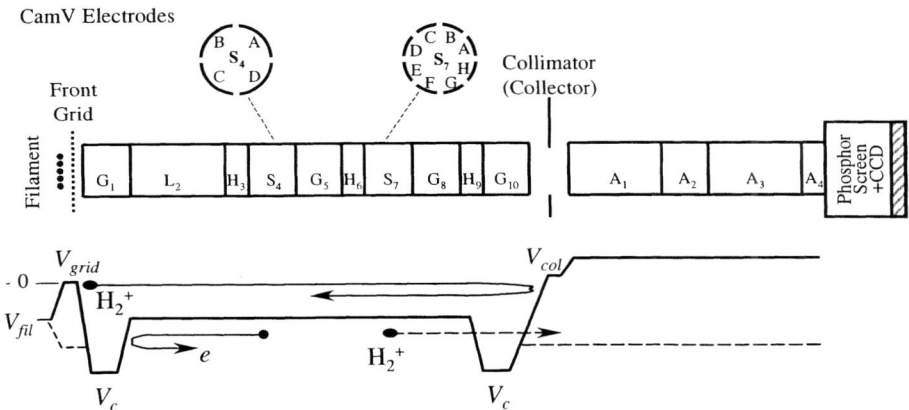

**FIGURE 1.** Cylindrical Penning-Malmberg trap and imaging diagnostics, with potential profiles for two configurations: a double-well configuration with axially *trapped* ions (solid); and a single-well configuration with *transient* ions (dashed).

# ENERGETIC ION INJECTION

Figure 2 shows the ion current $i_+$ transiting through the electron column as a function of the neutral pressure when ions are not trapped at the dump end of the trap. In this particular case, the front grid was positively biased, $V_{grid} = +10$ V, and the filament was at $V_{fil} = -40$ V. Thus, the maximum energy acquired by emitted electrons near the grid, $\mathcal{E}_{max} = V_{grid} - V_{fil} \approx 50$ eV, was close to the energy providing the maximum of the ionization cross-section for molecular hydrogen. The ion current is measured at the collector plate put temporarily behind the end confining cylinder $G_{10}$. One can see a linear dependence $i_+(P_{H2})$, with an offset probably due to ionization of gases desorbed from the grid surface; the pressure is measured 1m distant from the confinement electrodes.

Figure 3 shows the ion current as a function of the electron energy $\mathcal{E}_{max}$ for two extreme pressures. Thus, we have an adjustable source of transient ions; the current can be easily controlled over a vast range by the front grid bias, or by the background neutral pressure.

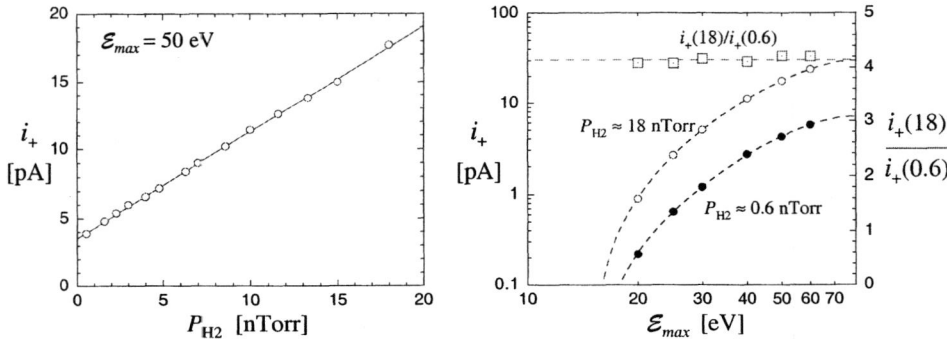

**FIGURE 2.** The ion current from ionization in the Grid-$G_1$ region versus neutral pressure.
**FIGURE 3.** The ion current versus the maximum beam energy $\mathcal{E}_{max}$ for two neutral pressures, and their ratios.

If we apply a potential to the collector plate, which is more positive than the potential at the front grid, then this continuously injected flow of ions gets trapped between the grid and the collector plate. Trapped ions bounce back and forth through the electron column, causing an even greater instability of the diocotron modes. Typical dependencies of the instability growth rates $\gamma_m$ on neutral pressure are shown in Fig. 4 for $m_\theta = 1,2$. The growth rates are linearly proportional to the neutral pressure (i.e., to the ion current) over two orders of magnitude. Here, we have subtracted the zero-current growth (or damping) rate $\gamma_m(0)$, arising from resistive instability and asymmetry-induced damping. For these conditions, $\gamma_1(0) = +0.24$ s$^{-1}$ and $\gamma_2(0) = -0.081$ s$^{-1}$. Higher $m_\theta$-modes show even greater instability rates, but they rapidly develop non-linear saturation due to spatial Landau damping in the radial edge of the plasma column. Hence, we focus predominantly on the basic ($m_\theta = 1$) diocotron mode.

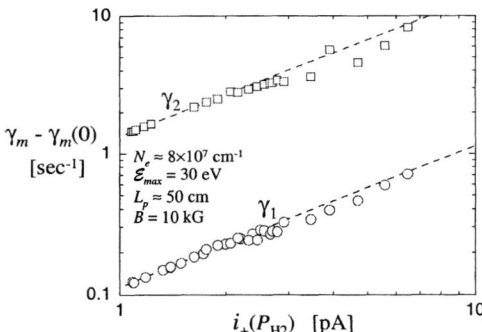

**FIGURE 4.** The instability growth rates versus (continuously injected) ion current in the double-well configuration.

Modulating the energy of the electron beam or the ion trapping potential modulates the flux of trapped ions, and this correspondingly modulates the growth rates $\gamma_m$, as shown in Figs. 5a,b. Opening a confinement gate for the trapped ions (by decreasing the bias $V_{col}$ on the collector plate below the plasma potential) causes an immediate and dramatic effect: the growth rate goes down more than 10 times. Re-establishing ion trapping causes linear growth in $\gamma_1$, due to increase in number of ions, limited by an "active time" $\tau_a \sim 0.1$ s. We believe that $\tau_a$ represents the time required for ions to move radially to the edge of electron column. This active lifetime shows an approximate empirical scaling $\tau_a \propto B/V_cL_c$, which suggests $\mathbf{E} \times \mathbf{B}$ $\theta$-drift nature. Here, $V_c$ is the electron confinement voltage applied to the end cylinders, and $L_c$ is the length of those cylinders at $V_c$. This probably arises from $\delta\theta \propto V_cL_c$ as the relative $\theta$-drift of the ions with respect to the phase of the electron column diocotron rotation.

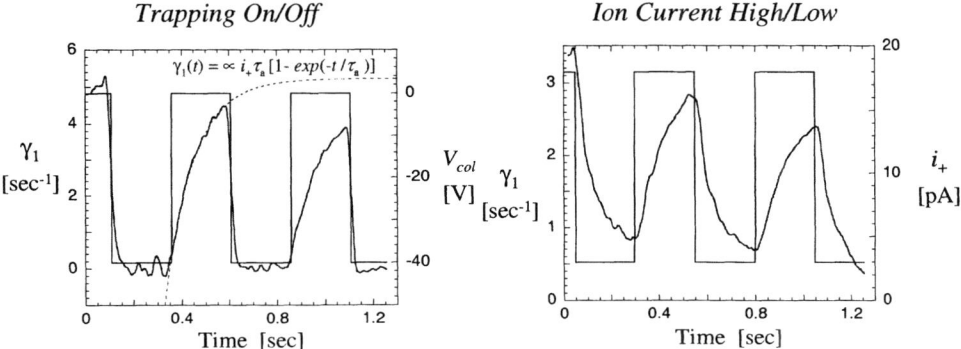

**FIGURE 5.** Growth rates with square-wave modulated trapping (left) and injection current (right) of the ions.

Note that while the growth rate $\gamma_1(t)$ shows an exponential saturation due to $\tau_a$ of the trapped ions, the fractional neutralization $\alpha(t) \equiv N_i/N_e = v_+t$ grows linearly due to continuous trapping of injected ions. Thus, at high enough values of $i_+(P_{H2})$, the "ionization rate" $v_+$ can be directly measured from the linear decrease in the diocotron frequency $f_1(t)$ due to continuous accumulation of the ion space charge (most likely, at the periphery of electron column), i.e., $f_1(t) \propto N_e(1 - v_+t)$, and hence, $v_+ \approx -(1/f_1)df_1/dt$.

Figure 6 demonstrates that the observed decrease in the diocotron frequency $f_1(t)$ is indeed due to the trapping and accumulation of positive ions: after letting them go away at $t = 6.6$ s, the diocotron frequency immediately rises up to its initial value simultaneously with a halt in the mode growth (the 5 Hz steps in $f_1$ are instrumental only).

Taking data similar to Fig. 6 at different values of $i_+(P_{H2})$, we can plot the corresponding growth rates $\gamma_1$ versus $v_+ = i_+/Q_e$. The best linear fit gives us the ratio $\gamma_1/v_+ \approx 400$. In general, we find that this non-resonant ion-induced instability shows exponential growth with a rate $\gamma_1$ directly proportional to the "ionization rate" $v_+ = i_+/Q_e$, and that $\gamma_1/v_+ \gg 1$ in the whole range of $\lambda(B) \gg 1$.

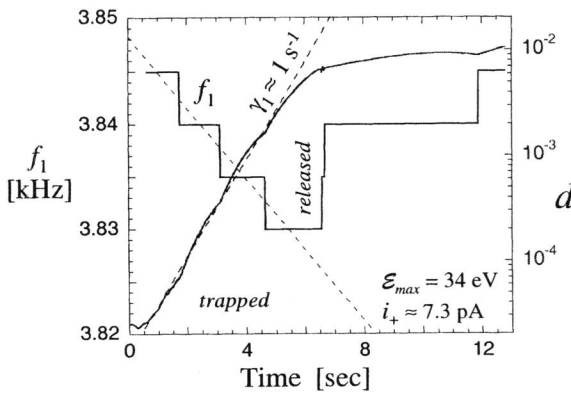

**FIGURE 6.** Temporal behavior of the diocotron mode frequency $f_1$ and its amplitude $d$ with trapping and release in double-well configuration.

## IN-PLACE IONIZATION

Another approach to this problem of the non-resonant ion-induced instabilities lies in ionization of background neutrals directly in an electron column. To get the ionization rate at measurable level we raise $T_e$ up to 7-9 eV with rf-heating burst. In this experimental configuration, each ion freely escapes from the trap in a time $\tau_i \sim L_p/v_i \leq 1$ ms, leaving its counterpart electron inside the trap. Therefore, the electron column image charge and the diocotron frequency have an increase proportional to $v_+(t)$, so we can obtain the ionization rate as $v_+(t) \equiv 1/f_1 \, df_1/dt$.

Figures 7a and 7b show typical evolutions of this calculated ionization rate $v_+(t)$ and the diocotron mode amplitude $d(t)$ after heating at $t_0 = 4.6$ s. At the low pressure of Fig. 7a, $v_+$ remains essentially constant after the heating burst, and $d$ growth exponentially. At the higher pressure of Fig. 7b, electron cooling causes $v_+$ to decrease with time, and a surprisingly *linear* growth of $d$ is observed. However, this represents the same exponential instability, i.e., $\partial d/\partial t = \kappa_1 v_+(t) \, d$, but with a rapidly decreasing $v_+$. This can be seen directly from the $d(t)$ and $f_1(t)$ data, since $\kappa_1 \equiv \gamma_1/v_+ = \partial \ln(d/d(t_0)) / \partial \ln(f_1/f_1(t_0))$. Figure 8 shows a slope $\kappa_1 \approx 20$ for all time, demonstrating that the surprising linear growth of $d$ is an insignificant consequence of electron cooling causing $v_+(t)$ to decrease.

Thus, we have got again a definitive evidence that $\gamma_1 = \kappa_1 v_+$, where $\kappa_1 \approx 20$. Adding double-well potential to the ends makes these ions temporarily trapped, and we then observe an additional twenty-fold increase in $\kappa_1$, which brings it close to the typical level for the ion injection case ($\kappa_1 \sim 400$). The ions have a finite residence time $\tau_i$ (depending on $v_{zi}$), and the growth rate depends on $\kappa_1(\tau_i)$. Note that a typical fractional neutralization for this case of in situ ionization is $\alpha(t) = v_+(t)\tau_i \ll 10^{-4}$.

The instability factor $\kappa_1$ shows a surprising sensitivity (factor of two) to the plasma column shape $n(r, z)$. It is hard to quantify this dependence experimentally. However, we have observed a good correlation between the $\kappa_1$ dependence on the radial density profile and a factor of coupling of the $m_\theta = 1$ diocotron mode to a wall sector.

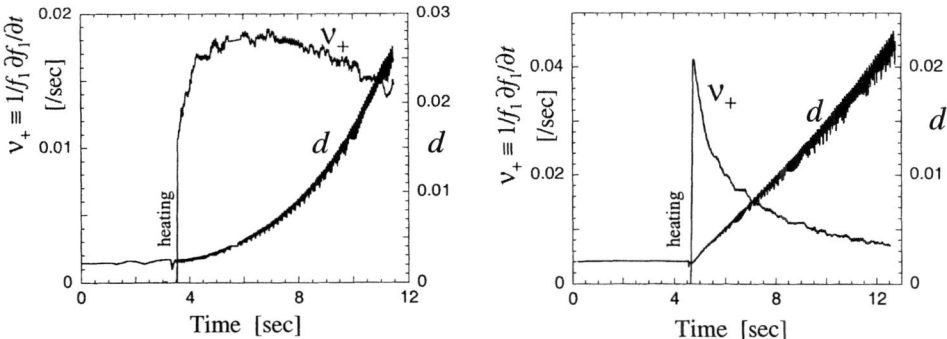

**Figure 7.** Temporal behavior of the ionization rate $v_+$ and the $m_\theta = 1$ diocotron mode amplitude $d$ in single-well configuration. Left figure shows simple exponential growth with nearly constant $v_+(t)$ at low pressure $P_{H2} \approx 0.5$ nTorr. Right figure shows *quasi-linear* growth due to strongly decreasing $v_+(t)$ caused by the fast cooling at higher pressure $P_{H2} \approx 2$ nTorr.

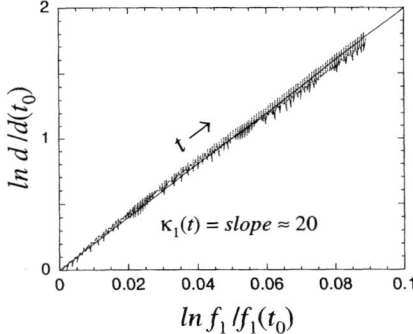

**Figure 8.** Exponential instability (Fig. 7b) with decreasing $\gamma_1(t) \approx 20\, v_+(t)$. The slope $\kappa_1 \equiv \gamma_1/v_+ = \partial\, ln(d/d(t_0))\, /\partial\, ln(f_1/f_1(t_0))$ is nearly constant for all time.

This plasma column shape is also sensitive to the magnetic field through transport processes. Due to these reasons, the growth rate scaling with magnetic field is not well established. Nevertheless, our preliminary measurements within similar plasma shapes show a scaling consistent with $\gamma_1 \propto B^{0 \pm 0.2}$ in the range $2 \leq B \leq 14$ kG. This result is also in apparent contradiction to the $1/B^2$ magnetic scalings predicted by the 2D-theories of the ion-induced instabilities [2,6].

## CONCLUSIONS

Diocotron instabilities are commonly observed when ions are present in pure electron plasmas [3-5,12]. Here, we establish an exponential growth of diocotron modes $m_\theta = 1,2,3$ with growth rates $\gamma_m$ that are directly proportional to the incoming ion rates $v_+$, and with corresponding coefficients $\kappa_m \equiv \gamma_m/v_+$ that are orders of magnitude greater than one. These effects may have strong implications for a variety of experiments that propel bunch of ions many times through an electron/positron cloud [7,8].

The present results show that the existing 2D-theory of the transient ion instability underestimates the growth rate by many orders of magnitude, and indicates that as-yet-unspecified non-2D effects play a dominant role in ion motion. One possible candidate for this effect is the difference between the bounce-averaged rotation frequencies of electrons and ions; this difference comes from the two species sampling somewhat different radial electric fields at the plasma ends (the so called the magnetron rotation). This basic effect has been incorporated in several theory approaches [12,13]; but these theories show little correspondence to the present experimental results.

Broadly, it appears that azimuthal drift of electrons and ions tends to polarize the diocotron mode density perturbations, thereby developing instability similar to the classical flute MHD-instability of neutral plasmas confined in non-uniform magnetic fields. Moreover, the observed dependencies of growth rate on the confinement voltage and on the plasma column length are generically consistent with this hypothesis.

## ACKNOWLEDGMENTS

This work was supported by NSF grant PHY-9876999.

## REFERENCES

1. Davidson, R.C., *Physics of Nonneutral Plasmas*, Redwood Addison-Wesley, 1990, pp. 260-264.
2. Levy, R. H., Daugherty, J.D., and Buneman, O., *Phys. Fluids* **12**, 2616-2629 (1969).
3. Daugherty, J.D., Eninger, J.E., and Janes, G.S., *Phys. Fluids* **12**, 2677-2693 (1969).
4. Clark, W., Korn, P., Mondelli, A., and Rostoker, N., *Phys. Rev. Letters* **37**, 592-595 (1976).
5. Perrung, A.J., Notte, J., and Fajans, J., *Phys. Rev. Letters* **70**, 295-298 (1993).
6. Fajans, J., *Phys. Fluids* **B 5**, 3127-3135 (1993).
7. Amoretti, M., Bonomi, G., Bouchta, A., *et al.*, *Phys. Plasmas* **10**, 3056-3064 (2003).
8. Gabrielse, G., Bowden, N.S., Oxley, P., *et al.*, *Phys. Rev. Letters* **89**, 213401(4) (2002).
9. Beck, B.R., Fajans, J., and Malmberg, J.H., *Phys. Rev. Letters* **68**, 317-320, 1992).
10. Fine, K.S., Flynn, W.G., Cass, A.C., and Driscoll, C.F., *Phys. Rev. Letters* **75**, 3277-3280 (1995).
11. Fine, K.S., Driscoll, C.F., and Malmberg, J.H., *Phys. Rev. Letters* **63**, 2232-2235 (1989).
12. Pasquini, T., and Fajans, J., *Bull. Am. Phys. Society* **47**(9), 127 (2002).
13. Hilsabeck, T.J., and O'Neil, T.M., *Workshop on NNP 2001*, Abstracts (2001).

# Damping of Trapped-Particle Asymmetry Modes in Non-Neutral Plasma Columns

## Grant W. Mason

*Department of Physics and Astronomy, Brigham Young University, Provo, Utah 84602*

**Abstract.** Asymmetry modes ($m = 1$, $k_z \neq 0$) are diocotron-like modes in finite-length plasma columns in Malmberg-Penning traps. We have investigated the modes with a detailed 3-d particle-in-cell (PIC) drift-kinetic computer simulation. Although PIC simulations do not employ realistic collisions, the simulations in this case reproduce many of the salient features of the data. Particle transport associated with the damping is seen not to be a direct collisional effect, but rather a feature of orbital dynamics associated with transitions from trapped-to-untrapped or untrapped-to-trapped state relative to the inversion plane of the asymmetry. In the simulations we observe a $B^{-1}$ dependence of the mode frequencies and a $B^{-0.5}$ dependence of the damping constant for large rigidity. We further observe a steepening of the dependence of the decay constant to $B^{-2}$ as the rigidity of the plasma falls below about 2.0. We have also used the simulations to investigate the modes at small seed amplitudes and observe linear flattening in the mode frequency as the seed amplitude becomes small. In contrast, the decay constant does not flatten for small seed amplitude.

Non-neutral plasmas, typically ions or electrons, can be confined for long periods of time in a cylindrical Malmberg-Penning trap. A stiff axial magnetic field confines the particles radially and charged rings at the ends of the otherwise grounded cylinder provide electrostatic longitudinal confinement. Diocotron modes are azimuthal drift waves in the cylindrical plasma that vary spatially as $\exp(im\theta)$. The theory of diocotron modes in non-neutral plasmas has its origins in seminal papers by Briggs, Daugherty and Levy [1] and the comprehensive treatment of non-neutral plasmas by Davidson [2]. Experimental and theoretical work at the University of California San Diego for more than a decade has also contributed particularly to the foundation of understanding of these modes [3].

Here we consider a modification of the diocotron mode when an applied "squeeze voltage" is applied to an additional ring installed at the longitudinal center of the trap and the plasma is offset from the symmetry axis in opposite directions on each side of the center ring such that the parity of $k_z \neq 0$ is odd. The squeeze voltage creates an energy barrier at the longitudinal median plane that gives rise to a population of trapped particles on either side of the divide as well as a population of untrapped particles with sufficient energy to traverse the entire length of the trap. The result is a new mode revolving at a frequency different from the wall value of the rotation frequency profile. These "trapped-particle asymmetry modes" have been experimentally observed and reported by Kabantsev *et al.* [4, 5, 6, 7]. The dependence of the decay constant on magnetic field has been particularly problematic. The most recent data are divided into two regimes, one at lower magnetic fields varying roughly as $B^{-1}$ eventually giving way to a less steep dependence ($B^{-0.5}$) at higher magnetic fields [6]. Hilsabeck and O'Neil have ascribed

CP692, *Non-Neutral Plasma Physics V*, edited by M. Schauer et al.
© 2003 American Institute of Physics 0-7354-0165-9/03/$20.00

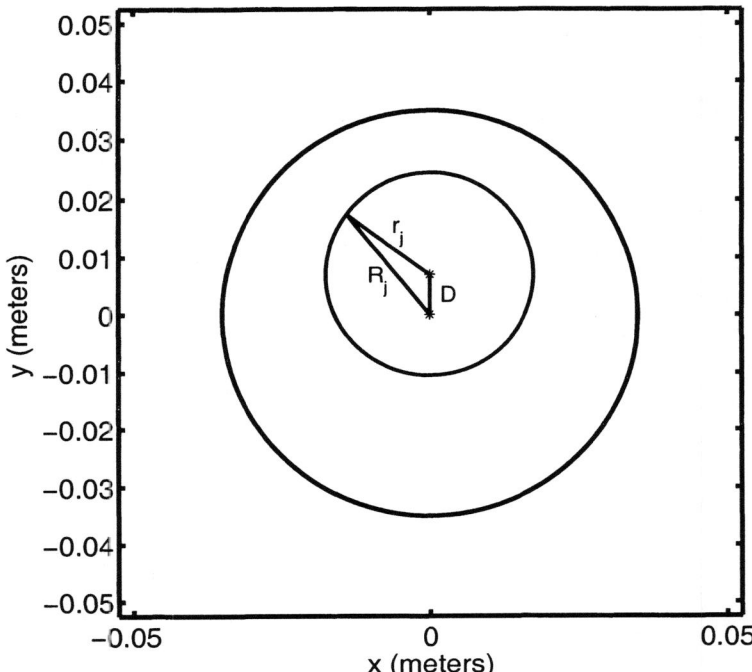

**FIGURE 1.** The $m = 1$ mode displaces the charge center of trapped particles relative to the symmetry center of the trap. Trapped particles revolve roughly in circles of radius $r_j$ around the displaced charge center. The orbit of a trapped particle about the charge center creates a radial oscillatory motion whose amplitude is the mode amplitude, $D$. $R_j$ is the radius of a particle relative to the symmetry center. Untrapped particles revolve roughly in circles around the symmetry center of the trap. Transitions from trapped to untrapped state or from untrapped to trapped state result in radial transport.

the damping mechanism to velocity scattering of marginally trapped particles. They have developed a theory based on this mechanism for which they report good agreement with measurements [5, 8].

We have reported elsewhere the results of detailed 3-dimensional computer simulations of the trapped-particle asymmetry modes [9]. In the simulations the decay mechanism is demonstrated to be a consequence of the orbital dynamics of particles moving back and forth between the trapped and untrapped populations. The interchange between the two populations is a result primarily of slow modulations (diffusion) of the longitudinal velocities of the particles. The simulations (and data) have demonstrated that the mode frequency increases with decreasing squeeze voltage, that the mode frequency varies as $B^{-1}$ and that the decay of the mode varies with magnetic field as $B^{-0.5}$ for a range of magnetic fields $0.02T \leq B \leq 0.32T$. Here we observe that the dependence of the decay constant steepens as the magnetic field decreases below this range (see Fig. 2).

We have not simulated any specific experiment, although we have chosen parameters

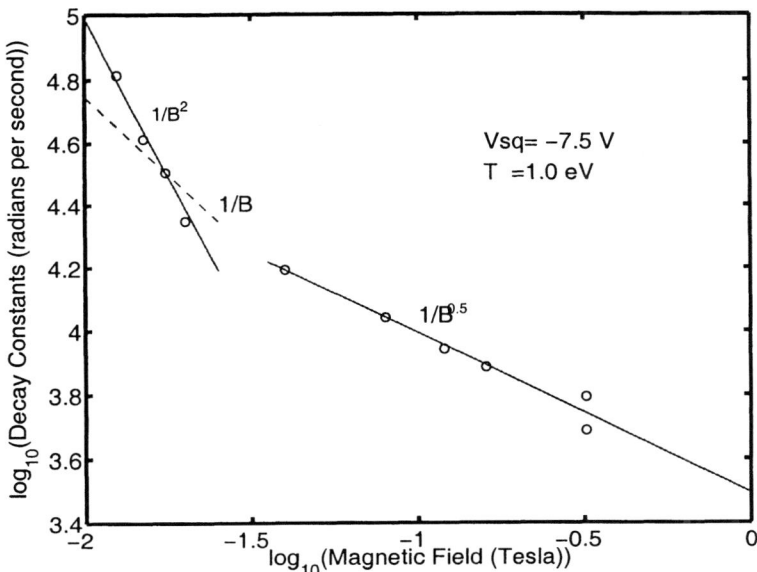

**FIGURE 2.** Dependence of the decay constant on magnetic field for a squeeze voltage that is about half the central potential of the plasma. The dependence steepens roughly where there are about two particle longitudinal bounce cycles for each mode cycle.

that lie in the general parameter space of the UCSD data [6]. Using Fig. 2 of Kabantsev *et al.* and the formula for rotation frequency on the same page of this latter reference, we get a typical decay constant of $10^3 s^{-1}$ for a typical density of $10^7 cm^{-3}$. The ratio of decay constant mode frequency is then about $10^{-3}$. This is for a 1 Tesla field. Taking numbers from our simulations [9] and using 0.8 Tesla (our closest magnetic field for the experimental 1 Tesla), we get a ratio of $25 \times 10^{-3}$, i.e. 25x too large. However, the decay constants from the simulations have a rough dependence on the number of simulation particles ($N^{-0.5}$, a collisional effect) that we have determined empirically. In this paper we use $N = 1.5 \times 10^6$ particles to represent $6.8 \times 10^8$ electrons. Thus we must divide our result by $\sqrt{450}$ to correct to a 1 : 1 simulation. The resulting value of the decay constant puts us very near the mark, other things being equal. The code accurately computes rotation frequencies where collisionality plays no role, so we conclude that the simulation roughly predicts the salient features of the data.

The simulation code being used is a drift-kinetic, particle-in-cell (PIC) code [9]. In such a code, all motion transverse to the longitudinal axis of the trap is handled as $\mathbf{E} \times \mathbf{B}$ drift. Thus, the only aspect of collisions that is handled in anything near realistic fashion is the longitudinal motion and even that has a built-in de-emphasis of small impact parameter collisions. Therefore, a conclusion to be drawn is that the decay of the asymmetry modes, insofar as the PIC code reproduces the features of the decay correctly, is virtually independent of transverse momentum transfer in the collisions.

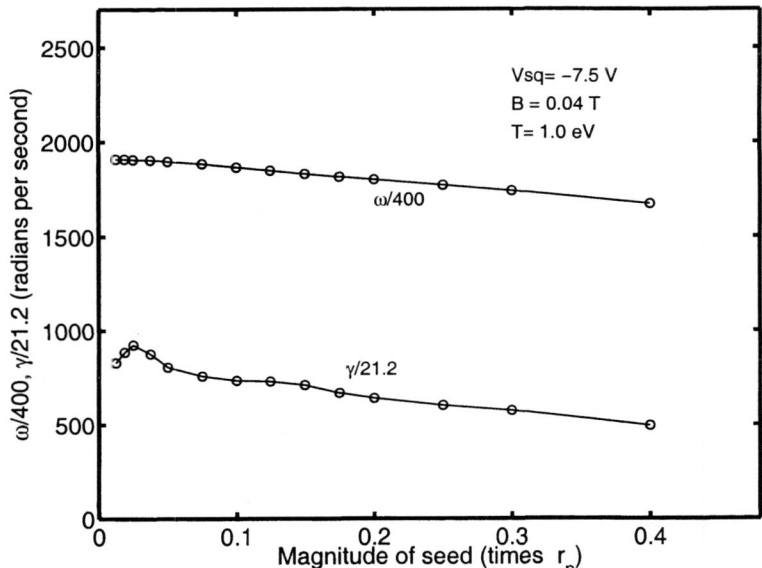

**FIGURE 3.** Both the mode frequency $\omega$ and the decay constant $\gamma$ depend on the magnitude of the seeded amplitude. However, $\gamma$ shows variation at small mode amplitude, indicating a nonlinearity in its behavior. The mode frequency has been divided by a factor of 400 merely to allow comparison of the dependences on the same graph. The decay constant has been divided by an empirically determined correction factor of $\sqrt{450}$ that is necessary to correct for the difference between the ratio of actual particles (electrons) to simulation particles in the simulated trap. The ratio is 450 in this case. A similar correction factor is necessary whenever comparisons to actual data are attempted.

The canonical angular momentum of the system is given approximately by [10]

$$P_\theta \approx \frac{eB}{2} \sum_{j=1}^{N} R_j^2 \approx \frac{eB}{2} N \left( D^2 + \frac{1}{N} \sum_{j=1}^{N} r_j^2 \right), \tag{1}$$

where $N$ is the number of particles, $\mathbf{R}_j$ is the position of the $j$th particle measured from the symmetry axis of the cylinder, $\mathbf{D}$ is the position of the $m = 1$ mode (charge) center, and $\mathbf{r}_j$ is the position of the particle relative to the mode center, i.e. $\mathbf{R}_j = \mathbf{D} + \mathbf{r}_j$. Since the simulation code conserves the angular momentum, any decrease in the mode amplitude $D$ must be accompanied by an overall adjustment in $\sum r_j^2$, i.e. some transport relative to the charge center of the mode. However, in the region of parameter space where we are working, $\sum r_j^2 \ll ND^2$. In the relatively short simulations times considered here (75 $\mu$sec), the net amount of radial transport needed to account for the mode decay is almost imperceptible.

The mode frequency $\omega$ and the decay constant $\gamma$ both depend mildly on the magnitude of the seed of the mode (created by moving a subset of particles radially) [9]. Linear behavior requires that the dependence of frequency and decay constant be independent of the (perturbed) amplitude of the mode for small amplitude. In Figure 3 we observe

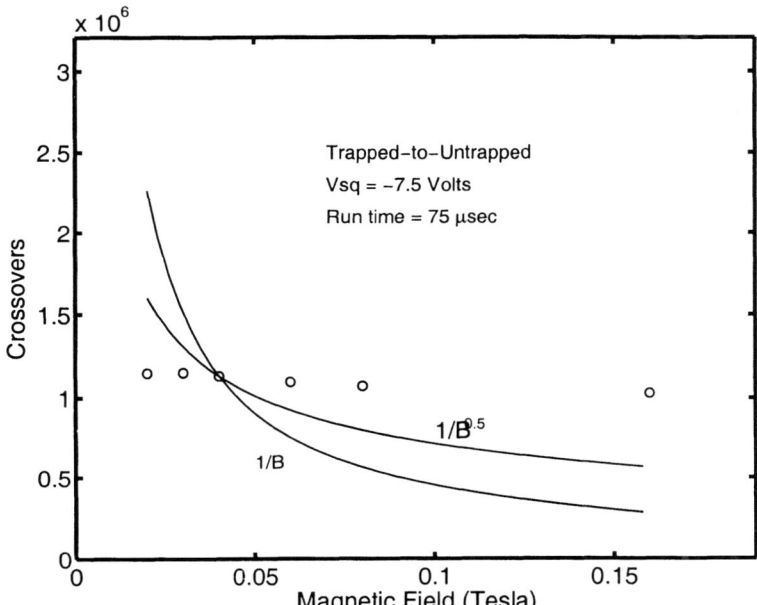

**FIGURE 4.** The number of simulation particles that change from trapped to untrapped status within a 75 μsec simulation as a function of magnetic field. The magnetic field has only a weak influence on the transition rate associated with the mode decay.

that the dependence of $\gamma$ violates this condition.

Within the parameter space considered here, the code reproduces the expected $B^{-1}$ behavior of the mode frequency (which takes its value from the $\mathbf{E} \times \mathbf{B}$ rotation profile). However, the decay constant varies as $B^{-0.5}$ for large rigidity. To understand why this is so, we have monitored the total number of particle transitions from trapped-to-untrapped state during a simulation as a function of magnetic field (see Fig. 4). The rather mild dependence on magnetic field suggests that the dependence of $\gamma$ on magnetic field must be sought elsewhere. Since the "radial diffusion" (small variation in $\Sigma r_j^2$) apparently does not depend on the transition rate, we look at the radial step size as a function of magnetic field (see Fig. 5). Although the connection to the $B^{-0.5}$ dependence of $\gamma$ is not readily evident, the marked dependence of the radial step size on magnetic field suggests that the answer is to be found in the step size.

In conclusion, our simulations indicate that the decay of asymmetry modes is accompanied by minimal net particle transport that is a consequence of adjustments in orbital motion when a trapped particle becomes untrapped (relative to the full length of the trap), or vice versa. The decay constant of the mode is seen to vary as $B^{-0.5}$, steepening to $B^{-2}$ for smaller rigidity. Unlike the mode frequency, the mode decay constant depends on amplitude, particularly at small amplitude. There is only a small dependence on magnetic field for the rate of transitions from trapped to untrapped status associated

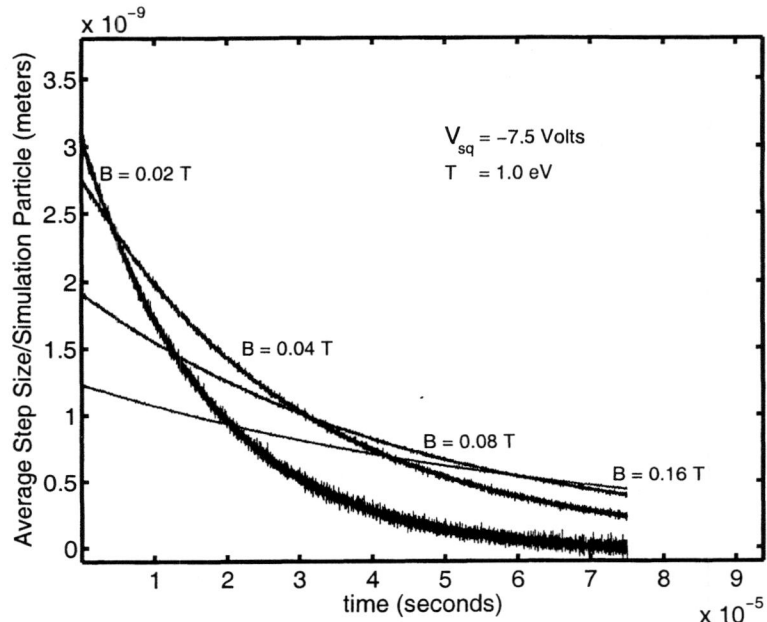

**FIGURE 5.** The average radial step size per simulation particle as a function of time for different choices of magnetic field. Since the transition rate for the diffusion associated with the mode decay is roughly independent of magnetic field, the $B^{-0.5}$ behavior of the decay constant likely stems from the dependence of the radial step size on magnetic field.

with the decay mechanism. The radial stepsize shows a much stronger dependence on magnetic field. Thus, it is likely that the dependence of the decay constant on magnetic field is to be found in the dependence of the radial stepsize on magnetic field.

# REFERENCES

1.  R. J. Briggs, J. D. Daugherty, and R. H. Levy, Phys. Fluids **13**, 421 (1970).
2.  R. C. Davidson, *Theory of Nonneutral Plasmas* (Benjamin, Reading, MA, 1974).
3.  K. S. Fine, Ph.D. thesis, Univ. of California at San Diego (1989).
4.  A. A. Kabantsev, C. F. Driscoll, T. J. Hilsabeck, T. M. O'Neil, and J. H. Yu, in *Non-Neutral Plasmas IV*, edited by F. Anderegg, L. Schweikhard and C. F. Driscoll (American Institute of Physics, New York, 2002), pp. 277-286.
5.  A. A. Kabantsev, C. F. Driscoll, T. J. Hilsabeck, T. M. O'Neil, and J. H. Yu, Phys. Rev. Lett **87**, 225002 (2001).
6.  A. A. Kabantsev and C. F. Driscoll, Phys. Rev. Lett. **89**, 245001 (2002).
7.  A. A. Kabantsev, J. H. Yu, R. B. Lyncy, and C. F. Driscoll, Phys. Plasmas **10** (5), 1628 (2003).
8.  T. J. Hilsabeck and T. M. O'Neil, Bull. Am. Phys. Soc. **47**(9), 125 (2002).
9.  G. W. Mason, Phys. Plasmas **10** (5), 1231 (2003).
10. S. M. Crooks and T. M. O'Neil, Phys. Plasmas **2** (2), 355 (1995).

# Measurements of Plasma Expansion due to Background Gas in the Electron Diffusion Gauge Experiment

Kyle A. Morrison, Stephen F. Paul, and Ronald C. Davidson

*Plasma Physics Laboratory, Princeton University, Princeton, NJ 08543*

**Abstract.** The expansion of pure electron plasmas due to collisions with background neutral gas atoms in the Electron Diffusion Gauge experiment device is observed. Measurements of plasma expansion with the new, phosphor-screen density diagnostic suggest that the expansion rates measured previously were observed during the plasma's relaxation to thermal quasi-equilibrium, making it even more remarkable that they scale classically with pressure. Measurements of the on-axis, parallel plasma temperature evolution support the conclusion.

## INTRODUCTION

Pure electron plasmas are trapped in the Electron Diffusion Gauge (EDG) experiment device [1-4], a cylindrically symmetric, Malmberg-Penning trap [5-12] with inside diameter I.D. $= 2 \cdot R_w = 5.08$ cm. Malmberg-Penning traps have a uniform magnetic field parallel to the common axis of several cylindrical electrodes, and particles with the same sign of charge can be confined by charging two nonadjacent electrodes to a sufficiently large voltage. Previously reported experimental results [2] from the EDG experiment indicate that the plasma expansion rates measured in the high-vacuum regime (where asymmetry-induced expansion is negligible) are in good agreement with with the predicted expansion rates [13] derived using a warm fluid treatment of the plasma. The evolution of the inferred perpendicular temperature during this expansion, however, did not account for the clear decrease in electrostatic potential energy, prompting improvements to the EDG diagnostic systems.

Previously, axially-integrated density profiles obtained from the EDG experiment were accumulated from a series of plasmas by measuring the number of electrons passing through a small hole in a radially-movable collimating plate. The particles moving along magnetic field lines aligned with the small hole in the plate would pass through to a Faraday cup, giving a radial profile as the collimating plate was scanned. By forming several (well-reproduced) plasmas in succession, a series of line-integrated, radial density profiles could be obtained and used to follow the expansion of the plasma. To determine the plasma behavior at different gas pressures, helium gas was fed into the chamber at different, controlled rates.

CP692, *Non-Neutral Plasma Physics V*, edited by M. Schauer et al.
© 2003 American Institute of Physics 0-7354-0165-9/03/$20.00

# ELECTRON DENSITY AND TEMPERATURE DIAGNOSTICS

The Faraday cup density diagnostic has been replaced with a CCD camera focused on a biased, phosphor-coated glass screen, a setup based on the diagnostics developed by other groups [14, 15]. The phosphor is coated with aluminum both to reflect excess light from the plasma source (a 1.27 cm-diameter spiral filament) and act as an additional electrode. The aluminum coating is biased to a few kilovolts to accelerate the plasma electrons to the point that they can pass through it and excite the phosphor molecules. The light emitted by the phosphor passes through the glass screen, a glass vacuum window, a notch filter tuned to the peak emission wavelength of the P-43 phosphor, a camera lens, and a separate image intensifier on its way to the CCD camera. A grounded (10 wires/inch) copper grid is attached to the end of the trap, about 1 inch away from the biased screen, to make the accelerating electric field more uniform. Figure 1 shows the improved resolution of the phosphor-screen diagnostic.

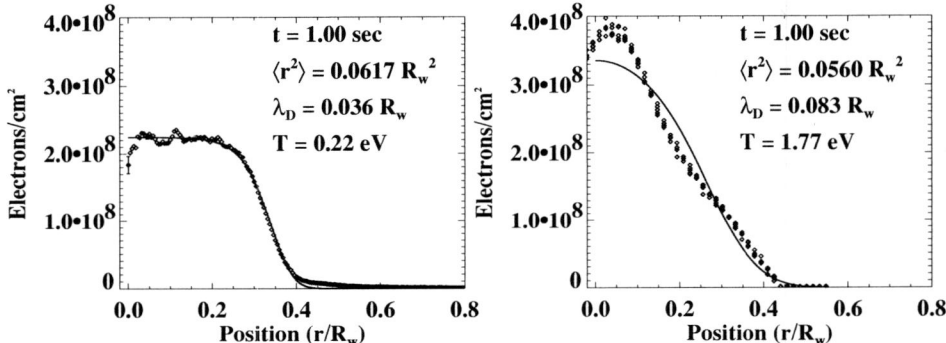

**FIGURE 1.** The figure on the left shows an example of the CCD-image-derived profiles, overlaid with a thermal quasi-equilibrium profile. The figure on the right shows an example of data obtained with the Faraday cup diagnostic, where each data point is from a different plasma.

The perpendicular electron temperatures displayed in Fig. 1 are estimated by fitting an ideal, thermal quasi-equilibrium density profile [13] to the measured, axially-integrated profile. The thermal quasi-equilibrium profile is

$$n(r,t) = \hat{n}(t)\exp\left\{\frac{e\phi(r,t) - e\hat{\phi}(t)}{T} - \frac{r^2}{\langle r^2 \rangle(t)}\left(1 + \frac{N_L e^2}{2T}\right)\right\}. \tag{1}$$

Here, $\hat{n}(t)$ is the central density as a function of time, $\phi(r,t)$ is the electrostatic potential, and $\hat{\phi}(t)$ is the electrostatic potential on axis ($r = 0$). This thermal quasi-equilibrium profile describes expanding, infinite-length, azimuthally symmetric plasmas that enjoy global energy conservation and elastic electron-neutral collisions, and have a spatially uniform temperature. Poisson's equation can be recast in a form that shows the underlying profile shape is dependent on only one parameter, $\gamma$, defined by $\gamma \equiv (\omega_r \omega_{ce} - \omega_r^2)/(\hat{\omega}_p^2/2) - 1$, where $\omega_{ce} = eB/m_e c$ is the electron cyclotron frequency, $\omega_r$ is the plasma rotation frequency, and $\hat{\omega}_p$ is the plasma frequency at $r = 0$. $\gamma$ is the only parameter necessary to describe the ideal density profiles. We also allow $\hat{n}(t)$ to vary in the fit for simplicity, though in principle it should be identifiable from the data.

On-axis parallel temperature measurements are performed as described by Eggleston [16]: the charge on one of the confining electrodes is slowly decreased, and the number of electrons escaping the trap as a function of time is recorded. The results are fit using the approximate relationship

$$\frac{d\ln(Q_{esc})}{d(e\phi_c)} = \frac{-1.05}{T_\parallel}, \tag{2}$$

where $Q_{esc}$ is the total amount of charge that has escaped, $-e$ is the charge of an electron, $\phi_c$ is the confining voltage on axis, and $T_\parallel$ is the parallel temperature in eV. In EDG, a charge-sensitive amplifier is capacitively coupled to the biased phosphor screen to measure the total charge that has escaped as a function of time. Typical plots of both $\ln(Q_{esc})$ versus $\phi_c$ and $Q_{esc}$ versus $\phi_c$ are displayed in Fig. 2.

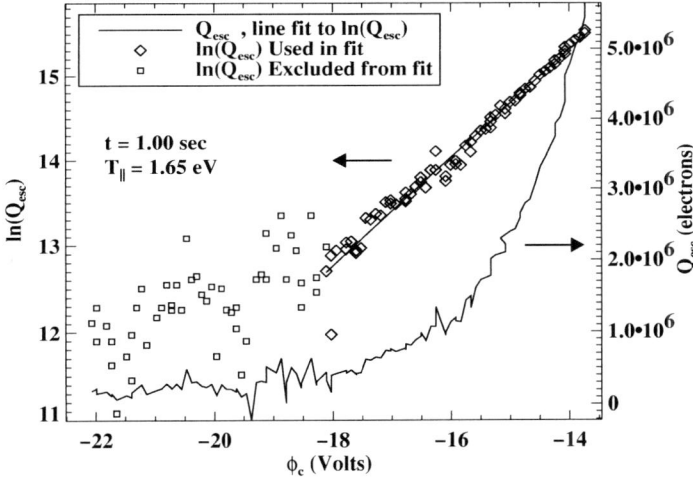

**FIGURE 2.** Total charge escaped versus confining voltage as a plasma is released from the trap. The diamonds denote the data used in the fit. This data was taken at magnetic field $B = 600G$, background gas pressure $P \sim 5 \times 10^{-9}$ Torr, filament heating voltage $V_h = 4.8$V, and filament bias voltage (the voltage at the center of the spiral filament) $V_b = -16.6$V. This plasma comprised $N \sim 5 \times 10^8$ electrons.

## MEASUREMENT OF PLASMA EXPANSION

The plasma's expansion rate is computed as the rate of change of its mean-square radius. The mean-square radius is approximated by

$$\langle r^2 \rangle = \frac{\int_0^{R_w} dr\, 2\pi r\, r^2\, Q(r)}{\int_0^{R_w} dr\, 2\pi r\, Q(r)}, \tag{3}$$

where $Q(r)$ is the axially-integrated density profile determined from one of the density diagnostics. For the Faraday cup diagnostic, $Q(r)$ corresponds to the axially-integrated density averaged over the collimating hole area (radius = 0.159 cm) at location $r$. For the phosphor-screen diagnostic, it corresponds to the axially integrated density averaged azimuthally between $r$ and $r + dr$ ($dr \sim .012$ cm, the width on the screen of one camera pixel's view).

Because many plasmas were needed to construct one measured density profile using the Faraday cup diagnostic and the trap conditions tend to drift with time, it was impossible to make profile measurements for plasmas held in the trap much longer than a second. The new, phosphor-screen density diagnostic measures the entire, axially-integrated density profile of a single plasma, allowing us to measure the evolution well past 1 second. The data in Fig. 3 show that the plasma behaves differently after it has been confined for about 3 seconds than it does initially.

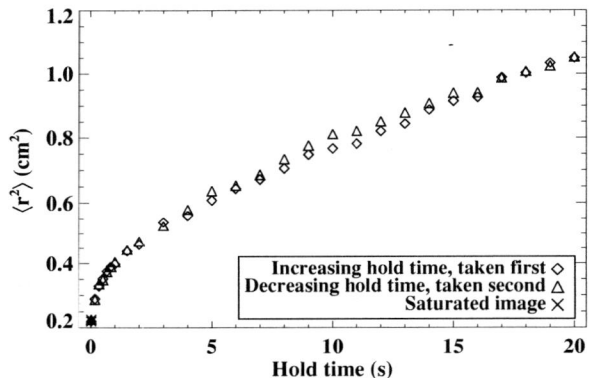

**FIGURE 3.** Plasma mean square radius as a function of time for the same experimental parameters as the data in Fig. 2. The agreement between the increasing hold time and decreasing hold time data indicate the plasma's reproducibility.

The density profiles for plasmas trapped longer than 3 seconds are fit somewhat better by the thermal quasi-equilibrium profiles than those for plasmas trapped for less time, but the inferred temperatures from the profile fits, shown in Fig. 4, hardly vary throughout the evolution. The parallel temperatures also shown in Fig. 4, however, rise dramatically at the beginning of the evolution. In addition, the initial evolutions of the plasma mean square radius and the parallel temperature appear to take the same amounts of time at several different filament conditions and background gas pressures below P $\sim 2 \times 10^{-7}$ Torr. The initial evolution of the plasmas in Fig. 3 may be due primarily to a transition to thermal quasi-equilibrium from an initially non-equilibrium state, rather than expansion due to background gas or trap asymmetries. The disagreement between the two temperature diagnostics is not presently understood.

The expansion rates measured previously on EDG were determined from plasmas trapped less than 1 second, meaning that they were computed from plasmas experiencing this apparent relaxation to thermal quasi-equilibrium. Figure 5 shows a comparison of

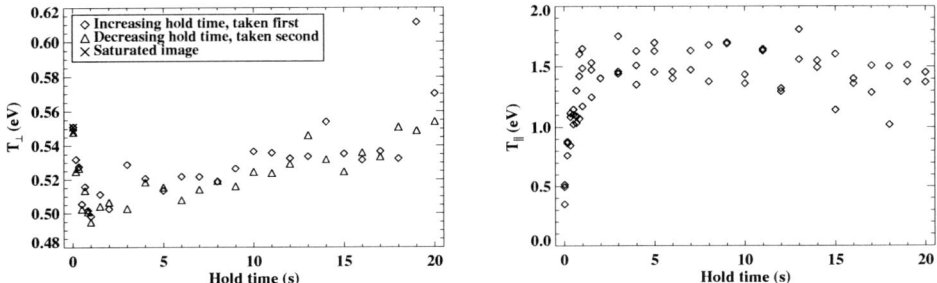

**FIGURE 4.** The inferred perpendicular temperature (left) and measured, on-axis parallel temperature evolution (right) for the same experimental parameters as the data in Fig. 2.

'late-time' expansion rates determined from the new profile data (excluding the initial relaxation, where possible) to the expansion rates measured previously. Note that while the older expansion rates start to level off to a value of about 0.1 cm²/s as the background gas pressure decreases, the late-time expansion rates are clearly smaller. Estimating expansion rates from the first 1 second of phosphor-screen profile data gives values that agree with the 0.1 cm²/s value measured previously at low pressures. The late-time expansion rates still level off at the lowest pressures, indicating that asymmetry-induced expansion is indeed affecting the measurements.

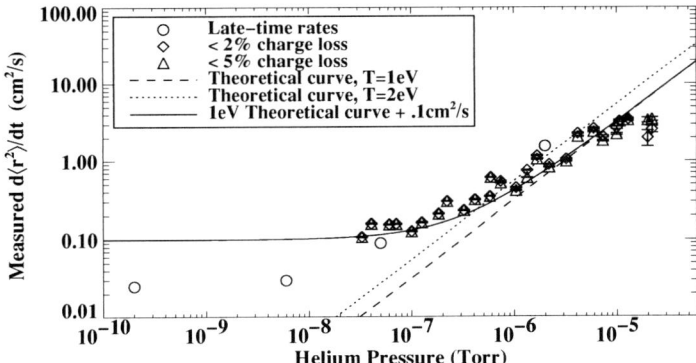

**FIGURE 5.** Comparison of expansion rates determined from the previous and present density diagnostics. The circles denote the new expansion rates computed by excluding the initial plasma relaxation at lower pressures.

The plasma expansion rate calculated [13] for thermal quasi-equilibrium profiles affected by background neutral gas is

$$\frac{d}{dt}\langle r^2 \rangle = \frac{2N_L e^2 \, v_{en}(T)}{m_e \omega_{ce}^2} \left(1 + \frac{2T}{N_L e^2}\right), \tag{4}$$

where $v_{en}(T) = n_R \sigma_{en} v_{Th}$ is the electron-neutral collision frequency, $T$ is the plasma temperature (in ergs), and $N_L$ is the line density of the plasma column. The theoretical curves in Fig. 5 agree with the data at higher pressures fairly well, despite the fact that the collisionally-induced expansion is superimposed upon the initial relaxation measured at lower pressures. Accordingly, we may infer that the fact that the plasma is not fully in thermal quasi-equilibrium does not prevent it from expanding at the same rate that a plasma in thermal quasi-equilibrium would. At high-vacuum pressures (above $P \sim 2 \times 10^{-7}$ Torr), the electron-neutral collision frequency is higher than the electron-electron collision frequency, so the temperature gradients that could exist in the plasma are also having a minimal effect on the classical expansion rate of the electron plasmas in EDG.

## ACKNOWLEDGMENTS

This research was supported by the Office of Naval Research, and in part supported by the U.S. Department of Energy.

## REFERENCES

1.  Chao, E. H., Davidson, R. C., Paul, S. F., and Morrison, K. A., *Phys. Plasmas*, **7**, 831–838 (2000).
2.  Morrison, K. A., Davidson, R. C., Paul, S. F., Belli, E. A., and Chao, E. H., *Phys. Plasmas*, **8**, 3506–3509 (2001).
3.  Chao, E. H., Davidson, R. C., Paul, S. F., and Morrison, K. A., "Effects of Background Gas Pressure on Pure Electron Plasma Dynamics in the Electron Diffusion Gauge Experiment," in *Proceedings of the 1999 Workshop on Nonneutral Plasmas: AIP Conference Proceedings 498*, edited by J. Bollinger, R. C. Davidson, and R. Spencer, American Institute of Physics, Melville, NY, 1999, pp. 278–289.
4.  Morrison, K. A., Davidson, R. C., Paul, S. F., and Jenkins, T. G., "Investigation of the Expansion Rate Scaling of Plasmas in the Electron Diffusion Gauge Experiment.," in *Non-Neutral Plasma Physics IV: Workshop on Non-Neutral Plasmas: AIP Conference Proceedings, Volume 606*, edited by F. Anderegg, L. Schweikhard, and C. F. Driscoll, American Institute of Physics, Melville, NY, 2002, pp. 416–421.
5.  deGrassie, J. S., and Malmberg, J. H., *Phys. Rev. Lett.*, **39**, 1077–1080 (1977).
6.  O'Neil, T. M., *Phys. Scripta*, **59**, 341–351 (1995).
7.  Gould, R. W., *Phys. Plasmas*, **2**, 2151–2163 (1995).
8.  Driscoll, C. F., Anderegg, F., Dubin, D. H. E., Jin, D. Z., Kriesel, J. M., Hollmann, E. M., and O'Neil, T. M., *Phys. Plasmas*, **9**, 1905–1914 (2002).
9.  Anderegg, F., Shiga, N., Dubin, D. H. E., Driscoll, C. F., and Gould, R. W., *Phys. Plasmas*, **10**, 1556–1562 (2003).
10. Bollinger, J. J., Kriesel, J. M., Mitchell, T. B., King, L. B., Jensen, M. J., Itano, W. M., and Dubin, D. H. E., *J. Phys. B*, **36**, 499–510 (2003).
11. Greaves, R. G., and Surko, C. M., *Nucl. Instrum. Meth. B*, **192**, 90–96 (2002).
12. Fajans, J., Gilson, E., and Backhaus, E. Y., *Phys. Plasmas*, **7**, 3929–3933 (2000).
13. Davidson, R. C., and Moore, D. A., *Phys. Plasmas*, **3**, 218–225 (1996).
14. Peurrung, A. J., and Fajans, J., *Rev. Sci. Instrum.*, **64**, 52–55 (1993).
15. Huang, X.-P., Fine, K. S., and Driscoll, C. F., *Phys. Rev. Lett.*, **74**, 4424–4427 (1995).
16. Eggleston, D. L., Driscoll, C. F., Beck, B. R., Hyatt, A. W., and Malmberg, J. H., *Phys. Fluids B*, **4**, 3432–3439 (1992).

# Experimental Investigation of m=1 Diocotron Mode Growth at Low Electron Densities

Stephen F. Paul, Kyle Morrison, Ronald C. Davidson

*Plasma Physics Laboratory, Princeton University, Princeton, New Jersey 08543, USA*

**Abstract.** Previous experiments on the Electron Diffusion Gauge showed that the diocotron mode damping increases with higher neutral gas filling pressure. Yet the energy dissipated from a rotating plasma by collisions with neutrals is predicted to excite the mode. To resolve this, experiments have been conducted to examine the coupling between expansion and the m=1 diocotron mode. Results from recent experiments have shown interesting phenomena: 1) The degree and sensitivity of mode growth is observed to be strongly dependent on filament conditions. Mode growth rates of nearly 20 sec$^{-1}$ have been observed even with negligible resistive drive. Specifically, at low filament bias voltages (and correspondingly low electron densities $\sim$ 1-2 $\times 10^7$ electrons/cm), the mode growth is very sensitive to the heating voltage across the filament, even though changes in filament heating voltage barely affect the plasma expansion, the plasma density profile, the filament emission, or the resulting electron density. 2) At low neutral gas pressure ($< 10^{-9}$ Torr), the diocotron mode growth rate increases with neutral pressure. However, the growth rate is several orders of magnitude larger than theoretical predictions.

## INTRODUCTION

Electron plasmas confined in Malmberg-Penning traps have been used to investigate important fundamental non-neutral plasma phenomena [1, 2, 3, 4]. The research emphasis for the Electron Diffusion Gauge (EDG) experiment has been the investigation of the effects of collisions between the confined pure electron plasma and a low-pressure neutral gas, with the goal of using the measured interaction to determine the neutral gas pressure over a wide range. The expansion of the plasma column has been related theoretically to the rate of collisions between the electrons and the neutrals [5, 6]. The expansion rate has also been measured experimentally on the EDG experiment [7, 8] using a Faraday cup collector and observed to scale classically [9] at relatively high neutral pressures ($P > 10^{-6}$), but this method requires hundreds of repeated discharges to determine the expansion rate. This approach was not only time-consuming, but required excellent plasma reproducibility and low measurement noise. The scatter in the data and the presence of asymmetry-induced expansion limited the detection of neutral pressure in the EDG via this technique to $P > 10^{-8}$ Torr.

As an alternative, the possibility of a non-destructive pressure measurement using the dependence of the $m = 1$ diocotron mode growth rate on background gas pressure has been explored more recently [8]. Here "diocotron mode" means the low-frequency, electrostatic oscillation with azimuthal mode number $m = 1$. As is typical in Malmberg-Penning traps, a uniform axial magnetic field provides the radial confinement, and applied electric potentials on cylindrical end-electrodes provide the axial confinement. For

CP692, *Non-Neutral Plasma Physics V*, edited by M. Schauer et al.

small-amplitude perturbations, the oscillations result from the plasma column being displaced from and precessing about the trap axis (the center of symmetry of the cylindrical electrodes that surround the plasma). The precession occurs at the diocotron mode frequency; the mode amplitude $A$ is the distance from the center of the plasma column to the trap axis; and the growth rate is $(1/A)dA/dt$. The advantage of this method is that the need to terminate the plasma and measure the density profile at successive times in the evolution is avoided because the entire mode evolution is recorded in one discharge. In the absence of electron-neutral collisions, the diocotron mode is predicted to be marginally stable. When collisions are present, they dissipate energy from the plasma column, forcing it to move nearer to the surrounding conducting wall, (i.e., the mode amplitude should grow). A calculation assuming that the expansion of the plasma is much slower than the growth time has predicted an instability growth rate that scales linearly with neutral pressure [10, 11],

$$ \gamma_n = \frac{\nu_{en}}{\omega_{ce}} \omega_\infty \tag{1} $$

where $\omega_\infty$ is the frequency of the diocotron mode for an infinite-length plasma column and $\nu_{en}$ is the electron-neutral collision frequency. For the plasma parameters in the EDG and a neutral gas pressure of $P = 5 \times 10^{-10}$ Torr, the mode growth rate is predicted to be small: about $10^{-4}\mathrm{sec}^{-1}$. In the early EDG experiments [8], the mode was observed to decay rather than grow in most cases, with the damping rate increasing with both pressure and electron density. Damping of trapped-particle modes (analogous to oppositely phased diocotron modes) has been observed in others' experiments and explained by associated particle transport resulting from applied electric asymmetries [12]. The damping results from velocity scattering of particles near the separatrix that are trapped as a result of the applied asymmetry. The damping of diocotron modes in EDG may be caused by trapped particles resulting from intrinsic magnetic asymmetries as well.

More recent experiments showed that at low pressures ($5 \times 10^{-11}$ Torr $< P < 4 \times 10^{-8}$ Torr) where asymmetry-induced transport dominates, only a weak, $P^{1/4}$ pressure dependence on damping is observed. However, in isolated cases ($P = 2$–$4 \times 10^{-9}$ Torr) where mode growth was resistively forced, the growth rate was seen to be very sensitive to changes in pressure. The purpose of this paper is to examine whether this sensitivity of the growth rate to neutral pressure can be observed, reproduced, and controlled at much lower pressures.

The EDG trap and its operation are similar to that in other experiments and has been described previously [7]. The co-linear, cylindrical copper trap electrodes have an inner radius of $R_w = 2.54$ cm, and the applied potentials at the end-electrodes are $-145$ V. The magnetic field, variable up to 1 kG, is generated with a solenoid whose axial current profile is adjusted so that the field in the trap region is constant to within 0.2% of the maximum field. The trap assembly is enclosed in an aluminum vacuum chamber that is evacuated to nearly $3 \times 10^{-11}$ Torr with a turbomolecular pump and a cryogenic pump. Helium gas is bled into the chamber with a precision metering valve and the fill pressure is measured with an Ionivac extractor gauge.

The trapped electron plasma has an initial density in the range $8 \times 10^6$ cm$^{-3} < n < 3 \times 10^7$ cm$^{-3}$, temperature $T \sim 1$–$2$ eV, radius $R_p \simeq 0.6$ cm, and length $L_p \simeq 15$ cm. For these parameters, the Debye length $\lambda_D = (T/4\pi n e^2)^{1/2}$ is smaller than the plasma radius ($R_p \simeq 6\lambda_D$) and $\omega_{pe}^2/\omega_{ce}^2 < 0.01$, where $\omega_{pe} = (4\pi n e^2/m_e)^{1/2}$ is the electron plasma frequency, and $\omega_{ce} = eB/m_e c$ is the electron cyclotron frequency. The $m = 1$ diocotron frequency ranges from 10 kHz to 100 kHz.

One of the copper cylinders that surrounds the plasma is divided into half-cylinders, and the mode amplitude is determined from the current induced by the precessing column to the half-cylinders [13]. Any odd-numbered diocotron mode ($m = 1, 3, \ldots$) can be measured with this diagnostic, but the $m = 1$ mode is dominant in the EDG.

## SENSITIVITY OF THE $M = 1$ DIOCOTRON MODE TO FILAMENT CONDITIONS AND BACKGROUND PRESSURE

As mentioned above, collisions with neutral particles can apply a torque, causing the column to expand symmetrically, or dissipate energy from the plasma column, displacing it towards the surrounding conducting wall. To isolate and improve the sensitivity of diocotron mode behavior to pressure, the EDG was operated at low electron density to minimize plasma expansion.

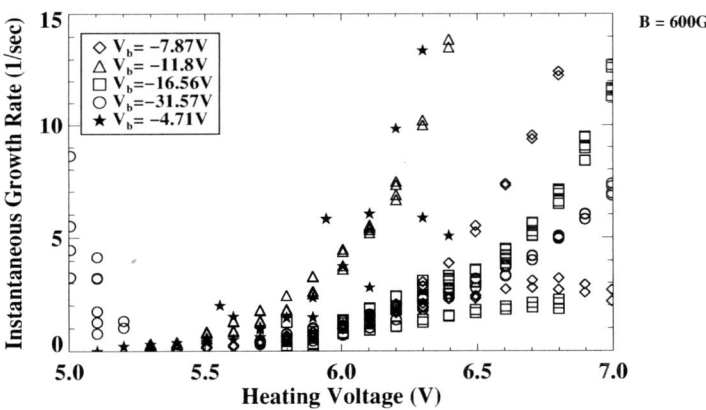

**FIGURE 1.** Plot of growth rate of the diocotron mode versus filament heating voltage for several values of filament bias voltage. The growth rate is measured at the point where the measured mode amplitude is 0.05 V.

The filament conditions were observed to have a profound effect on the growth rate. Even in the absence of resistive-wall mode growth, the diocotron mode was made to grow at a substantial rate by increasing the filament heating voltage, even by modest amounts. Growth rates exceeding 20 sec$^{-1}$ are observed with heating voltages of 6–7 volts. Figure 1 shows the mode growth increasing by factors of 50 or more with a change in heating voltage of less than 1 Volt (a 15% variation). Figure 1 also shows a dependence on bias voltage, with a distinct peak in sensitivity at about $-12$ Volts and falling substantially by $V_b = -8$ or $-16$ Volts.

To examine this further, a dedicated bias voltage scan was conducted $-4 > V_b > -15$ V at the base pressure. As shown in Figure 2, a narrow peak in mode sensitivity at a filament bias of about $-5$ V is evident. A number of measurements were made to detect whether the emission current, electron line density, temperature, or the initial column density profiles were similarly sensitive to the filament voltage. Neither the filament emission nor the electron density was found to be a function of heating voltage in the range 4.8 V $< V_h <$ 6.5 V. This is a result of the fact that the filament emission is space-charge limited for $V_h > 4.8$ V and this range of filament bias voltage. The emission and density increase with $V_b$ as expected because the plasma potential follows $V_b$, but no peak in density or emission occurs at $-5$ V that is coincident with the strong diocotron mode growth.

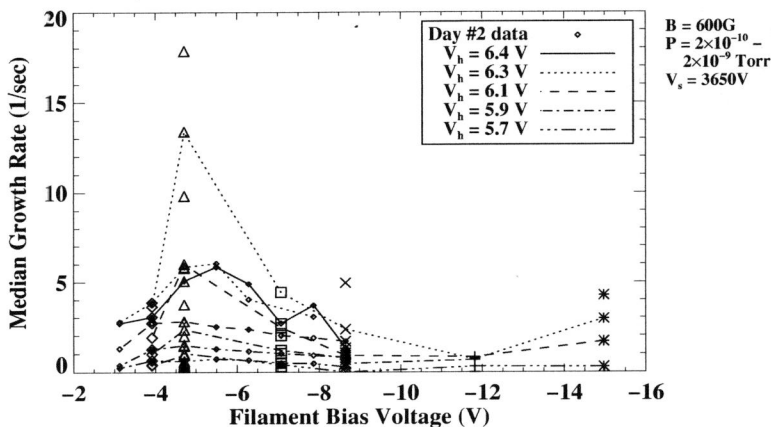

**FIGURE 2.** Plot of median growth rate of the diocotron mode versus filament bias voltage. Contours of filament heating voltage are indicated. The median growth rate is evaluated between A = 0.02 V and the point where $dA/dt$ is a maximum.

To determine whether temperature and profile effects might explain this behavior, filament heating voltage scans for 4.2 V $< V_h <$ 6.7 V at $V_b = -4.7$ V were performed. The perpendicular temperature $T_\perp$ is inferred by fitting the measured axially-integrated density profile with a calculated equilibrium and rises linearly from 0.3 to 0.6 eV. The parallel temperature $T_\parallel$ was determined by measuring the fast electrons escaping from the trap as the discharge is terminated and assuming a Maxwellian parallel velocity distribution. $T_\parallel$ is measured to be $1.7 \pm 0.3$ eV—independent of heating voltage. The estimated electrostatic energy per electron is found to range only from 2.1 to $2.2 \pm 0.1$ eV per electron.

The only obvious difference is that the predicted initial plasma radius changes with $V_h$. The bias voltage $V_b$ is applied to the center of the filament, and the heating voltage $V_h$ is applied across the filament. At low filament bias potentials, $V_h$ can exceed $V_b$, forcing the bias at the edge of the filament to become positive and suppressing electron emission. As $V_b$ is increased, the area of the filament that is positively biased becomes larger and the emitting area smaller. Throughout these measurements at the base pressure, the momentum transport rate due to expansion is less than 1.5 sec$^{-1}$, ten times less than the peak diocotron mode growth rate.

The previous data do not indicate how the introduction of neutral atoms affect the mode growth or the filament conditions where the mode growth was observed, so a pressure scan was conducted at $V_h = 6$ V. As shown in Figure 3, the $V_b$ envelope for strong mode growth between 4 V $< V_b <$ 9 V remains. Figure 3 also illustrates that strong mode growth can appear under other conditions; diocotron mode growth is observed for $V_b > 10$ V as well. The mode growth at the highest pressure in this scan is often non-exponential, and the growth at higher $V_b$ is also often delayed for several seconds or preceded by the damping of a small initial mode.

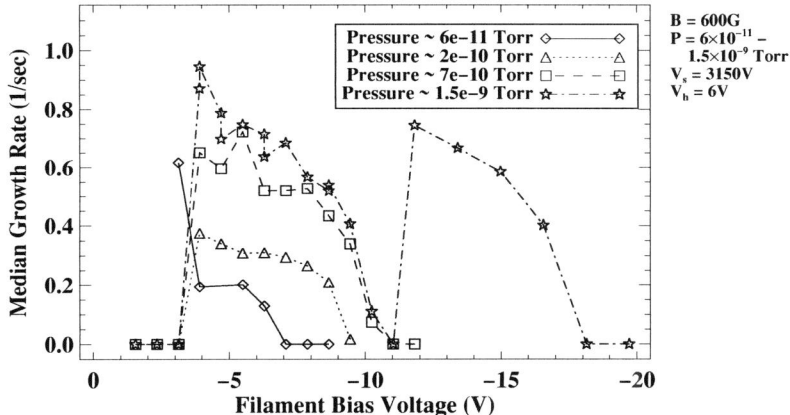

**FIGURE 3.** Plot of median diocotron mode growth rate versus $V_b$ for several values of helium filling pressure. The growth rate is the median value between A = 0.02 V and where dA/dt is a maximum.

A more detailed measurement of pressure dependence was conducted by simply allowing the base pressure to rise over time as the filament heated the trap. The pressure was allowed to rise from $6 \times 10^{-11}$ Torr to $2 \times 10^{-8}$ Torr. In Figure 4, the mode growth is seen to increase almost linearly with pressure up to $6 \times 10^{-10}$ Torr, a factor of six above the base pressure, above which the mode growth levels off. Also plotted is the scaled value for mode growth that would result from electron-neutral collisions, assuming that the base pressure residual gas is hydrogen and the electron temperature is 1 eV and isotropic. Although the scaling is consistent with the theory of collision-based transport, the scale factor is six orders of magnitude larger, clearly indicating that some additional, strong effect is present.

In summary, it is found that:

- The diocotron mode can be made to grow in response to electron-neutral collisions without resistive forcing by employing low filament bias voltages that produce a low-density ($N_L = 1 - 2 \times 10^7$cm$^{-1}$), small-diameter plasma column.
- The mode growth rate increases strongly with the filament heating voltage, with measured growth rates as high as 20 sec$^{-1}$.

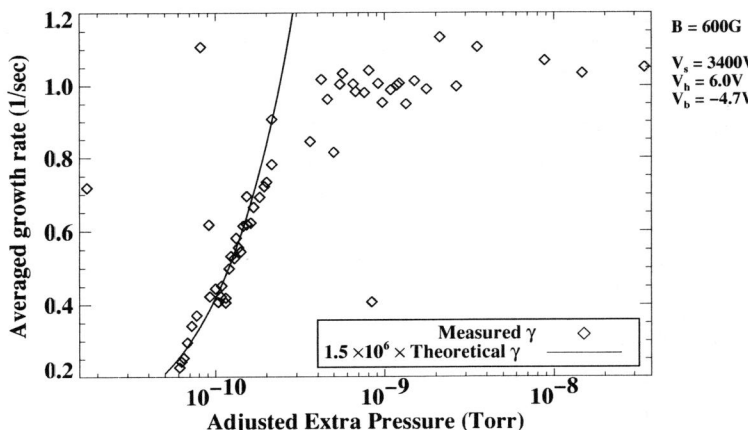

**FIGURE 4.** Plot of average growth rate to maximum $dV/dt$ of the diocotron mode versus $V_b$ for several values of helium filling pressure. The solid line is the mode growth rate predicted from electron-neutral collisions in Eq. (1) multiplied by a factor of 1.5 million.

- The sensitivity of mode growth to pressure for $P < 10^{-9}$ Torr is much greater than that previously observed for mode damping.
- The scaling of growth rate with pressure is in agreement with theory (both are approximately linearly proportional), but the growth rates are orders of magnitude larger than predicted for electron-neutral collisions.

This research was supported by the Office of Naval Research and in part by the U.S. Department of Energy.

## REFERENCES

1. Anderegg, F., Schweikhard, L., and Driscoll, C. F., editors, *Non-Neutral Plasma Physics IV: Workshop on Non-Neutral Plasmas: AIP Conference Proceedings, Volume 606*, American Institute of Physics, Melville, NY, 2002.
2. Bollinger, J. J., Spencer, R. L., and Davidson, R. C., editors, *Non-Neutral Plasma Physics III: AIP Conference Proceedings 498*, American Institute of Physics, Melville, NY, 1999.
3. Davidson, R. C., *Physics of Nonneutral Plasmas*, Addison-Wesley, Redwood City, 1990.
4. Fajans, J., and Dubin, D. H. E., editors, *Non-Neutral Plasma Physics II*, vol. AIP 331, AIP Conf. Proc., New York, 1995.
5. Davidson, R., and Moore, D., *Phys. Plasmas*, **3**, 218–225 (1996).
6. Davidson, R. C., and Chao, E. H., *Phys. Plasmas*, **3**, 2615 (1996).
7. Chao, E. H., Davidson, R. C., and Paul, S. F., *J. Vac. Sci. Technol. A*, **17**, 2050 (1999).
8. Chao, E. H., Davidson, R. C., Paul, S. F., and Morrison, K. A., *Phys. Plasmas*, **7**, 831–838 (2000).
9. Morrison, K. A., Davidson, R. C., Paul, S. F., Belli, E. A., and Chao, E. H., *Phys. Plasmas*, **8**, 3506–3509 (2001).
10. Davidson, R. C., and Chao, E. H., *Phys. Lett. A*, **219**, 95–101 (1996).
11. Davidson, R. C., and Chao, E. H., *Phys. Plasmas*, **3**, 3279 (1996).
12. Kabantsev, A. A., Yu, J. H., Lynch, R. B., and Driscoll, C. F., *Phys. Plasmas*, **10**, 1628–1635 (2003).
13. Kapetanakos, C., and Trivelpiece, A., *J. Appl. Phys.*, **42**, 4841–4847 (1971).

# Numerical Study of Diocotron Instability with MDGRAPE-2

Yuichi Yatsuyanagi*, Yasuhito Kiwamoto* and Toshikazu Ebisuzaki†

*Graduate Scool of Human and Environmental Studies, Kyoto University, Sakyo, Kyoto 606-8501,
Japan
†RIKEN (The Institute of Physical and Chemical Research), Wako, Saitama 351-0198, Japan

**Abstract.** The diocotron instability in a low-density non-neutral electron plasma is examined via numerical simulations. The vortex method is used in the simulations. A special-purpose computer, MDGRAPE-2, accelerates the calculations of the Biot-Savart integral in the vortex method. The diocotron modes reproduced by the simulations agree with the results obtained by the linear theory. This indicates that the simulation results are qualitatively correct. The linear growth rates of the diocotron instability in the simulations also agree with the theoretical ones. This implies that MDGRAPE-2 gives the sufficiently accurate results for the calculations of the Biot-Savart integral. A boundary effect from the conducting wall is discussed. It is concluded that the electric field induced by the conducting wall makes the nonlinear stage unstable and causes the clumps to merge.

## INTRODUCTION

Many features of non-neutral electron plasmas have been investigated both experimentally and theoretically [1, 2, 3, 4, 5, 6]. Above all, one of the most ubiquitous phenomena is the diocotron instability observed in a low-density ($\omega_{pe}^2 \ll \omega_{ce}^2$) non-neutral electron plasma column confined radially by a uniform axial magnetic field. The diocotron instability was first examined theoretically by MacFarlane *et al.* [7], and Levy*et al.* [8, 9, 10], and observed experimentally by Webster [11], Kapetanakos *et al.* [12], and Peurrung *et al.* [13]. The linear theory for the diocotron instability has been developed and well understood [14, 6]. Thus we have chosen this phenomenon for the qualitative and quantitative benchmark of our simulation model and method [15].

As a simulation model, we use a vortex method [16, 17, 18]. As a simulation method, we use a special-purpose computer, MDGRAPE-2. To improve performance of vortex simulations, one may needs special methods, such as the vortex-in-cell code. However, we took a way to use a special-purpose computer, MDGRAPE-2. It was originally designed for molecular dynamics simulations, and accelerates calculations of the Coulomb interactions, the van der Waals interactions, and so on. We find that MDGRAPE-2 can accelerate calculations of the Biot-Savart integral [19]. You will see that MDGRAPE-2 is an "accelerator" not only for molecular dynamics simulations but also for plasma simulations.

The diocotron modes obtained by the simulations agree with the results obtained by the linear theory. This indicates that the simulation results are qualitatively correct. The linear growth rates of the diocotron instability in the simulations also agree with the theoretical ones. This implies that MDGRAPE-2 gives the sufficiently accurate results

CP692, *Non-Neutral Plasma Physics V*, edited by M. Schauer et al.

for the calculations of the Biot-Savart integral.

The angular velocities of the particles are influenced by the radius of the conducting wall. If the conducting wall radius is small, the nonlinear stage is unstable. It is concluded that the electric field induced by the conducting wall makes the nonlinear stage unstable and causes the clumps to merge.

## BASIC EQUATIONS

We use the following basic equations:

$$n_e m_e \left( \frac{\partial}{\partial t} + (\mathbf{u} \cdot \nabla) \right) \mathbf{u} = -\nabla p - e n_e (\mathbf{E} + \mathbf{u} \times \mathbf{B}), \tag{1}$$

$$\frac{\partial n_e}{\partial t} + \nabla \cdot (n_e \mathbf{u}) = 0, \tag{2}$$

$$\nabla \cdot \mathbf{E} = \frac{e n_e}{\varepsilon_0}, \tag{3}$$

$$\omega_z = \hat{\mathbf{z}} \cdot \nabla \times \mathbf{u}, \tag{4}$$

$$\mathbf{B} = B_0 \hat{\mathbf{z}}, \tag{5}$$

where $n_e$, $m_e$, $e$, $p$, and $B_0$ are the number density of electrons, the electron mass, the electron charge, the kinetic pressure, and the uniform magnetic field in the $z$ direction, respectively. Notation $\omega_z$ is the $z$-component of the vorticity. Notations $\mathbf{u}$, $\mathbf{B}$, and $\mathbf{E}$ are the flow velocity, the magnetic field, and the electric field on the $x$-$y$ plane. A unit vector in $z$ direction is denoted by $\hat{\mathbf{z}}$. Equations (1), (2), and (3) are the equation of motion, the equation of continuity, and Gauss' theorem, respectively.

In the present analysis, a cold-fluid guiding-center approximation is adopted. In this limit, two-dimensional electron fluid motion can be determined by the Euler equation. Thus we use the vortex method in the simulations to calculate the time evolution of the electron fluid.

$$\frac{d\mathbf{r}_k(t)}{dt} = \mathbf{u}(\mathbf{r}_k(t), t), \tag{6}$$

$$\frac{d\Omega_k(t)}{dt} = 0, \tag{7}$$

$$\mathbf{u}(\mathbf{r}_k(t), t) = \sum_{i \neq k} \Omega_i(t) \nabla G(\mathbf{r}_k(t) - \mathbf{r}_i(t)) \times \hat{\mathbf{z}}, \tag{8}$$

where $\mathbf{r}_k(t)$, $\Omega_k(t)$ are the vecocity and the vorticity of the $k$th point vortex. The function $G(\mathbf{r})$ is the two-dimensional Green function for the Poisson equation. Velocity at $\mathbf{r}_k(t)$ is determined by the Biot-Savart integral (8).

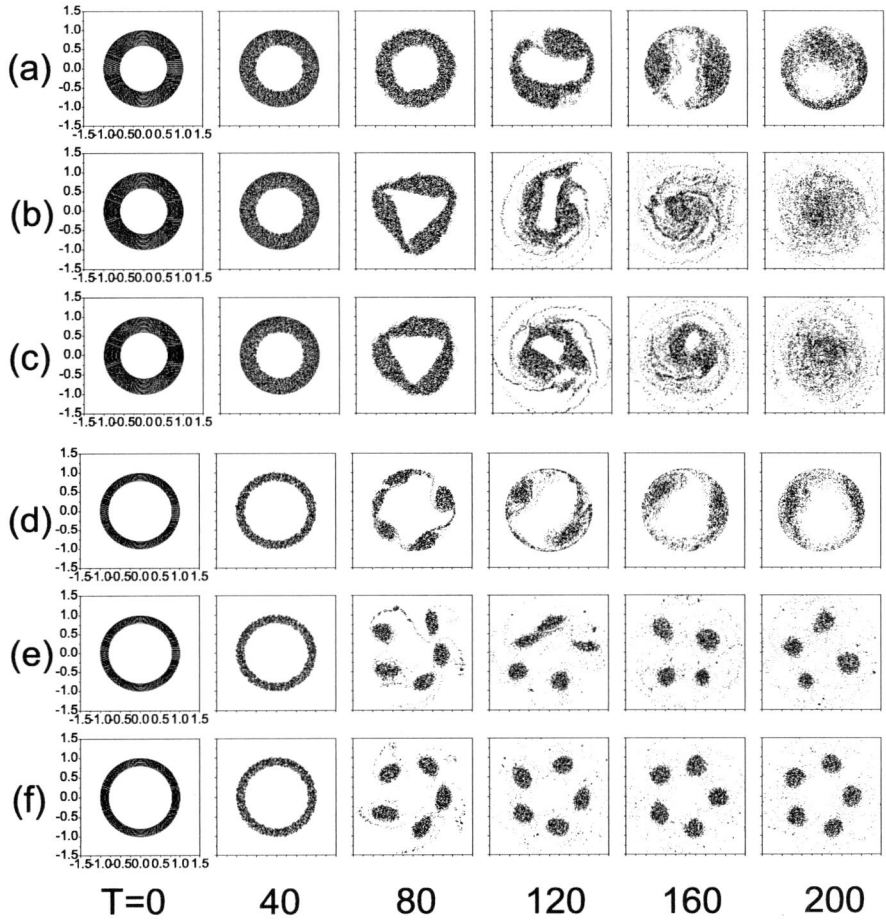

**FIGURE 1.** Time evolutions of the electron distributions at $T = 0, 40, 80, 120, 160$ and $200$ are shown. Initial outer radii of the electron distributions are denoted by $R_0$, which are all the same in (a) $-$ (f). Initial inner radii of the electron distributions are $R_1 = 0.6R_0$ for (a), (b), and (c), $R_1 = 0.8R_0$ for (d), (e), and (f). Conducting walls are located at $R_w = 1.1R_0$ for (a) and (d), $1.6R_0$ for (b) and (e), and $\infty$ for (c) and (f).

## SIMULATION RESULT

In Fig. 1, time evolutions of the electron distributions at $T = 0, 40, 80, 120, 160$, and $200$ are shown. The linearly most unstable modes $m$ obtained by the linear theory are $m = 2$ for (a), $m = 3$ for (b), $m = 3$ for (c), $m = 4$ for (d), $m = 5$ for (e), and $m = 5$ for (f). Thus, the unstable modes obtained by the simulations agree with the modes obtained by the linear theory.

To check the accuracy of the simulations quantitatively, we compare the linear growth

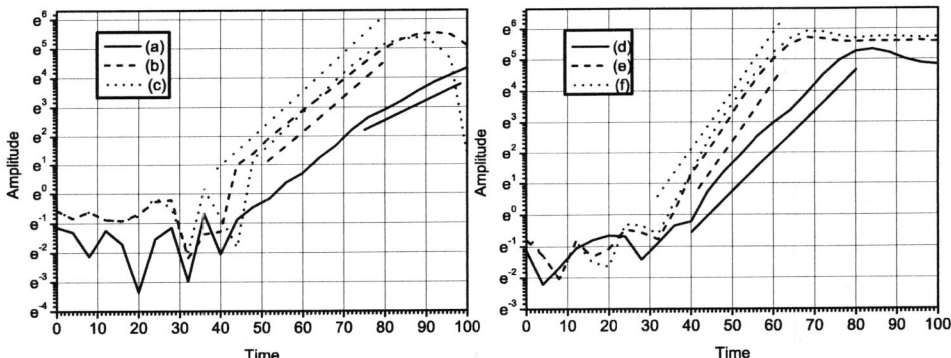

**FIGURE 2.** Time evolutions of the linearly most unstable Fourier coefficients corresponding to (a) − (f) in Fig. 1 are plotted.

rates obtained by the simulations to the rates obtained by the linear theory. A dispersion relation for complex eigenfrequency $\omega$ is given by [6],

$$\left(\frac{\omega}{\omega_D}\right)^2 - b_\ell \left(\frac{\omega}{\omega_D}\right) + c_\ell = 0, \tag{9}$$

where

$$b_\ell = \ell \left[1 - \left(\frac{R_1}{R_0}\right)^2\right] + \left[1 - \left(\frac{R_1}{R_0}\right) - 2\ell\right] \left(\frac{R_0}{R_w}\right)^{2\ell}, \tag{10}$$

$$c_\ell = \ell \left[1 - \left(\frac{R_1}{R_0}\right)^2\right] \left[1 - \left(\frac{R_1}{R_0}\right)^{2\ell}\right] - \left[1 - \left(\frac{R_0}{R_w}\right)^{2\ell}\right] \left[1 - \left(\frac{R_1}{R_0}\right)^{2\ell}\right]. \tag{11}$$

Time evolutions of the Fourier coefficients for Fig. 1 are given in Fig. 2. The curved lines indicate the values obtained by the simulations. The straight lines indicate the growth rates obtained by the linear theory. We conclude that the growth rates obtained by the simulations agree with the rates obtained by the linear theory. This indicates that MDGRAPE-2 gives sufficient precision for the vortex method.

We can see that the final state strongly depends on the wall radius in Fig. 1. In Fig. 1 (d) − (f), if the wall radius is small, the merging of the clumps occurs. The most unstable modes do not survive in (d) and (e), while the most unstable mode 5 survives in the nonlinear stage in (f). So, we focus on the results in Fig. 1 (d) − (f) to reveal the mechanism of the merging. In Figs. 3 and 4, we plot time evolutions of azimuthal positions of point vortices. In these figures, a bundle of lines indicates an azimuthal bunching. This means that the clumps are produced. Slope of a line indicates an angular velocity of a point vortex. If the slopes of the lines during the nonlinear stage is constant, the nonlinear stage is stable, which is shown in Fig. 3 (f). On the other hand, in the case of smaller wall radius, the slopes are still varying during the nonlinear stage and mergings occur, which is shown in Fig. 3 (d). It shows that the nonlinear stage is unstable. It is

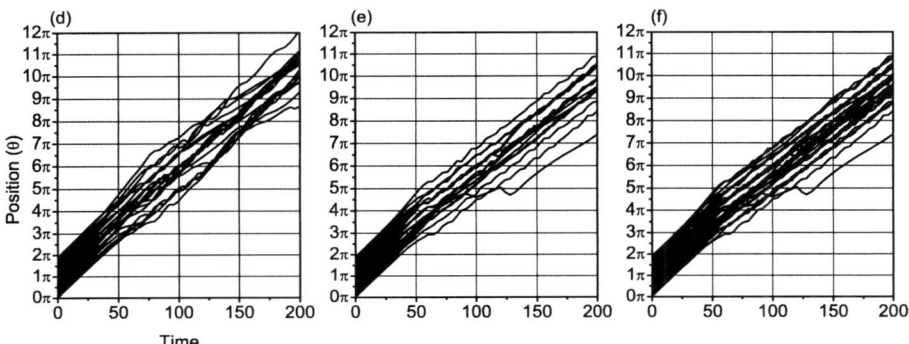

**FIGURE 3.** Time evolutions of azimuthal positions of point vortices in Fig. 1 (d), (e) and (f) are plotted. Initial point vortices at $\theta = n\pi/20$, $r = 0.93R_0$ ($n = 0, 1, 2, \cdots, 19$) are traced.

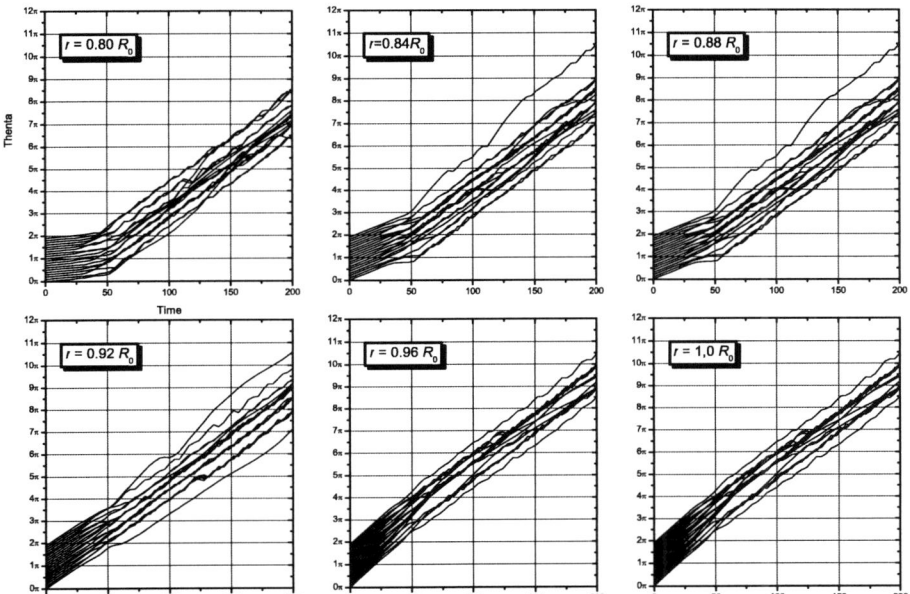

**FIGURE 4.** Time evolutions of azimuthal positions of point vortices in Fig. 1 (f) are plotted. Initial point vortices at $\theta = n\pi/20$, $r = 0.8R_0$, $0.84R_0$, $0.88R_0$, $0.92R_0$ and $0.93R_0$ ($n = 0, 1, 2, \cdots, 19$) are traced.

due to the electric field induced by the conducting wall. We conclude that the electric field induced by the conducting wall makes the nonlinear stage unstable and causes the clumps to merge. In Fig. 4, it is found that the transition from the linear growth stage to the nonlinear stage yields the change of the slopes. At that time, the momentum transfer from outer to inner regions occurs.

# CONCLUSIONS

The diocotron instability in a low-density non-neutral electron plasmas is examined via numerical simulations. In the present analysis, a cold-fluid guiding-center approximation is adopted. In this limit, two-dimensional electron fluid motion can be determined by the Euler equation. Thus we use the vortex method in the simulations to calculate the time evolution of the electron fluid. A special-purpose computer, MDGRAPE-2, is used to accelerate the calculations of the Biot-Savart integral in the vortex method. The diocotron modes and the linear growth rates obtained by the simulations agree with the results obtained by the linear theory. If the conducting wall radius is small, the nonlinear stage is unstable. The electric field induced by the conducting wall makes the nonlinear stage unstabe and causes the clumps to merge.

# ACKNOWLEDGMENTS

This work was supported by the Research Fellowships of the Japan Society for the Promotion of Science for Young Scientists.

# REFERENCES

1. Kiwamoto, Y., Ito, K., Sanpei, A., and Mohri, A., *Phys. Rev. Lett.*, **85**, 3173 (2000).
2. Kiwamoto, Y., Ito, K., Sanpei, A., Mohri, A., Yuyama, T., and Michishita, T., *J. Phys. Soc. Jpn.*, **68**, 3766 (1999).
3. Schecter, D. A., Dubin, D. H. E., Fine, K. S., and Driscoll, C. F., *Phys. Fluids*, **11**, 905 (1999).
4. Schecter, D. A., Dubin, D. H. E., Cass, A. C., Driscoll, C. F., Lansky, I. M., and O'Neil, T. M., *Phys. Fluids*, **12**, 2397 (2000).
5. Schecter, D. A., and Dubin, D. H. E., *Phys. Fluids*, **13**, 1704 (2001).
6. Davidson, R. C., *An Introduction to the Physics of Nonneutral Plasmas*, Addison-Wesley, California, 1990, chap. 6.
7. MacFarlane, C. C., and Hay, H. G., *Proc. Phys. Soc. London, Sect. B*, **63**, 409 (1950).
8. Levy, R. H., *Phys. Fluids*, **8**, 1288 (1965).
9. Levy, R. H., *Phys. Fluids*, **11**, 920 (1968).
10. Levy, R. H., Daugherty, J. D., and Buneman, O., *Phys. Fluids*, **12**, 2616 (1969).
11. Webster, H. F., *J. Appl. Phys.*, **26**, 1386 (1955).
12. Kapetanakos, C. A., Hammer, D. A., Striffler, C., and Davidson, R. C., *Phys. Rev. Lett.*, **30**, 1303 (1973).
13. Peurrung, A. J., and Fajans, J., *Phys. Fluids*, **5**, 493 (1993).
14. Davidson, R. C., and Felice, G. M., *Phys. Plasmas*, **5**, 3497 (1998).
15. Yatsuyanagi, Y., Kiwamoto, Y., Ebisuzaki, T., Hatori, T., and Kato, T., *Phys. Plasmas*, **10**, 3188 (2003).
16. Christiansen, J. P., *J. Comput. Phys.*, **135**, 189 (1997).
17. Koumoutsakos, P., *J. Comput. Phys.*, **138**, 821 (1997).
18. Leonard, A., *J. Comput. Phys.*, **37**, 289 (1980).
19. Yatsuyanagi, Y., Ebisuzaki, T., Hatori, T., and Kato, T., *Phys. Plasmas*, **10**, 3181 (2003).

# Two-Fluid Model for the Evaluation of the $m_\theta=1$ Diocotron Instability

Gianni Coppa, Gian Luca Delzanno, Federico Peinetti

*INFM (Istituto Nazionale per la Fisica della Materia) and Dipartimento di Energetica, Politecnico di Torino, Corso Duca degli Abruzzi 24, 10129 Torino (Italy)*

**Abstract.** The evolution of the $m_\theta=1$ diocotron mode for hollow density profiles has been an open problem of non-neutral plasma physics for more than a decade. In fact, the classic 2D fluid model leads to results that are in substantial disagreement with the ones provided by experiments. In the present work, a simplified kinetic model is proposed, in order to provide a quantitative estimate of the role played by kinetic effects on the evolution of the $m_\theta=1$ diocotron instability. A two-fluid model is deduced and a linear analysis of stability is performed, showing that kinetic effects can play an important part in the description of the evolution of the $m_\theta=1$ diocotron mode.

## INTRODUCTION

The work represents an attempt of evaluating the effect of the kinetic energy spread of the particles on the diocotron instability. In fact, as already pointed out by Peurrung and Fajans [1], electrons with different kinetic energy have different trajectories and different penetration depths in the region of the side electrodes; for that reason, they experience a different electric field and a different drift velocity. To evaluate quantitatively this effect, the motion of a single electron in the $z$ direction has been considered. The equilibrium 3D electrostatic potential $\phi(r,z)$ is calculated by solving the non-linear Poisson equation [2, 3]

$$\frac{1}{r}\frac{\partial}{\partial r}\left(r\frac{\partial \phi}{\partial r}\right) + \frac{\partial^2 \phi}{\partial z^2} = \frac{e}{\varepsilon_0}N(r)\exp\left(\frac{e\phi(r,z,t)}{kT}\right) \tag{1}$$

by supposing a Maxwellian velocity distribution for the plasma. The $z$-trajectory of an electron has been calculated by solving numerically the equation of motion

$$\begin{cases} m\dfrac{dv_z}{dt} = -eE_z(r_0,z) \\[2mm] z(0)=0, \quad v_z(0)=v_c \end{cases} \tag{2}$$

being $r_0$ the radial coordinate (supposed constant) of the guiding center of the electron, while $v_c$ is the velocity of the particle in the center ($z=0$) of the trap. Once $z(t; r_0, v_c)$ is known, the averaged angular velocity, $\omega_{av}$, is evaluated as

$$\omega_{av}(r_0,v_c) = \frac{1}{r_0 B_0}\frac{1}{\tau}\int_0^\tau E_r(r_0,z(t))\,dt \tag{3}$$

CP692, *Non-Neutral Plasma Physics V*, edited by M. Schauer et al.
© 2003 American Institute of Physics 0-7354-0165-9/03/$20.00

being $\tau$ the bouncing period of the particle.

## THE TWO-FLUID MODEL

For a Penning trap where the central region (in which the particle density does not depend on $z$) is much longer than the border region, the relation between the $z$-averaged angular velocity, $\omega_{av}$, and the classical 2D E x B drift, $\omega_{2D}$, is approximately linear:

$$\omega_{av}(v_c) = a_1(v_c) + b_1(v_c)\omega_{2D} \tag{4}$$

where the coefficients $a_1(v_c)$ and $b_1(v_c)$ have been calculated using a the best fit procedure. According to the previous analysis, a simplified 2D kinetic model can be written in the form:

$$\begin{cases} \dfrac{\partial f(r,\theta,v_c,t)}{\partial t} + \left[a(v_c)\,r\hat{e}_\theta + b(v_c)\dfrac{\hat{e}_z \times \nabla\phi(r,\theta,t)}{B_0}\right] \cdot \nabla f(r,\theta,v_c,t) = 0 \\ \\ \nabla^2\phi(r,\theta,t) = \dfrac{e}{\varepsilon_0}\displaystyle\int f(r,\theta,v_c,t)\,dv_c \end{cases} \tag{5}$$

The coefficients $a$ and $b$ of Eq (5) can be rewritten as

$$a(v_c) = \langle a\rangle(1 + v(v_c)) \qquad\qquad b(v_c) = \langle b\rangle(1 + \mu(v_c)) \tag{6}$$

where the mean values are evaluated by supposing the equilibrium distribution function to be Maxwellian:

$$f_{eq}(r,v_c) = f_0(r)\exp\left(-\frac{mv_c^2}{2kT}\right) \tag{7}$$

By supposing $v(v_c) \approx k\cdot\mu(v_c)$, a set of equations governing the evolution of the first two moments of $f(r,\theta,v_c,t)$, namely

$$n(r,\theta,t) = \int f(r,\theta,v_c,t)\,dv_c$$

$$j(r,\theta,t) = \int \mu(v_c)f(r,\theta,v_c,t)\,dv_c \tag{8}$$

is deduced: in fact, by considering a reference frame rotating with angular velocity $\langle a\rangle$, one obtains:

$$\frac{\partial n}{\partial t} + kr\langle a\rangle\,\hat{e}_\theta\cdot\nabla j + \langle b\rangle\,\mathbf{v}_\perp\cdot[\nabla n + \nabla j] = 0 \tag{9}$$

$$\frac{\partial j}{\partial t} + kr\langle a\rangle\,\hat{e}_\theta\cdot\nabla\left(\langle\mu^2\rangle n\right) + \langle b\rangle\,\mathbf{v}_\perp\cdot\left[\nabla\left(\langle\mu^2\rangle n\right) + \nabla j\right] = 0$$

where the system is closed by assuming

$$\langle\mu^2\rangle \simeq \frac{\displaystyle\int \mu^2 f_{eq}(r,v_c)\,dv_c}{\displaystyle\int f_{eq}(r,v_c)\,dv_c} \tag{10}$$

The same equations can be deduced starting from a two-fluid model of the form

$$\begin{cases} \dfrac{\partial f_+}{\partial t} + [k\mu_0 r \langle a \rangle \,\widehat{e}_\theta + \langle b \rangle \,(1+\mu_0)\,\mathbf{v}_\perp] \cdot \nabla f_+ = 0 \\[4mm] \dfrac{\partial f_-}{\partial t} + [-k\mu_0 r \langle a \rangle \,\widehat{e}_\theta + \langle b \rangle \,(1-\mu_0)\,\mathbf{v}_\perp] \cdot \nabla f_- = 0 \end{cases} \tag{11}$$

with $f_+ + f_- = n$, $\mu_0(f_+ - f_-) = j$ and by defining $\mu_0 = \sqrt{\langle \mu^2 \rangle}$: these equations describe the motion of two fluids (with densities $f_+$ and $f_-$) having different $\mathbf{E} \times \mathbf{B}$ drifts.

## LINEAR STABILITY ANALYSIS

In order to evaluate the growth rates of the diocotron modes, the equations of the model have been linearized in the case of a plasma ring of constant electron density, $n_0$, between $r = r_1$ and $r = r_2$. A dispersion relation has been obtained, along with a system of ODEs governing the early stage of the evolution of the perturbation. The equilibrium configuration is defined as:

$$f_+ (r, t = 0) = f_- (r, t = 0) = f_0(r) = \frac{n_0}{2} \tag{12}$$

and the corresponding equilibrium potential is denoted by $\phi_0$. In the linear analysis, a single azimuthal mode is considered and the perturbed quantities are indicated as follows:

$$f_\pm^{(1)}(r, \theta, t) = \widehat{f}_\pm^{(1)}(r, t)\, e^{im\theta} \tag{13}$$

$$\phi^{(1)}(r, \theta, t) = \widehat{\phi}^{(1)}(r, t')\, e^{im\theta}$$

By defining

$$\omega_0^\pm (r) = im \left[\pm k \langle a \rangle \mu_0 + \omega_D (1 \pm \mu_0)\left(1 - r_1^2/r^2\right)\right] \tag{14}$$

the solution of the linear equations for $f_+$ and $f_-$ can be written as

$$\widehat{f}_\pm^{(1)}(r, t) = \widehat{f}_\pm^{(1)}(r, 0)\, e^{-\omega_0^\pm(r)t} + (1 \pm \mu_0)\,\frac{im \langle b \rangle\, f_0'(r)}{r B_0}\, \varphi_\pm^{(1)}(r, t) \tag{15}$$

being

$$\varphi_\pm^{(1)}(r, t) = \int_0^t \widehat{\phi}^{(1)}(r, t')\, e^{-\omega_0^\pm(r)(t - t')}\, dt' \tag{16a}$$

$$f_0'(r) = \frac{n_0}{2}\, [\delta(r - r_1) - \delta(r - r_2)] \tag{16b}$$

As $\widehat{\phi}^{(1)}(r, t)$ can be evaluated as

$$\widehat{\phi}^{(1)}(r, t) = \frac{e}{\varepsilon_0} \int_0^{R_w} \left[\widehat{f}_+^{(1)}(r', t) + \widehat{f}_-^{(1)}(r', t)\right] G_m(r' \to r)\, r'\, dr' \tag{17}$$

(being $G_m$ the $m$-th Fourier component of the Green function for the Poisson equation), the evolution of the initial perturbation can be studied by a system of differential equations having the form:

$$\frac{d}{dt}\begin{pmatrix} \varphi_+^{(1)}(r_2,t) \\ \varphi_+^{(1)}(r_1,t) \\ \varphi_-^{(1)}(r_2,t) \\ \varphi_-^{(1)}(r_1,t) \end{pmatrix} = \hat{\Lambda}_m \begin{pmatrix} \varphi_+^{(1)}(r_2,t) \\ \varphi_+^{(1)}(r_1,t) \\ \varphi_-^{(1)}(r_2,t) \\ \varphi_-^{(1)}(r_1,t) \end{pmatrix} + \Gamma_m(t) \qquad (18)$$

being $\hat{\Lambda}_m$ a suitable 4x4 matrix whose elements depend on the parameters of the unperturbed plasma distribution, while $\Gamma_m(t)$ is a vector depending upon the initial perturbation.

## RESULTS

The linear analysis of stability provides a quantitative estimate of the role played by the kinetic effects in destabilizing the $m_\theta = 1$ diocotron mode: in general, the growth rate of the $m_\theta = 1$ mode is approximately proportional to $\mu_0$; on the other hand, the effect on higher azimuthal modes is less relevant. In Fig. 1, the growth rate for some diocotron modes are reported as functions of $\mu_0$.

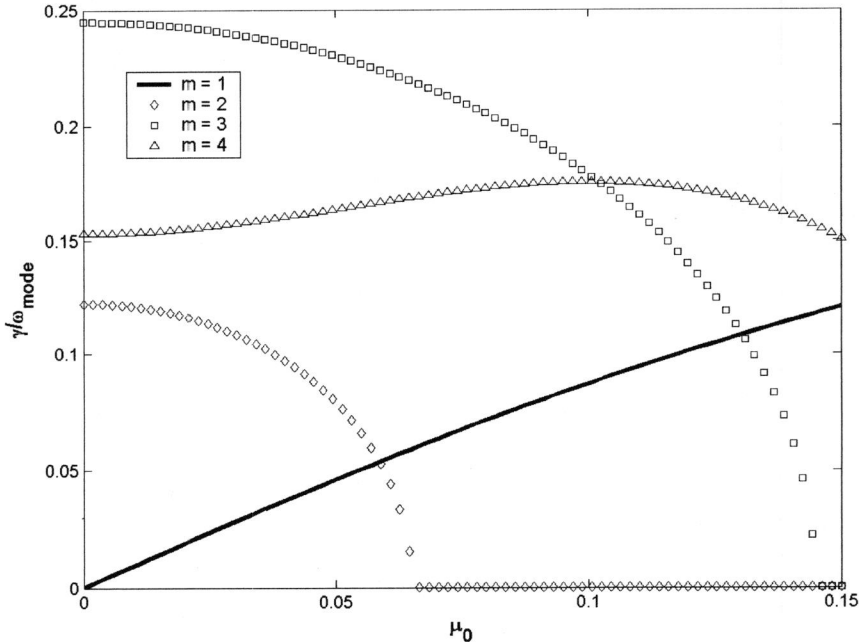

**FIGURE 1.** Behavior of normalized growth rates for different diocotron modes, as functions of $\mu_0$; $\langle a \rangle = 0.05 \omega_D$, $\langle b \rangle = 0.9$ and k = 8.

A plasma ring of radii $r_1 = 0.4R_W$ and $r_2 = 0.6R_W$ is considered; in this case, the parameters of the model ($\langle a \rangle$, $\langle b \rangle$ and $k$) are set to typical values.

The results of the two-fluid model have been compared with the ones obtained with the classical model, for different values of the radii of the plasma ring and for different lengths of the trap. In Fig. 2 the evolution in time of the density perturbation for the inner and outer contour of the ring is reported. The plasma ring is initially perturbed according to the $m_\theta = 1$ mode (the relative amplitude of the perturbation is 0.1%): in the framework of the classical model, the perturbed configuration is linearly stable. Figure 2a reports the results of the two-fluid model, showing the instability of the plasma equilibrium with respect to $m_\theta = 1$ azimuthal perturbations, while Fig. 2b reports the well-known results of the classical model.

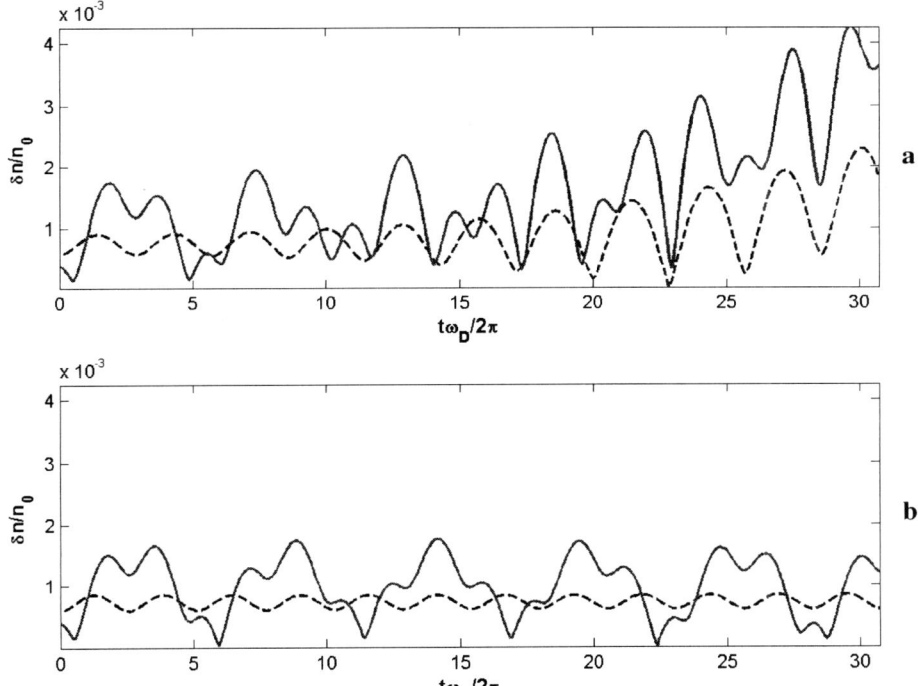

**FIGURE 2.** Evolution in time of the perturbation of the density on the inner contour (full line) and on the outer contour (dashed line) of the plasma ring, as predicted by the two-fluid model (a)) and by the classic 2D model (b). Normalized radii of the plasma ring: $r_1 = 0.4R_W$ and $r_2 = 0.6R_W$. Model parameters: $\mu = 0.025$, $\langle a \rangle = 0.004\omega_D$, $\langle b \rangle = 0.98$ and $k=39.7$.

The same results are reported in Fig. 3, for a different plasma ring ($r_1 = 0.2R_W$ and $r_2 = 0.8R_W$) in a short trap. In these calculations, all the parameters of the model have been evaluated consistently with the solution of the 3D Poisson equation for the equilibrium configuration.

The results show the role played by kinetic effects in destabilizing the $m_\theta = 1$ mode. In fact, to obtain quantitative results for the growth one should start from a self-consistent kinetic model, as the ones proposed in Ref. [4] and [5]. However, the present simplified model proves the importance of taking into account the energy distribution in the $z$-direction to study the diocotron instability.

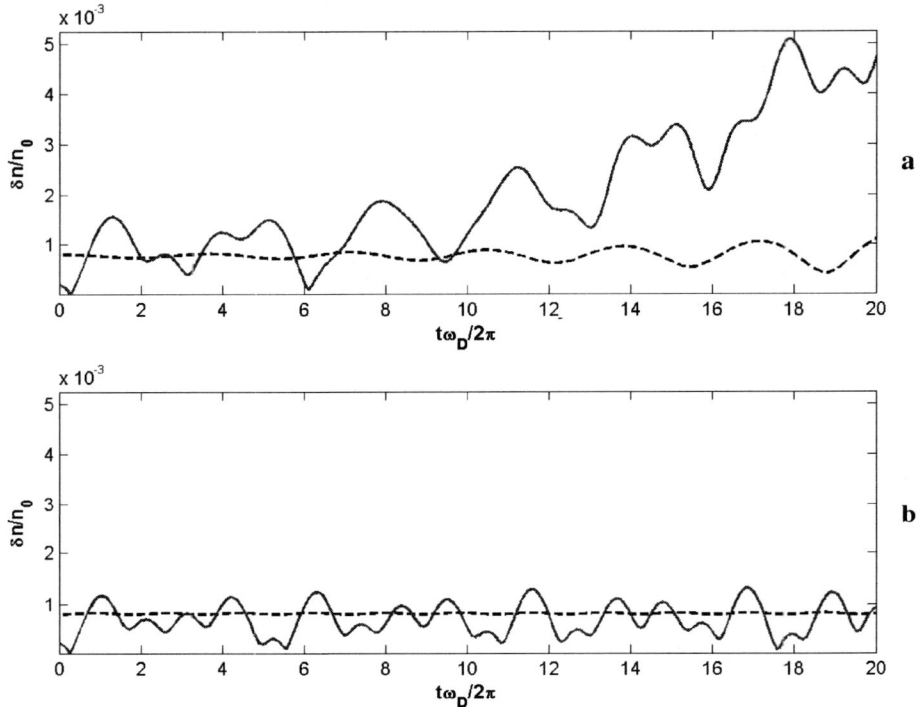

**FIGURE 3.** Evolution in time of the perturbation of the density on the inner contour (full line) and on the outer contour (dashed line) of the plasma ring, as predicted by the two-fluid model (a) and by the classic 2D model (b). Normalized radii of the plasma ring: $r_1 = 0.2R_W$ and $r_2 = 0.8R_W$. Model parameters: $\mu = 0.090$. $\langle a \rangle = 0.1\omega_D$, $\langle b \rangle = 0.78$ and $k=3.97$.

# REFERENCES

1. A.J. Peurrung, J. Fajans, *Phys. Fluids* B, **5**, 4295 (1993).
2. J.M. Finn, D. Del-Castillo-Negrete, D.C. Barnes, *Phys. Plasmas* **6**, 3744 (1999).
3. G.G.M. Coppa, A. D'Angola, G.L. Delzanno, G.L. Lapenta, *Phys. Plasmas* **8**, 1133 (2001).
4. T.J. Hilsabeck, T.M. O'Neil, *Phys. Plasmas* **8**, 407 (2001).
5. G.G.M. Coppa, P. Ricci, *Phys. Rev. E* **66**, 046409 (2002).

# Linear and Non-linear Interaction of Plasma Vortices

Gianni G.M. Coppa, Fabio Peano, Federico Peinetti

*INFM (Istituto Nazionale per la Fisica della Materia)*
*and Politecnico di Torino, Dipartimento di Energetica*
*Corso Duca degli Abruzzi 24, 10129 Torino (Italy)*

**Abstract**: The present work deals with two important theoretical aspects of the 2D fluid theory for non-neutral plasmas: the study of the merger between plasma vortices and the non-linear study of the evolution of the diocotron instability. In both cases, analytical and semi-analytical models have been deduced and the results have been tested by means of a numerical code based on the contour dynamics technique developed by the Authors.

## INTRODUCTION

As the velocity field due to the $\mathbf{E} \times \mathbf{B}$ drift is solenoidal, the dynamics of regions of constant charge density (vortices) in a Penning trap can be studied by considering the evolution of their contours.

In the present work, this idea is employed to study two different kinds of phenomena. The first analysis concerns the interaction between a finite-size vortex (of constant charge density and placed in the center of the trap) and a weak, pointlike vortex. By using a proper harmonic representation of the contour of the finite vortex, a linear model for the Fourier coefficients of the surface waves is deduced, allowing one to describe the interactions between the two vortices when small surface waves are induced. The resonance condition between the surface modes and the point vortex, along with a sufficient condition for the merger not to occur, is obtained, so generalizing the results obtained by O'Neil *et Al.* [1] for an infinite domain.

The second study concerns the non-linear interaction of surface modes in a plasma ring. Starting from the equations governing the evolution of the outer and inner contours (derived by supposing the charge density to be constant within the ring), a second-order model (with respect to the surface-wave amplitudes) is developed, in which the coupling-mode coefficients, determined in a fully analytical way, are evaluated numerically. The results presented concern the non-linear excitation of linearly-instable diocotron modes by means of one or a couple of linearly-stable modes. These non-linear effects can become relevant, for instance, when studying the evolution of the azimuthal $m=1$ diocotron mode. The couplings between modes cannot be described within the framework of a classic, first-order analysis of stability and, for this reason, their semi-analytical evaluation represents an innovative aspect in the study of the dynamics of the non-neutral plasmas dynamics.

The validity of all the results presented has been checked by means of a non-linear contour dynamics code developed by the Authors [2].

CP692, *Non-Neutral Plasma Physics V*, edited by M. Schauer et al.
© 2003 American Institute of Physics 0-7354-0165-9/03/$20.00

# LINEAR ANALYSIS OF VORTEX MERGER IN A PENNING TRAP

The merger process between plasma vortices has been extensively studied in the literature and important results have been obtained. In general, analytical results cannot be easily achieved, expecially when the vortices have comparable size and/or total charge: nevertheless, accurate results for these distributions have been produced by means of non-linear numerical codes. The present work, which extends the results of a previous work by O'Neil et al [2], shows that, when the system is made up of a finite-size vortex (of constant charge density and placed in the center of the trap) and a weak, pointlike vortex (i.e., in which the ratio, $\gamma$, between the charge of the two vortices is <<1), interesting analytical results can be achieved, along with correct predictions on the stability properties of the system.

Due to the divergence free condition on the velocity field, the charge density inside the extended vortex remains constant in time: for this reason, its dynamics can be studied just by considering the evolution of its contour. By doing so, the contour of a vortex can be represented by the function $r_c(\theta,t)$, defined as the radial coordinate of the vortex contour, at the angular coordinate $\theta$ and at time $t$. The study of the evolution in time of $r_c(\theta,t)$, along with the evolution of the coordinates $(r_v, \theta_v)$ of the point vortex, gives full informations on the evolution of the plasma distribution inside the trap. In the present work, a Fourier expansion of $r_c(\theta,t)$ with respect to $\theta$ is used ($r_0$ being the unperturbed radius of the vortex):

$$r_c(\theta,t)=r_0 + \sum_{m=-\infty}^{+\infty} C_m(t)e^{im\theta} \tag{1}$$

By means of a Green-function approach, the electrostatic potential can be expressed in a similar way, as

$$\phi(r,\theta,t)= \phi^{(0)}(r)+\frac{2\pi e n_0}{\varepsilon_0}r_0 \sum_{m=-\infty}^{+\infty} G_m(r_0 \to r)C_m e^{im\theta} +\frac{q}{\varepsilon_0} \sum_{m=-\infty}^{+\infty} G_m(r_v \to r)e^{im(\theta-\theta_v)} \tag{2}$$

being $G_m(r' \to r)$ the $m$-th Fourier component of the Green function for the Poisson equation with proper boundary conditions.

By doing so, a linear model (with respect to the Fourier coefficients of the surface waves) is deduced in the non-dimensional form

$$\begin{cases} \dfrac{d\hat{C}_m}{d\hat{t}} = -im\left[\left(1+\dfrac{G_m(\hat{r}_0 \to \hat{r}_0)}{\hat{r}_0}\right)\hat{C}_m +\dfrac{\gamma}{2}G_m(\hat{r}_v \to \hat{r}_0)e^{-im\theta_v}\right] \\[2ex] \dfrac{d\hat{r}_v}{d\hat{t}} = -\dfrac{i}{\hat{r}_v} \sum_{m=-\infty}^{+\infty} mG_m(\hat{r}_0 \to \hat{r}_v)\hat{C}_m e^{im\theta_v} \\[2ex] \dfrac{d\theta_v}{d\hat{t}} = \dfrac{\hat{r}_0^2}{\hat{r}_v^2}+\dfrac{1}{\hat{r}_v}\sum_{m=-\infty}^{+\infty} m\dfrac{\partial G_m(\hat{r}_0 \to \hat{r}_v)}{\partial \hat{r}_v}\hat{C}_m e^{im\theta_v} +\gamma\dfrac{\hat{r}_0^2}{1-\hat{r}_v^2} \end{cases} \tag{3}$$

where all the lengths are normalized to the radius of the trap and where the time is normalized to the inverse of the diocotron frequency. The model describes the interaction between the two vortices when small surface waves (i.e., having amplitude << $r_0$) are induced on the contour of the finite vortex.

100

If the interaction is sufficiently weak, the finite vortex remains nearly circular and the point vortex spins around it on a circumference, with angular velocity ω that can be assumed as constant. By inserting these hypothesis into Eqs. (3), the equations for the contour coefficients have the form of a driven harmonic oscillator, thus revealing the existence of resonant configurations for the contour modes. The resonance condition has the form

$$\left|m\right| + \hat{r}_0^{2|m|} - 1 = \left|m\right|\hat{r}_0^{\,2}\left[\frac{1}{\hat{r}_v^{\,2}} + \frac{\gamma}{1 - \hat{r}_v^{\,2}}\right] \tag{4}$$

where $m$ is the azimuthal mode number. This result suggests that, if the initial configuration is resonant, the first stage of the evolution of the distribution can be described by means of a simplified, one-harmonic model, taking into account only the resonant contour mode. This assumption allows one to obtain a self-consistent equation for the evolution of the radial coordinate of the point vortex and to the determination of an "oscillation layer" in which the point vortex is trapped. In Fig. 1, the behavior of the oscillation layers for different modes, as a function of the charge ratio γ, is shown for three different values of the unperturbed radius of the finite vortex: $\hat{r}_0 = 0.2$, 0.5 and 0.8.

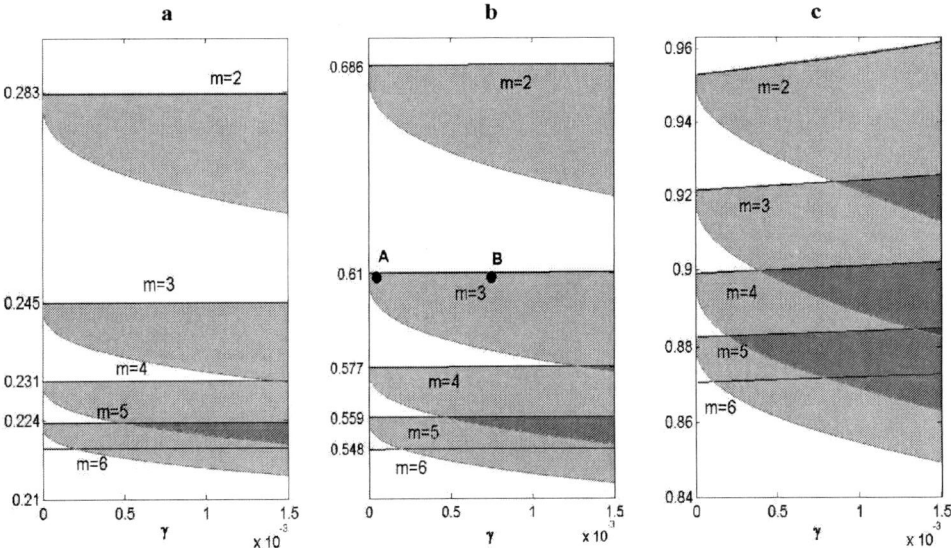

**FIGURE 1.** Resonant radii, along with their respective oscillation layers, as functions of γ, for $\hat{r}_0 = 0.2$, 0.5 and 0.8.

The one-harmonic model provides a sufficient condition for the merger not to occur. Furthermore, a simple criterion for the stability of the system can be deduced, only by comparing the width of the oscillation layer to the distances between the resonant radius and the resonant radii for the adjacent modes.

In the following, two figures are reported, showing the evolution in time of the radial coordinate of the point vortex, for a stable configuration marked by the point A in Fig. 1b and for an unstable configuration (point B). In the first case, the simplified, one-harmonic model is in excellent agreement with the reference solution obtained with a numerical code based on the contour dynamics technique developed by the Authors. In the second case, after a few bounces inside the oscillation layer, the point vortex eventually spirals towards the center of the trap (leading to the merger of the two vortices), as shown by the CD code (curve (b) of Fig. 2b) and by the full linear model (curve (c)).

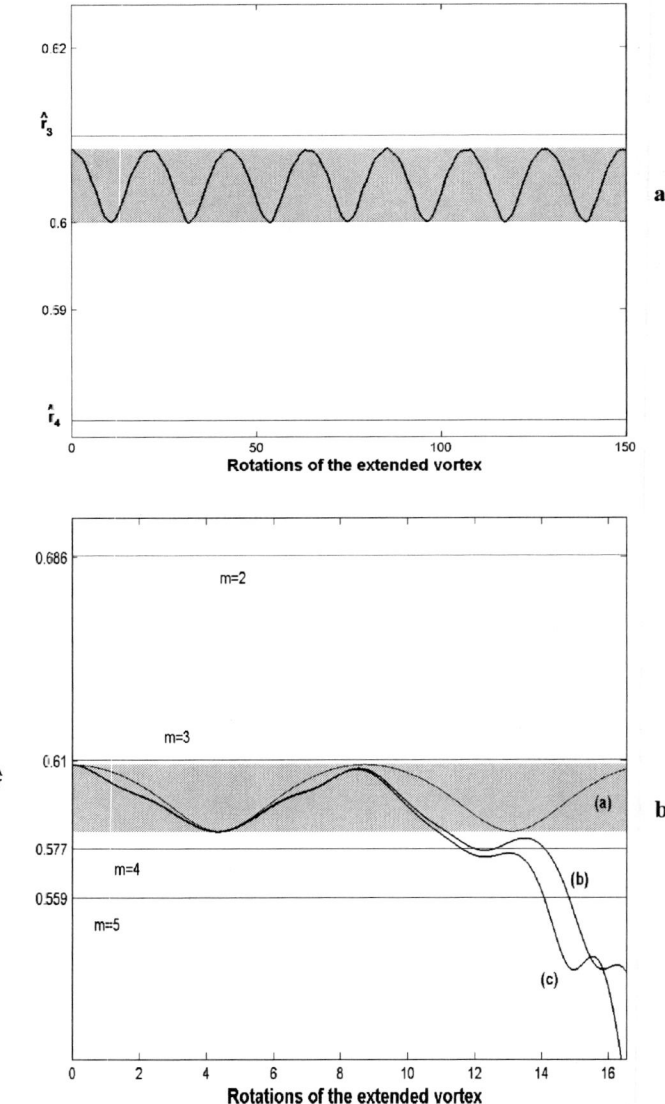

**FIGURE 2.** Evolution in time of the radial coordinate of the point vortex, for the configurations marked by "A" (a) and "B" (b) in Fig. 1b, respectively.

# NON-LINEAR STUDY OF THE EVOLUTION OF A PERTURBED PLASMA RING

The evolution of a plasma ring of constant charge density has been extensively studied in the literature, both analytically and numerically. In particular, great attention has been given to the investigation of the stability properties of this configuration by means of a linear stability analysis. In the present work, a semi-analytical model to study the non-linear part of the evolution of the diocotron instability is proposed. The analysis is carried out in the transformed Fourier space, in order to study the coupling mechanisms between different azimuthal modes that cannot be investigated by a first-order perturbative analysis.

The starting point is given by the analytical expression for the velocity field, valid for a generic, piece-wise constant plasma distribution inside the trap [2]. Following this approach, the generic $i$-th contour of the plasma distribution evolves in time according to an equation of the following general form:

$$\frac{\partial r_c^{(i)}(\theta,t)}{\partial t} = f\left(\left\{r_c^{(j)}(\theta,t)\right\}_{j=1,...N}\right) \tag{5}$$

where $N$ is the total number of contours and $f$ is a suitable functional. Nevertheless, due to the complexity of the functional $f$, analytical results cannot be achieved from Eq. (5) and a numerical solution has to be used. Considering the case of a plasma ring, a uniform azimuthal grid of $N_g$ steps has been used and the generic function $r_c^{(j)}(\theta,t)$ has been replaced by a set of $N_g$ functions of time, $\left\{\varepsilon_i^{(j)}(t)\right\}_{i=1,...N_g}$, being

$$\varepsilon_i^{(j)}(t) = r^{(j)}(\theta_i,t) - r_0^{(j)} \tag{6}$$

and being $r_0^{(j)}$ the unperturbed radius of the $j$-th contour of the ring ($j = 1,2$). In this way, a system of $2N_g$ ODEs can be obtained, as

$$\frac{d\varepsilon_i^{(j)}}{dt} = f_i^{(j)}\left(\varepsilon^{(1)}, \varepsilon^{(2)}\right) \qquad i = 1,..., N_g \ , \ j = 1, 2 \tag{7}$$

The function $f_i^{(j)}$ in Eq. (7) can be expanded with respect to its arguments $\left\{\varepsilon^{(1)}, \varepsilon^{(2)}\right\}$ using a second-order Taylor polynomial (the coefficients of the expansion being evaluated numerically), as

$$\frac{d\varepsilon_i^{(j)}}{dt} = \sum_{k,l} \sigma^{(j,l)}_{k-i} \varepsilon_k^{(l)} + \sum_{l,m} \sum_{k,h} \gamma^{(j,l,m)}_{k-i,h-i} \varepsilon_k^{(l)} \varepsilon_h^{(m)} \tag{8}$$

and the analysis is carried out in the transformed Fourier space by introducing the Fourier components $\left\{C_\alpha^{(j)}\right\}_{j=1,2}$ defined as

$$C_\alpha^{(j)} = \frac{1}{N_g} \sum_k \varepsilon_k^{(j)} e^{-i\alpha\theta_k} \tag{9}$$

By doing so, a new set of ODEs is obtained, of the form:

$$\frac{dC_\alpha^{(j)}}{dt} = -\sum_{i=1,2} \Lambda_\alpha^{(j,i)} C_\alpha^{(i)} + \sum_{l,m=1,2} \sum_\beta C_\beta^{(l)} C_{\alpha-\beta}^{(m)} \Gamma^{(j,l,m)}_{\beta,\alpha-\beta} \qquad (10)$$

showing non-linear couplings between modes.

The results presented here are focused on the non-linear excitation of linearly-unstable diocotron modes by means of one or a couple of linearly-stable modes. These non-linear effects can become relevant, for instance, when studying the evolution of the azimuthal $m = 1$ diocotron mode. In the following figures, the results of the semi-analytical model are compared with the reference results provided by the CD code. The figures show the excellent agreement between the two different sets of results.

**Case 1:** Unperturbed radii of the ring: $r_{1,0} = 0.3R_w$ and $r_{2,0} = 0.7\ R_w$; mode initially excited: $m = 1$; relative amplitude of the perturbation (on each radius): 2 %.

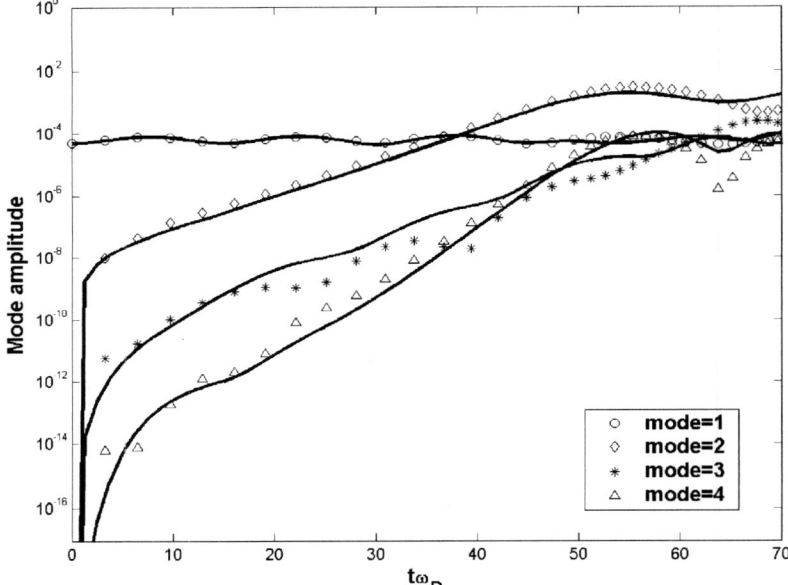

**FIGURE 3.** Evolution in time of the mode amplitudes, as predicted by the semi-analytical model (full line) and by the CD code for the case 1.

**Case 2:** Unperturbed radii of the ring: $r_{1,0} = 0.3R_w$ and $r_{2,0} = 0.7\ R_w$; modes initially excited: $m = 3, 4$; relative amplitude of the perturbation (on each radius): 0.5 %.

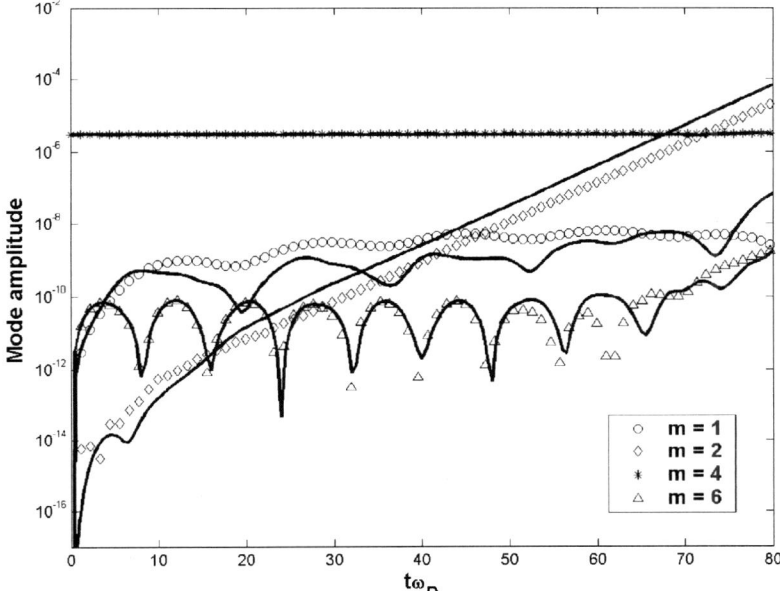

**FIGURE 4.** Evolution in time of the mode amplitudes, as predicted by the semi-analytical model (full line) and by the CD code.

# REFERENCES

1. G.G.M. Coppa, F. Peano, F. Peinetti, *Image-charge method for Contour Dynamics in Systems with Cylindrical Boundaries*, J. Comput. Phys. **182**, 392-417 (2002).

2. I.M. Lansky, T.M. O'Neil, D.A. Schecter, *A theory of vortex merger*, Phys. Rev. Lett. **79**, **1479**, (1997).

# Formation of a Triangle Vortex Configuration Assisted by a Background Vorticity Distribution

A. Sanpei, Y. Kiwamoto, Y. Soga and J. Aoki

*Graduate School of Human and Environmental Studies, Kyoto University, Sakyo-ku, Kyoto 606-8501, Japan*

**Abstract.** A vortex crystal is a quasi-stationary, symmetric array of intense vortices (clumps). A low-level of background vorticity is experimentally observed to assist three clumps in forming an equilateral triangle starting from initial positions on a linear array. The triangle constitutes a unit cell of a crystal in a many-vortex system. The background vortex curbs the orbital motion of the clumps with unequal strengths to arrest them at the vertices of an equilateral triangle by wrapping them with different sized belts of depleted vorticity (ring holes). We characterize the contributions of a low-level background vorticity distribution on the formation of ordered states of clumps.

## INTRODUCTION

The crystallization of intense vortices (clumps) is a complex self-organization phenomenon. The number of clumps decreases intermittently on merging between themselves at a rate as provided phenomenologically in terms of punctuated scaling law [1, 2, 3]. During relatively long periods punctuated by the merging the clumps form metaequilibrium configurations with high symmetry [1]. The contribution of low level distribution of background vorticity (BGV) is important as a remover of energy or entropy of the clumps' motion [4, 5]. The theoretical models, however, cannot fully describe the observed evolutionary paths of vorticity distributions, and further experimental examinations focused on relevant processes are required [6].

The purpose of this paper is to evaluate the contribution of the background vortices to the formation of the meta-equilibrium distribution of clumps starting from the initial distribution immersed in a continuous background vortex controlled at varied levels of distribution $\zeta_b(x,y)$. To focus on the equilibration of the vortex distribution we limit the number of the clumps to three as the minimum number that determines the unit cell of the 2D lattice, and the merging between clumps is not in the scope of this paper. The dynamics of mutually interacting point vortices has been the subject of theoretical and simulational studies of 2D turbulence [7, 8, 9]. The experiment by Fine *et al.* [1] and a subsequent simulational study by Schecter *et al.* [4] were the first to reveal the important role of the interaction between the strong patches of vorticity (clumps) and the low level vorticity filling the space around the clumps in the vortical relaxation processes toward the crystal structures. In addition, the background vorticity was called for in the statistical model that successfully reproduced the observed crystal arrays of electron vortices [5].

CP692, *Non-Neutral Plasma Physics V*, edited by M. Schauer et al.

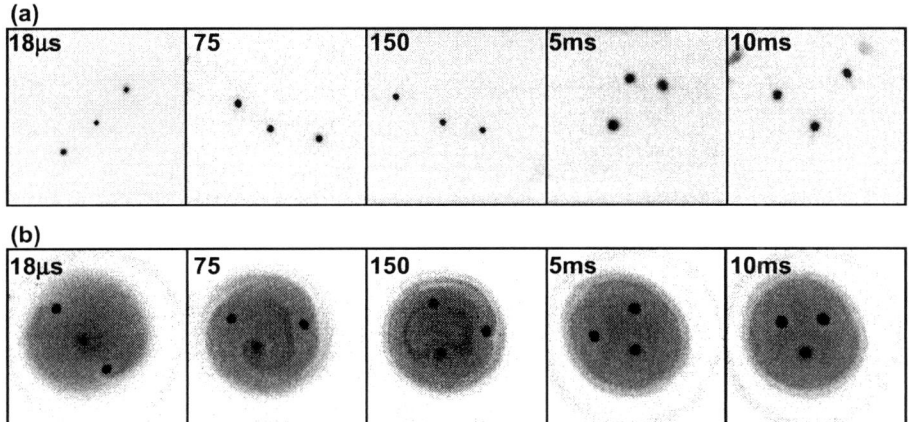

**FIGURE 1.** Snapshots of the clumps' dynamics in vacuum (a) and in a background vortex with $\Gamma_b \propto N_b/10^7 = 9.4$ (b). These snapshots have a linear luminosity scale with saturation at high level. The clumps with equal strength of $\Gamma_c \propto N_c/10^7 = 0.74$ are placed in line as the initial condition. The panel size is 37.4 mm $\times$ 37.4 mm.

The experiment is carried out in a Malmberg-type trap with $B_0 = 0.048$ T [10]. The details of the configuration and diagnostics have been reported elsewhere [10, 11].

## BGV-ASSISTED FORMATION OF A SYMMETRIC CELL

In Fig. 1 we compare the time evolution of a set of three clumps in vacuum (a) and in a BGV distribution (BGVD) (b). The clumps are injected at $t = 10 \ \mu$sec, and start to move freely on disconnection from the cathode at $t = 18 \ \mu$s from the initial distribution aligned in a straight line with the same circulation $\Gamma_c = \int \int dx dy \zeta_c(x,y) = eN_c/\varepsilon_0 B_0 L = 1.64 \times 10^{-6} N_c$. Here, $N_c$ is the total number of electrons constituting each clump with the length of $L = 235 \pm 5$ mm, and the electron density $n(x,y)$ is related to the vorticity $\zeta(x,y) = \nabla \times \mathbf{v}$ by $\zeta = en/\varepsilon_0 B_0$. The time of observation is indicated at the upper left corner. In vacuum, the clumps continue orbital motion with the period of $\tau_R \approx 50 \ \mu$s without showing any stationary configuration until non-ideal effects set in at $t > 100$ ms [9].

In the presence of BGVD as shown in Fig. 1(b), however, the clumps move in a limited region to go into an ordered state. Here the initial BGVD is a continuous distribution with the maximum height of $\zeta_b = \zeta_{c0}/60$, where $\zeta_{c0}$ is the height of a clump's vorticity and the ratio of the circulation is $\Gamma_b/\Gamma_c = 12.7$. At first, clumps show orbital motion with an averaged period of $\tau_R = 25 \ \mu$s locally wrapping the background vortex around them. ($t = 75, 150 \ \mu$s) Then BGVD becomes increasingly fine-structured in the differential rotation around the clumps [10]. The perturbations surrounding the clumps eventually evolve into ring holes [4, 5, 12, 13]. In the later stage, the clumps settle down to form a symmetric triangular array and the fine structures in BGV are smoothed out in our coarse-grained observation. ($t = 5, 10$ ms) In vacuum such a symmetric configuration

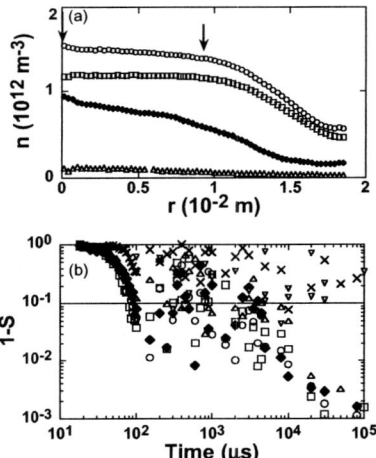

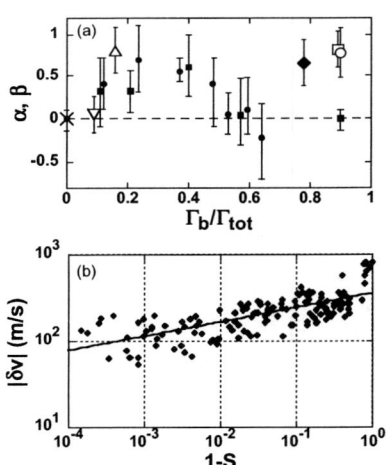

FIGURE 2. (a) Radial profiles of the electron density distribution $n_b\ (x,y)$ constituting the background vorticity. The initial radial locations of the clumps are indicated by arrows. (b) Deviation from the symmetry 1-S is plotted as a function of time.

FIGURE 3. (a) The exponent $\beta$ is plotted against $\Gamma_b\ /\Gamma_{tot}$. (b) $|\delta\ v_{rms}\ |$ is plotted against 1-S determined from observed data at $t > 200\ \mu$s.

can be established only when the three clumps are of equal circulation and placed initially in symmetric locations [9, 14, 15].

For a quantitative analysis of the formation of the symmetric cell, we introduce a symmetry parameter defined as

$$S = 12\sqrt{3}\frac{A}{l^2}. \tag{1}$$

Here $A$ is the area, and $l$ is the peripheral length of a triangle that has three clumps at its vertices. $S$ is maximized at 1 when the clumps form an equilateral triangle. We examine the clumps' dynamics in different shapes and levels of BGVD as described in Fig. 2(a). Initially the clumps are placed in a straight line at the points indicated by arrows with equal circulations $\Gamma_c \propto N_c/10^7 = 0.74$. The time evolution as illustrated in Fig. 1(b) is summarized in Fig. 2(b) in terms of $1 - S$, representing the degree of deviation from symmetry. The symbols employed in Figs. 2(a) and (b) correspond to each other. Two symbols added in Fig. 2(b) represent the cases with $N_b/10^7 = 0.25$ ($\triangledown$) and 0 ($\times$), i.e. vacuum. The density distribution for $\triangledown$ is close to the sensitivity limit of the diagnostics and cannot be determined.

In the initial stage within 100 $\mu$s, all symbols for $1 - S$ show a common feature of rapid reduction. Here the clumps show kinetic motion weakly modified by the presence of BGVD. Then comes the next stage of decaying oscillations. The frequency and decay rate of $1 - S$ depend on BGVD. In the vacuum case ($\times$), $1 - S$ continues undamped oscillation for $\sim 10^5$ $\mu$s until one of the clumps disappears and $S$ becomes undetermined. With the addition of a low level BGVD ($\triangle: N_b/10^7 = 0.52$, $\zeta_b = \zeta_{c0}/400$), the excursion of the clumps is reduced and the oscillation of $1 - S$ damps in the time scale of 1 ms. These features are more evident for higher levels of BGVD, $N_b/10^7 = 9.4$(closed $\diamond$), 21.0($\square$), 24.8($\bigcirc$), but less clear for the case with $N_b/10^7 = 0.25$ ($\triangledown$). In the third stage of $t \geq 10$ ms, $1 - S$ further decreases below $10^{-2}$ without showing strong oscillations. For more than 100 ms in the last stage, we observe stationary configuration in the shape of an equilateral triangle. For quantitative examination of the relaxation speed of $S$, we introduce the settling time $\tau_S$ as the time after which $S$ remains above 0.9. The $\tau_S$ decreases as the circulation of BGVD, $\Gamma_b \propto N_b/10^7$, increases. Almost the same dependence is observed for $\tau_S$ vs. $\zeta_b$, the height of BGVD at the initial location of the outermost clump. If $\tau_S$ is normalized by the initial turn-over time $\tau_R$, $\tau_S/\tau_R$ decreases from 120 to 55 as $N_b$ increases. These observations indicate that BGVD assists the formation of the unit of a vortex crystal.

## CONTRIBUTION OF BGVD TO COOLING

The BGVD has been called for as a cooler of randomly moving clumps [1, 4]. Fine et al. have observed that the average chaotic velocity decreases as $|\delta v| \propto t^{-\alpha}$ with $\alpha = 0.5 - 0.6$ between 10 $\tau_R$ and 100 $\tau_R$ [1]. Schecter et al. have evaluated that the exponent $\alpha$ lies between $-0.2$ and $+0.6$ depending on $\Gamma_b/\Gamma_{tot}$ where $\Gamma_{tot} = \Gamma_b + \Gamma_c$ [4].

From Fig. 2(b) we can evaluate the exponent $\beta$ for the symmetrization by fitting to $1 - S \propto t^{-\beta}$. Restricting our analysis to the data at $t \geq 200$ $\mu$s, i.e. excluding the initial orbital motion, we obtain $\beta = 0.05 \pm 0.21$ for $\triangledown$, $0.81 \pm 0.28$ for $\triangle$, $0.66 \pm 0.28$ for closed $\diamond$, $0.82 \pm 0.22$ for $\square$, $0.78 \pm 0.21$ for $\bigcirc$. If we further limit our analysis for $t > 10^4 \mu$s, the asymptotic value of $\beta$ is $\sim 1$. Figure 3 (a) summarizes the cooling exponent $\beta$ as a function of $\Gamma_b/\Gamma_{tot}$. Starting from zero at $\Gamma_b/\Gamma_{tot} = 0$, $\beta$ approaches $\sim 0.8$ and remains there for $0.15 \leq \Gamma_b/\Gamma_{tot} < 0.9$. The exponent $\alpha$ quoted from Fig. 10 of Ref. [4] are shown with solid symbols. The observation that $\beta$ lies around the upper limit of $\alpha$ suggests that $\beta$ indicates the contribution of BGVD to the cooling of clumps' motion.

We examine the correspondence between $\alpha$ and $\beta$ in more detail. We evaluate the root-mean-square value of the velocity fluctuation $\delta v_{rms}$ as defined by $\delta v_{rms} = (\sum_{i=1}^M \delta v_i^2/M)^{1/2}$ for $M$ clumps [4]. Here, $\delta \mathbf{v}_i = \mathbf{v}_i - r_i \bar{\Omega}(t)\hat{\theta}$ is velocity fluctuation about the radial-weighted average rotation $\bar{\Omega}(t) = \sum_{i=1}^M v_{i,\theta}/\sum_{i=1}^M r_i$ at the position $\mathbf{r}_i$ of the $i$th clump. The vectors, $\mathbf{r}_i$ and $\mathbf{v}_i$, are evaluated relative to the flow's center of vorticity. The velocity $\mathbf{v}_i$ is calculated from the potential distribution $\phi(x,y)$ as determined from the measured density distribution $n(x,y)$. In Fig. 3 (b) we directly compare $|\delta v_{rms}|$ with $1 - S$ for a typical data representing a similar case to closed $\diamond$ ($N_b/10^7 = 9.4$). We can observe a correlation given as $|\delta v_{rms}| = (1 - S)^\gamma$, and obtain a linear relation $\alpha = \gamma\beta$ with $\gamma = 0.166 \pm 0.043$. Applying this method to other cases, we obtain $\gamma = 0.239 \pm 0.062$ for $\triangle$, $0.198 \pm 0.055$ for closed $\diamond$, $0.189 \pm 0.041$ for $\square$, $0.182 \pm 0.059$ for $\bigcirc$.

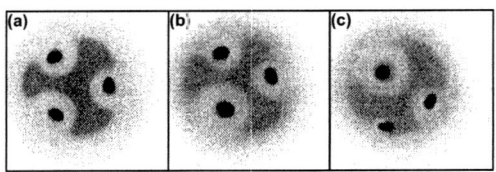

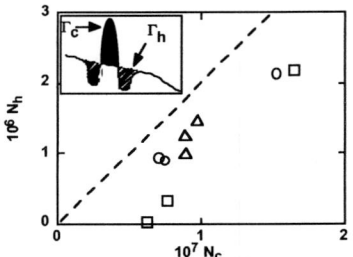

**FIGURE 4.** Quasi-stationary configurations of three clumps immersed in the background vortex : $(N_{c1}, N_{c2}, N_{c3}, N_b)/10^7$ = (a) $(1.1, 1, 0.93, 12)$; (b) $(1.4, 0.84, 0.74, 12)$; (c) $(1.6, 0.93, 0.42, 11)$. Clumps with unequal circulations form an equilateral triangle only in the presence of BGVD.

**FIGURE 5.** The circulation of ring holes is plotted as a function of the circulation of the associated clump. The symbols represent different sets of clumps in Fig. 4.

# EQUILIBRIUM STATE SUPPORTED BY RING HOLES

An important feature characterizing the structures in BGVD is a ring hole, a depleted zone of vorticity surrounding a clump. A ring hole is generated when a part of the distribution of an inhomogeneous BGV is advected by a clump [12, 13]. Figure 4 shows the vorticity distributions recorded in a quasi-stationary state at $t \geq 5$ ms for different combinations of clumps immersed in the almost identical initial distributions of BGV. Different sizes of ring holes are clearly observed around the clumps. The BGVD enables a set of extremely unbalanced clumps to form an equilateral triangle with their center of mass keeping the orbital motion.

Modification of BGVD in the shape of ring holes has been observed in experiments [1] and in a simulation [4]. It was called for in a statistical model [5]. We evaluate the circulation $\Gamma_h$ of the ring holes by integrating the deficit of the vorticity $\delta\zeta_b(x,y)$ from the smoothly interpolated distribution $\bar{\zeta}_b(x,y)$ of BGVD *i.e.* $\Gamma_h = \iint dxdy\, \delta\zeta_b(x,y) = \iint dxdy\, \{\bar{\zeta}_b(x,y) - \zeta_b(x,y)\}$. Here $\delta\zeta_b(x,y)$ is schematically illustrated by the hatch in the inset of Fig. 5. Figure 5 plots $N_h \propto \Gamma_h$ as a function of $N_c \propto \Gamma_c$ for each clump shown in Fig. 4. It indicates that $\Gamma_h$ increases with $\Gamma_c$ for each combination of the clumps and that all the data points lie within a narrow belt demonstrating a good positive correlation between $\Gamma_h$ and $\Gamma_c$. It should be noted, however, that $\Gamma_h$ remains less than 20 % of $\Gamma_c$. Therefore the effective circulations, $\Gamma_{eff} = \Gamma_c - \Gamma_h \propto (N_c - N_h)/10^6$, are (7.9, 7.5, 8.2) for $\triangle$, (13, 6.1, 6.6) for $\bigcirc$ and (14, 7.3, 6.2) for $\square$. In a statistical model it is requested for crystallization of vortices that the circulations of the clumps are not fully cancelled by the holes in BGVD [5].

On the basis of these experimental data, the role of BGVD on symmetrization of clumps is discussed in the Ref [16] in light of Aref's theorem [9].

# SUMMARY

In summary, we have reported an observation revealing a strong influence of a low-level BGVD on the formation of ordered states of clumps. The BGVD curbs the motion of three clumps to arrest them at the vertices of an equilateral triangle constituting a unit cell of vortex crystal. The settling process quantified by the symmetry parameter $S$ is related to the reduction of random velocities $\delta v$ of the clumps. The clumps generate ring holes (depleted zone of vorticity) around them and tend to partially compensate the imbalance among the circulations.

# ACKNOWLEDGMENTS

This work was supported by a Grant-in-Aid from the Ministry of Education, Culture, Sports, Science and Technology and by the collaborative research program of the National Institute for Fusion Science.

# REFERENCES

1. K. S. Fine, A. C. Cass, W. G. Flynn and C. F. Driscoll, Phys. Rev. Lett. **75**, 3277 (1995).
2. D. Z. Jin and D. H. E. Dubin, Phys. Rev. Lett. **84**, 1443 (2000).
3. J. C. MacWilliams, J. Fluid Mech **146**, 21 (1984).
4. D. A. Schecter, D. H. E. Dubin, K. S .Fine and C. F. Driscoll, Phys. Fluids. **11**, 905 (1999).
5. D. Z. Jin and D. H. E. Dubin, Phys. Rev. Lett. **80**, 4434 (1998).
6. C. F. Driscoll, D. Z. Jin, D. A. Schecter, E. J. Moreau and D. H. E. Dubin, Physica Scripta **T84**, 76 (2000).
7. L. Onsager, Nuovo Cinmeto **6**, 276 (1949).
8. G. Joyce and D. Montgomery, J. Plasma Phys. **10**, 107 (1973).
9. H. Aref, Phys. Fluids. **22**, 393 (1979).
10. Y. Kiwamoto, K. Ito, A. Sanpei and A. Mohri, Phys. Rev. Lett. **85**, 3173 (2000).
11. K. Ito, Y. Kiwamoto and A. Sanpei, Jpn. J. Appl. Phys. **40**, 2558 (2001).
12. A. Sanpei, Y. Kiwamoto and K. Ito, J. Phys. Soc. Jpn. (Lett.) **70**, 2813 (2001).
13. D. Durkin and J. Fajans, Phys. Rev. Lett. **85**, 4052 (2000).
14. Y. Kiwamoto, A. Mohri, K. Ito, A. Sanpei and T. Yuyama, *Non-neutral Plasma Physics* III (AIP1999) pp.99-105.
15. Y. Kiwamoto, K. Ito, A. Sanpei, A. Mohri, T. Yuyama and T. Michishita, J. Phys. Soc. Jpn. (Lett.) **68**, 3766 (1999).
16. A. Sanpei, Y. Kiwamoto, K. Ito and Y. Soga, Phys. Rev. E. **68**, 016404 (2003).

# Mechanisms of Merger and Binary Structure Formation of Two Discrete Vortices in a Non-Neutral Plasma

Y. Soga, Y. Kiwamoto, A. Sanpei and J. Aoki

*Graduate School of Human and Environmental Studies, Kyoto University, Sakyo-ku, Kyoto 606-8501, Japan*

**Abstract.** Observations have shown that two discrete vortices (clumps) immersed in a broad profile of background vorticity merge quickly or form a binary state that lasts for a long period. The different paths of the vortical evolution critically depend on slight differences in the initial background vorticity distribution (BGVD). By a fine control of the BGVD the asymptotic inter-clump distance is found to be a two-valued function of $\nabla \zeta_b / \zeta_b{}^2$, where $\zeta_b$ is the BGV at the initial location of the clumps, corresponding to the merger and the binary state. The degeneracy is removed by considering the degree of depletion of BGVD between two clumps at the time of their proximity in the initial phase.

## INTRODUCTION

A highlight among recent experimental achievements of two-dimensional (2D) vortex dynamics using a strongly magnetized pure electron plasma is the observation of vortex crystals generated in the relaxation processes from unstable initial vorticity distributions [1]. The role of the background vorticity distribution (BGVD) filling the space among prominent patches of vortices (clumps) was pointed out in the experiment and subsequently evaluated in its simulational analyses [2]. BGVD was called for also in statistical derivation of the crystal distribution [3].

This paper reports an experimental study focused on interaction between two clumps in BGVD. Merger between clumps is a key process in vortical relaxation accompanied by reduction of number of clumps or turbulent cascade toward long wavelengths. On the other hand, processes keeping the clumps separate at a certain distance are also essential to the formation of crystal structures in a quasi-steady state.

Key factors causing the branching into the two states are interesting as an example of bifurcation processes in the paths of relaxation toward ordered structures, though they have not been systematically examined so far. Since the bifurcation depends sensitively on the BGVD structure, for a systematic analysis precise control is necessary in preparation of BGVD with a medium strength of circulation. This report is a first step aiming at this objective with the introduction of a new scheme of fine controlling the initial shape of BGVD into which two test clumps are immersed. The narrow range of BGVD parameters available for this experimental examination may impose some limitations to the generalization of the results.

This experiment is carried out in a Malmberg-type trap with $B_0 = 0.048$T. The BGV

CP692, *Non-Neutral Plasma Physics V*, edited by M. Schauer et al.
© 2003 American Institute of Physics 0-7354-0165-9/03/$20.00

distribution under test is generated in two stages. The first stage of BGV distribution is created in a smooth shape after mixing and relaxation of about 300 strings of electrons which accumulate in the trap through multiple injection-hold-mixing cycle [4, 5, 6]. The second stage of BGV shaping consists of active spilling a fraction of the trapped electrons through the injection-side plug by reducing the potential barrier. We select the time-width of the reduced barrier as a useful knob for fine-control of the final BGV distribution. The time-width is varied from $100\mu s$ to 10ms for a pre-selected height of the reduced potential barrier, and the electrons remaining in the trap are left isolated again for additional 70ms to form a second stage of BGVD with a smooth profile. This distribution serves as the initial BGVD for the study of dynamics of a pair of electron vortices (clumps). The details of the configuration and diagnostics have been reported elsewhere [4, 5, 6].

## EXPERIMENTAL OBSERVATION

Figure 1 shows typical examples of time evolution of a vortex system consisting of BGVD and two clumps. Brightness in each snapshot corresponds to the height of vorticity, and the white dots represent the clumps. The ratio of the circulations of the BGV and the clump is $\Gamma_b/\Gamma_c = N_b/N_c = 59(a)$ and $57(b)$. The total circulation of the vortex is evaluated as $\Gamma = eN/\varepsilon_0 B_0 L = 1.64 \times 10^{-6} N [m^2/s]$. Here, $N$ is the total number of electrons consisting each vortex with the length of $L \approx 235$mm, and the electron density $n(x,y)$ is related to the vorticity $\zeta(x,y) = \nabla \times v$ by $\zeta = en(x,y)/\varepsilon_0 B_0$. The two clumps injected in the BGV start free motion at $t = 20\mu s$ on electrical disconnection from the cathodes associated with negative biasing of the left plug potential.

The upper panels show that clumps come up to the center of the BGV within $100\mu s$ to merge by $t = 150\mu s$. The merging is achieved in less than a few turnover time. The lower panels, on the other hand, show that the clumps stay apart for more than 10ms (about 200

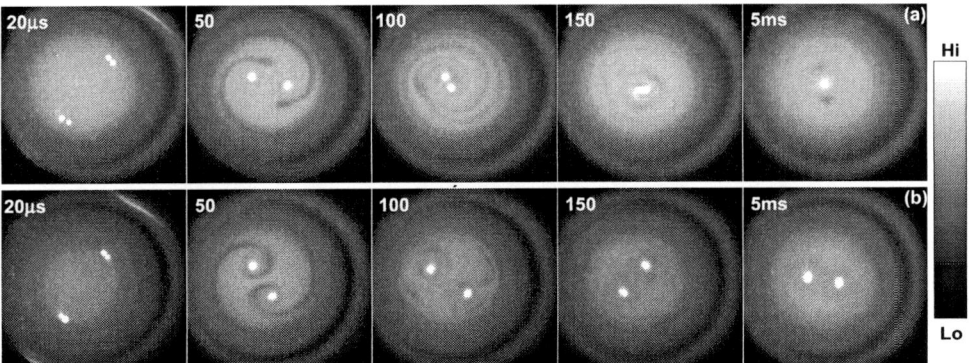

**FIGURE 1.** Snapshots describing clumps' dynamics in BGVD with $\Gamma_b \propto N_b/10^8 = 3.55$(a) and $\Gamma_b \propto N_b/10^8 = 3.39$(b). Two clumps with equal circulation $\Gamma_c \propto N_c/10^7 = 0.6$ are placed symmetrically in the initial BGVD. The panel size is 37.4 mm × 37.4 mm at the mid-plane of the trap.

turnover times) and form a binary structure, when the initial profile of BGV is modified slightly. The fate of the vortical evolution critically depends on slight differences in the initial distribution of BGV.

The radial profiles of the electron density for the BGV are plotted in Fig.2 $(a)$. Here the zero levels are successively shifted by $5 \times 10^{11} [m^{-3}]$ to distinguish closely resembling distributions. The total number $N_b$ of the BGV electrons lies in the range $(3.16 - 3.55) \times 10^8$. The clumps as shown in Fig.1 ($t = 20\mu s$) are injected at two symmetric locations indicated by the arrow.

The inter-clump distance is plotted in Fig.2(b) as a function of the time after the clumps start free motion in different profiles of the BGV. Common symbols are used for Figs.2(a) and (b). Each data point represents the average over five shots. The vertical bars stand for the full dispersion of the data points. The narrow dispersion indicates that the evolutionary path of the vortical dynamics is well defined by the initial condition under the fine control.

A general trend in the vortical evolution is that the clumps move inward during the initial phase of $\sim 50\mu s$ at nearly the same velocity (within a factor of 2). For the initial conditions with $N_b/10^8 \geq 3.46$, the clumps continue to move inward then merge within a few hundred $\mu s$. On the other hand the clumps immersed in a BGV, with $N_b/10^8 = 3.16, 3.17, 3.13, 3.08$, cease inward motion around the radial point of $r = (4-6)mm$ then move outward finally to form a binary structure that lasts more than 10ms characterized by the coulomb collision time between electrons.

The time scale of the merging process is characterized by the period of vortex rotation $\leq 100\mu s$, while the decay time of the inter-clump distance in the binary state is $\sim$ 10ms. We expect that the latter process is related to collisional dissipation in the fine spatial structure. A new feature obtained in this experiment is that the two states are not separated by an abrupt transition if a narrow parameter range is carefully expanded by a fine control of the initial BGV distribution.

## INITIAL PARAMETERS AS KEY ELEMENTS

For quantitative discussion we try to relate the inter-clump distances $D(1ms)$ at $t = 1ms$ to combinations of parameters characterizing the initial BGVD's. The selected time 1ms corresponds to the geometrical mean of the characteristic time ($\sim 100\mu s$) of vortex dynamics and a typical electron-electron collision time ($\sim$ 10ms). Therefore it belongs to the asymptotic time scale of collisionless vortex dynamics.

The parameters we have examined for the correlational analyses with respect to $D(1ms)$ include the total electron number $N_b$, the local density of electrons $n_b$, its derivative $\nabla n_b$, the rotational shear, and their combinations. It is noted that the circulations, $\Gamma_c \propto N_c$ and $\Gamma_b \propto N_b$, of the clumps and BGV do not show a good correlation with $D(1ms)$ contrary to the results obtained in cruder but more extensive surveys [1, 2, 7]. The parameter that provides the most coherent display of $D(1ms)$'s has turned out to be $\nabla n_b/n_b{}^\alpha$ with $\alpha = 2 \pm 0.5$. Here we take the parametric values at the initial locations of the clumps. Figure 3 plots $D(1ms)$'s as a function of $\nabla n_b/n_b{}^\alpha$ with $\alpha = 2$. The symbols correspond to those used in Fig.2(a).

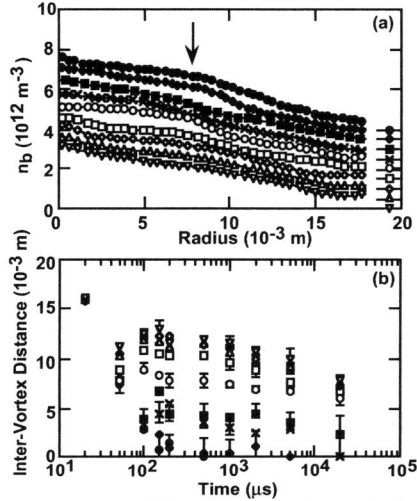

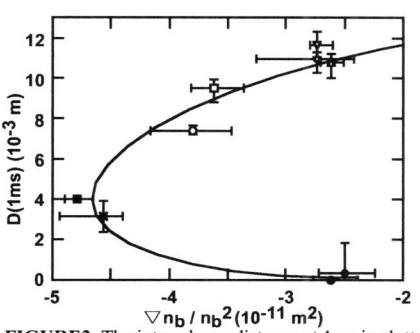

**FIGURE3.** The inter-clump distance at 1ms is plotted against $\nabla n_b/n_b^2$ evaluated at the initial locations of the clumps. The full width of five shots attached to each symbol.

**FIGURE2.** (a) Radial profiles of the electron density distribution $n_b(x,y)=\varepsilon_o B_o \zeta_b(x,y)/e$ constituting the background vortices with $Nb/10^8$ =3.55(●), 3.46(◆), 3.25(+), 3.14(×), 3.39(○), 3.08(□), 3.13(◇), 3.17(△), and 3.16(▽).
(b) Distance between two clumps is plotted as a function of time for differnent values of $N_b$.

An interesting point is that $D(1\text{ms})$ is a multivalued function of $\nabla n_b/n_b^2$. The power index $\alpha$ is determined so that the two branches connect smoothly. The upper branch suggests that the two clumps climb the vorticity hill of BGVD with increased speed for larger $|\nabla n_b|$ so that the inter-clump distance decreases with $|\nabla n_b|$ [4, 8, 9]. But the upper branch indicates that the inward drive by the density gradient is not strong enough to let the clumps merge at the hill top of BGVD.

As far as $|\nabla n_b|$ is concerned, the lower branch shows a trend contradictory to the Schecter's model [8, 9]. This may be attributed to the contribution of the factor $1/n_b^2$. As the density ratio $n_b/n_c$ between the BGV and the clump decreases, local perturbations $\delta n_b/n_b$ in the BGV increases, and the clumps can be driven by the strong perturbations in a manner quite different from the linear theory. Mutual interaction between the clumps may also be equally influential.

The dynamics of the clumps is consistent with the $\mathbf{E} \times \mathbf{B}$ drift calculated numerically from the observed electron density distribution. In spite of the experimental consistency, a theoretical model that can describe the dynamics is not available.

## HOLE STRUCTURES IN BGV

From the observed facts that the initial distributions of BGV shown in Fig.2 are close in shape and height and that the data points in Fig.3 lie on a single continuous curve we expect that there should be another parameter that varies monotonically along the

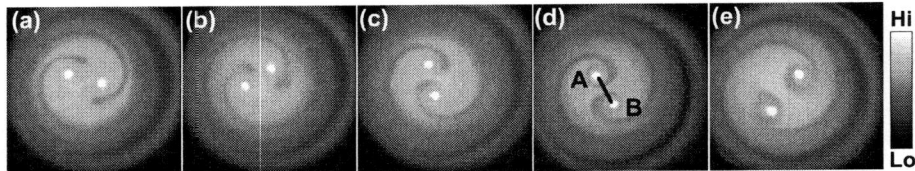

**FIGURE4.** Snapshots of the vorticity distributions at 50μs starting from the initial BGVD's with $N_b/10^8 = 3.55$(a), 3.46(b), 3.25(c), 3.39(d), and 3.13(e).

curve. The parameter, if exists, should form a single-valued function for $D(1\text{ms})$ which connects the two branches and eliminates the multiplicity of $D(1\text{ms})$ against $\nabla n_b/n_b^2$.

We have not been successful in finding the second parameter directly connected to the initial distribution, so that we turn our attention to patterns of the BGVD in the evolutionary stage of vortical dynamics. In Fig.4 we compare the vorticity distributions at 50μs starting from different initial BGVD's as given in Fig.2(a). In the selected time-domain the clumps show stagnation in the radial motion, some rebound outward, some stay there and some regain inward motion.

In the panels we focus our attention to characteristic structures generated around the clumps. A noticeable feature is that there is a belt of depleted density around each clump and the angle spanned by the arc varies monotonically from a panel to panel. We observe that clumps ($N_b/10^8 = 3.55$(a) and 3.46(b)) with a depletion arc which does not reach the line connecting their peaks merge within the vortical time scale ( < 1ms). On the other hand clumps ($N_b/10^8 = 3.25$(c) and 3.39(d)) that are enclosed by the depletion remain separated to form binary vortices. These observations indicate a good correlation between the degree of completion of the ring hole and the asymptotic distance $D(1\text{ms})$.

We therefore examine the correlation more quantitatively. Figure 5(a) plots the density distribution along the chord AB connecting the peaks of the clumps shown in Fig.4(d). The ordinate is enlarged to clearly show the density profile around the feet of the clumps. In the distribution we observe two dips corresponding to the belts of depletion surrounding the clumps. We introduce $\Delta n_b$ representing the amount of the density dip and $\Delta r$ corresponding to the width of the depletion belt, both along the chord AB. Analyses have led us to a good correlation between $D(1\text{ms})$ and the product $\Delta n_b \Delta r$ as plotted in Fig.5(b). Here the same symbols are used as in Fig.2(b). The almost linear dependence of $D(1\text{ms})$ on $\Delta n_b \Delta r$ indicates that the deficit of circulation in the space between two clumps tends to prevent further convergence and merger of the clumps. Detail discussion about mechanism of bifurcation has been reported elsewhere [10].

## CONCLUSION

In conclusion we have carefully examined the mechanism of bifurcation leading to merger or to the formation of binary vortices by introducing a fine control of the initial profile of BGVD. The inter-clump distance at a vortically asymptotic time of 1 ms has been related to the initial parameters only in terms of $\nabla n_b/n_b^2$ at the initial positions

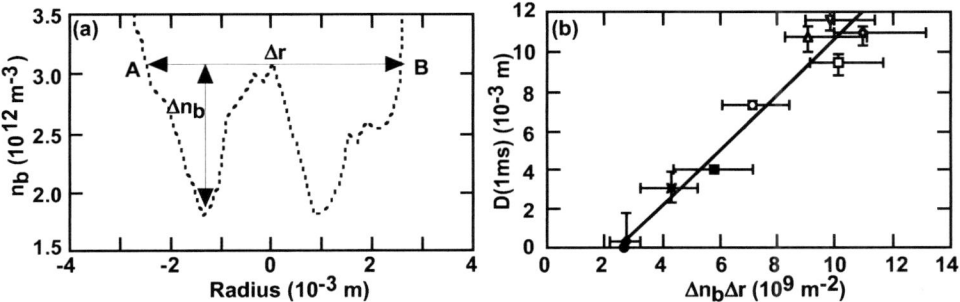

**FIGURE5.** (a) The profile of electron density distribution along the straight line AB in Fig.4(d). (b) The inter-clump distance at 1ms is plotted against $\Delta n_b \Delta r$ at 50ms starting from different distributions of initial BGV. Symbols correspond to the profiles in Fig. 2 (a).

of the clumps. The empirical relation is double valued. The degeneracy can be removed if the distance is related to a shaping parameter $\Delta n_b \Delta r$ at the critical time period of clumps' approaching in the initial phase. The observations stress a dominant role which structures in BGVD play in determining the clumps' state, merger or a binary state.

## ACKNOWLEDGMENTS

This work was supported by a Grant-in-Aid from the Ministry of Education, Culture, Sports, Science and Technology and by the collaborative research program of the National Institute for Fusion Science. The authors appreciate stimulating discussion with Prof. Kimitaka Itoh of NIFS.

## REFERENCES

1.  K. S. Fine, A. C. Cass, W. G. Flynn and C. F. Driscoll, Phys. Rev. Lett. **75**, 3277 (1995).
2.  D. A. Schecter, D. H. E. Dubin, K. S .Fine and C. F. Driscoll, Phys. Fluids. **11**, 905 (1999).
3.  D. Z. Jin and D. H. E. Dubin, Phys. Rev. Lett. **80**, 4434 (1998).
4.  Y. Kiwamoto, K. Ito, A. Sanpei and A. Mohri, Phys. Rev. Lett. **85**, 3173 (2000).
5.  Y. Kiwamoto, K. Ito, A. Sanpei, A. Mohri, T. Yuyama and T. Michishita, J. Phys. Soc. Jpn. (Lett.) **68**, 3766 (1999).
6.  K. Ito, Y. Kiwamoto and A. Sanpei, Jpn. J. Appl. Phys. **40**, 2558 (2001).
7.  A. Sanpei, Y. Kiwamoto, K. Ito and Y. Soga, Phys. Rev. E. bf 68, 016404 (2003).
8.  D. A. Schecter and D. H. E. Dubin, Phys. Rev. Lett. **83**, 2191 (1999).
9.  D. A. Schecter and D. H. E. Dubin, Phys. Fluids. **13**, 1704 (2001).
10. Y. Soga, Y. Kiwamoto, A. Sanpei and J. Aoki, "Merger and binary structure formation of two discrete vortices in a background vorticity distribution of a pure electron plasma" accepted for publication in Phys. Plasmas (2003)

# SECTION 2

# ANTIMATTER PLASMAS
# AND RECOMBINATION

# Non-Destructive Positron Plasma Diagnostics for Antihydrogen Production

M. Amoretti[1,*], C. Amsler[†], G. Bonomi[**], A. Bouchta[**], P. D. Bowe[‡],
C. Carraro[*], C. L. Cesar[§], M. Charlton[‡], M. Doser[**], V. Filippini[¶],
A. Fontana[¶], M. C. Fujiwara[‖], R. Funakoshi[‖], P. Genova[¶], J. S. Hangst[††],
R. S. Hayano[‖], L. V. Jørgensen[‡], V. Lagomarsino[*], R. Landua[**],
D. Lindelöf[†], E. Lodi Rizzini[‡‡], M. Macri[*], N. Madsen[††], G. Manuzio[*],
P. Montagna[¶], H. Pruys[†], C. Regenfus[†], A. Rotondi[§§], G. Testera[*],
A. Variola[*], D. P. van der Werf[‡],

(ATHENA COLLABORATION)
and
R. L. Spencer[¶¶]

[*]*INFN and Dipartimento di Fisica, Università di Genova, 16146 Genova, Italy*
[†]*Physik-Institut, Zürich University, CH-8057 Zürich, Switzerland*
[**]*EP Division, CERN, CH-1211 Geneva 23, Switzerland*
[‡]*Department of Physics, University of Wales Swansea, Swansea SA2 8PP, UK*
[§]*Instituto de Fisica, Universidade Federal do Rio de Janeiro, Rio de Janeiro 21945-970, and
Centro Federal de Educação Tecnologica do Ceara, Fortaleza 60040-531, Brazil*
[¶]*INFN and Dipartimento di Fisica Nucleare e Teorica, Università di Pavia, 27100 Pavia, Italy*
[‖]*Department of Physics, University of Tokyo, Tokyo 113-0033, Japan*
[††]*Department of Physics and Astronomy, University of Aarhus, DK-8000 Aarhus C, Denmark*
[‡‡]*Dipartimento di Chimica e Fisica per l'Ingegneria e per i Materiali, Università di Brescia and
INFN (Gruppo collegato di Brescia), 25123 Brescia, Italy*
[§§]*INFN and Dipartimento di Fica Nucleare e Teorica, Università di Pavia, 27100 Pavia, Italy*
[¶¶]*Department of Physics and Astronomy, Brigham Young University, Provo, Utah 84602*

**Abstract.** Production of antihydrogen atoms by mixing antiprotons with a cold, confined, positron plasma depends on parameters such as the plasma density and temperature. We discuss a non-destructive diagnostic, based on an analysis of excited, low-order plasma modes, that provides comprehensive characterization of the positron plasma in the ATHENA antihydrogen apparatus. The dipole and quadrupole modes of a spheroidal positron plasma are interpreted in the framework of a cold fluid theory. In particular, the excitation and detection of the dipole mode are analytically modeled considering the response of the center-of-mass to a resonant driving perturbation. The model is compared to, and validated by, numerical simulations with a particle-in-cell code. Measurements of the positron plasma properties are discussed.

---

[1] Corresponding author (marco.amoretti@ge.infn.it)

CP692, *Non-Neutral Plasma Physics V*, edited by M. Schauer et al.
© 2003 American Institute of Physics 0-7354-0165-9/03/$20.00

# INTRODUCTION

Recently cold antihydrogen atoms where produced in the ATHENA experiment (ApparaTus for High precision Experiments on Neutral Antimatter, or shortly AnTiHydrogEN Apparatus) at CERN (European Organization for Nuclear Research) by mixing low energy antiprotons with a cold dense positron plasma inside an electromagnetic trap [1].

Very low positron and antiproton temperatures (a few K) and high positron density ($\simeq 10^8$ particles/cm$^3$) are the two key ingredients necessary to enhance the recombination rate in ATHENA. Under these conditions the positron cloud is in the plasma regime. In order to understand and control the recombination process, several parameters describing the positron plasma in thermal equilibrium should be measured in a non-destructive way to avoid perturbing the system.

Harmonically confined one component plasmas at temperatures close to absolute zero take the shape of an ellipsoid characterized by an aspect ratio $\alpha = z_p/r_p$, where $z_p$ and $r_p$ are the semi-major axis and semi-minor axis respectively. A simple analytic model for low-order axisymmetric plasma modes (Trivelpiece-Gould modes) in a spheroidal plasma has been used as a diagnostic in the ATHENA experiment [2, 3], where these modes were excited and detected to gain information about the positron plasma. The two lowest-order (dipole and quadrupole) modes were interpreted in the framework of a cold fluid theory [4]. The mode frequencies depend only on the plasma density $n$ and aspect ratio $\alpha$. Corrections due to finite temperature have also been calculated [5, 6]. Previous studies [6, 7, 8, 9, 10, 11, 12, 13] have extracted the information contained in the mode frequencies themselves. The plasma density and aspect ratio can be derived by comparing the measured frequencies of the dipole and quadrupole modes with those predicted by theory. But the actual plasma length and radius (or, equivalently, the number of particles) cannot be ascertained by using only frequency data. However, in Ref. [3], we demonstrated that the plasma length can be extracted from a detailed analysis of the power transmitted through the plasma near the resonance of the center-of-mass mode.

The model described in Ref. [3] can be numerically validated using a non-neutral plasma equilibrium code EQUILSOR [14] coupled with a two-dimensional ($r - z$) particle-in-cell (PIC) simulation RATTLE [6]. The first code (EQUILSOR) is dedicated to the evaluation of the plasma thermal equilibrium. The code solves the Poisson-Boltzmann equilibrium equation assuming axisymmetry [15]. The second code (RATTLE) uses the computed equilibria to create initial distributions of particles in ($r, z, v_z$)-space and then simulates the motion of the particles via a standard PIC technique. The same numerical codes were used by Surko and co-workers [6] to investigate the applicability of Dubin's mode theory [4] to their electron plasma.

This paper is organized in the following manner. The following section briefly describes the experimental setup. The third section is a concise review of the model used to describe the center-of-mass (dipole) mode and confirms its validity by means of a comparison with numerical simulations. In the fourth section a possible extension of the model in order to study the plasma response for the quadrupole mode is discussed. The simulation results show that the extension developed in Ref. [3] also works for the quadrupole mode. The concluding section presents an example of the application of the diagnostic.

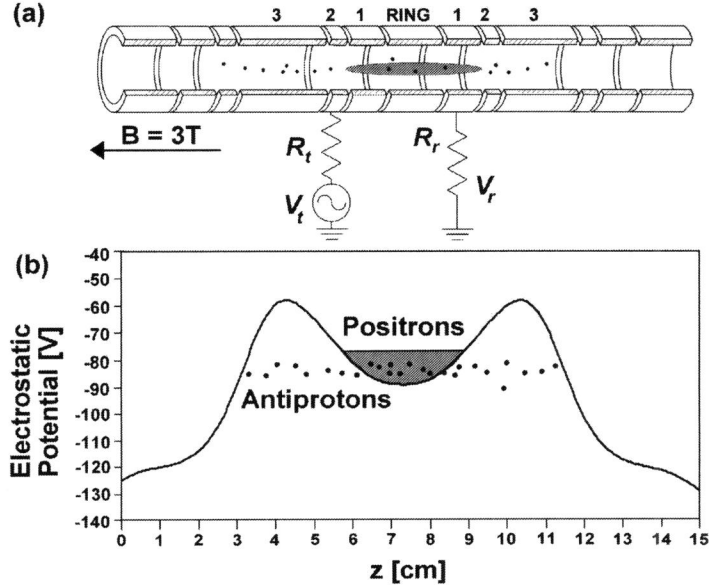

**FIGURE 1.** (a) Schematic of the mixing trap electrodes. The positrons are confined in the center group designed to create a harmonic electrostatic field. The central electrode is called the ring electrode and the other 3 pairs are labeled 1, 2 and 3. The driving signal applied to one electrode is shown together with the resistances on the transmitting and receiving electrodes. The shape of the prolate positron ellipsoid is shown schematically. (b) The axial potential of the ATHENA nested trap is shown and the ranges of axial motion of the positrons and of the antiprotons is indicated schematically.

# EXPERIMENTAL SETUP

Electromagnetic traps of the Penning-Malmberg type are used in the ATHENA experiment to confine charged particles. The traps are realized by placing a series of cylindrical electrodes of various lengths and with an inner radius of 1.25 cm inside a uniform 3-Tesla magnetic field parallel to the trap axis and applying static voltages to them. A potential well along the trap (or $z-$) axis is thereby produced which provides axial confinement for particles having energies lower than the top of the well. The magnetic field ensures radial confinement. The trap structure is installed inside a cryogenic bore and can be cooled to about 15 K. The voltages on these electrodes are manipulated to perform various procedures.

The mixing trap [Fig. 1(a)] is composed of 3 groups of electrodes which produce the nested trap configuration [Fig. 1(b)]. Thus, the simultaneous confinement of particles having opposite signs of charge is achieved. The positron confining region is comprised of seven electrodes. The lengths of these electrodes have been chosen, according to Ref. [16], in order to create a harmonic potential when the ratios between the applied voltages are suitably chosen.

The axial modes are excited by applying a sinusoidal perturbation to one electrode with an electromotive force $V_t = v_t e^{j\omega t}$. The resulting oscillation of the plasma induces a current in the pick-up electrode [17, 18, 19] and a voltage $V_r = v_r(\omega)e^{j\omega t}$ is detected across the resistance $R_r$ [Fig. 1(a)].

## DIPOLE MODE

In cold fluid theory [4], the lowest-order mode is a coherent oscillation of the whole plasma along the $z$ axis with a frequency $\omega_1$ equal to that of single particle motion inside the trap,

$$\omega_1 = \omega_z = \sqrt{\frac{qV_0}{md^2}}. \tag{1}$$

In Eq. (1) $q$ is the particle electric charge and $m$ is its mass, $V_0$ is the potential difference between the ring and the type 3 electrode, and the length $d$ is related to the trap radius $r_w$ ($d = 1.74\, r_w$ for the mixing trap design).

A simple model, based on the observation that the dipole mode can be described as a damped oscillation of the center-of-mass of the positron cloud, enables study of the excitation and detection processes. The center-of-mass equation of motion along the trap axis can be written as [3]:

$$\ddot{z}_{cm} + \gamma\dot{z}_{cm} + \omega_1^2 z_{cm} = \frac{q}{m}\langle E_{zi}\rangle, \tag{2}$$

where $z_{cm}$ is the axial position of the cloud center-of-mass and $\gamma$ describes the damping of the oscillations. In this equation the driving term $\langle E_{zi}\rangle$ is an effective axial electric field acting on the center-of-mass when a potential $V_i$ is applied on the electrode labeled $i$ ($i$ indicates the electrode type, see also Fig. 1(a)). It can be approximated by

$$\langle E_{zi}\rangle = g_i(\alpha, z_p)\frac{V_i}{2r_w}, \tag{3}$$

where the characteristic function $g_i(\alpha, z_p)$, defined by

$$g_i(\alpha, z_p) = \frac{3\alpha^2 r_w}{z_p^3} \int_{-z_p}^{z_p} dz \int_0^{z_p\alpha^{-1}\sqrt{1-z^2/z_p^2}} r\,dr\,F_{zi}(r,z), \tag{4}$$

has been introduced to describe the coupling between the perturbation signal and the center-of-mass response. In Eq. (4), $F_{zi}(r,z)$ represents the axial component of the electric field at the position $(r,z)$ when a unit potential is applied on the electrode $i$ while the rest of the electrodes are grounded. The factor $g_i(\alpha, z_p)$ can be numerically evaluated using a truncated Fourier-Bessel series (as in Ref. [3]) or by using directly the EQUILSOR-RATTLE numerical Poisson solver. This coupling function depends not only on the trap geometry and on the type of the electrode used to drive the mode, but also on the size and shape of the plasma. It has a strong dependence on the plasma length, but only weakly depends on the aspect ratio (or, equivalently, on the plasma radius; see Fig. 2).

124

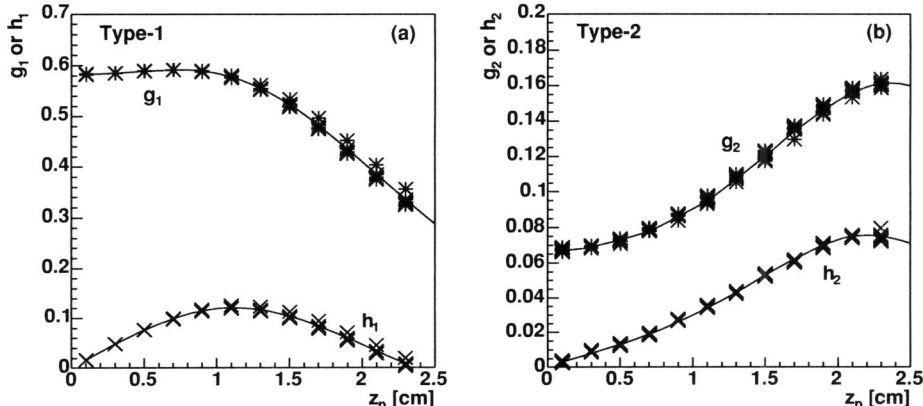

**FIGURE 2.** Dependence of the characteristic functions describing the coupling between the perturbation signal and dipole ($g_i$) and quadrupole ($h_i$) modes on the plasma semi-length for (a) type-1 and (b) type-2 electrodes (see the discussion in the text). The values are evaluated using the EQUILSOR-RATTLE numerical Poisson solver. The scattering of the points reflects the dependence on the aspect ratio. The solid curves are for aspect ratio $\alpha = 6$.

This model allows estimation of the growth of the mode due to the application of a sinusoidal perturbation to one transmitting electrode [$i = t$, see also Fig. 1(a)]. If we suppose that the driving frequency is in perfect resonance with the natural frequency of this mode, i.e., $V_t \sin \omega_1 t$, then the velocity of the center-of-mass, $\dot{z}_{cm}$, is characterized by a linear growth with an acceleration

$$a_{cm} = g_t(\alpha, z_p) \frac{qV_t}{4mr_w}. \tag{5}$$

To test the accuracy of this simple model, the excitation process was directly simulated using EQUILSOR-RATTLE. First, a series of thermal equilibria characterized by the same plasma radius, but with different numbers of particles $N$ and plasma half-lengths $z_p$, were generated. The plasmas were confined in the central part of the nested trap configuration with electrode potential values equal to those used in the mixing studies in the ATHENA experiment. The exciting perturbation was applied in separate runs to both type-1 and type-2 electrodes. The simulated growth rates were then compared with the rates predicted by Eq. (5). Figure 3 shows the good agreement between the simulation and the model.

The coupling function $g_r(\alpha, z_p)$ also allows estimation of the induced current on a single electrode due to the center-of-mass oscillation. Following the model, the current induced on the receiving electrode [labeled by $r$, see also Fig. 1(a)] is given by Eq. (19) of Ref. [3], i.e.,

$$I_r = g_r(\alpha, z_p) \frac{qN}{2r_w} \dot{z}_{cm}. \tag{6}$$

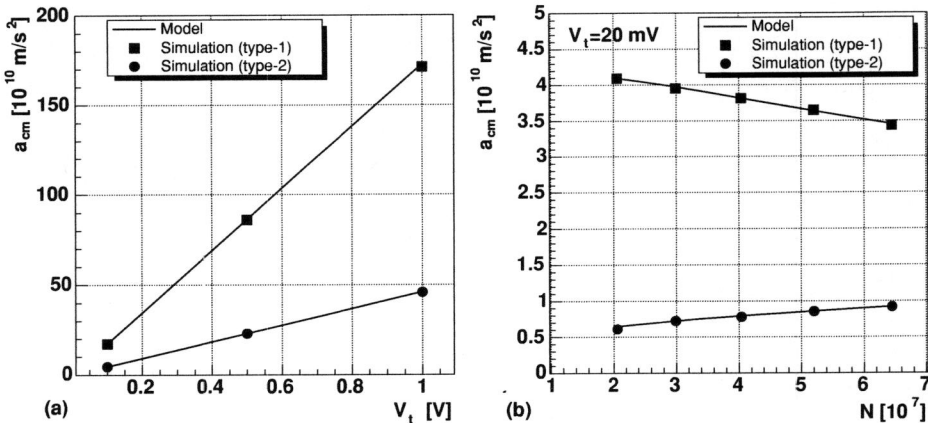

**FIGURE 3.**  (a) Dependence of the acceleration amplitude $a_{cm}$ of the center-of-mass oscillation on the amplitude of the perturbation $V_t$ applied in resonance with $\omega_1$. The time duration of the perturbation is limited to a few cycles of the dipole oscillation. Otherwise, for $V_t > 0.1$ V saturation effects are soon visible. The plasma was characterized by $z_p = 1.63$ cm, $r_p = 0.22$ cm, $n = 2 \times 10^8$ cm$^{-3}$, and temperature $T = 0.1$ meV. (b) Acceleration amplitudes $a_{cm}$ of the center-of-mass oscillation for different plasmas subjected to an external sinusoidal perturbation $V_t \sin \omega_1 t$ applied on a type-1 or type-2 electrode. The simulated plasmas are characterized by different values of particle number $N$ and axial semi-length $z_p$ but they have all the same radius $r_p = 0.22$ cm. In both (a) and (b) the solid lines are the growth rates evaluated using Eq. (5).

If the maximum of the oscillation is $z_{\text{offset}}$ then

$$I_r = g_r(\alpha, z_p) \frac{qN}{2r_w} \omega_1 z_{\text{offset}}. \tag{7}$$

The simplest way to test the validity of this relation is to simulate the oscillation of the center-of-mass by giving an axial offset $z_{\text{offset}}$ to all of the plasma particles in RATTLE. The simulated current is obtained by numerically integrating the normal component of the electric field over the electrode under consideration, then numerically differentiating this charge as a function of time to find the current. Figure 4 shows the comparison between the analytic model and the simulation and, again, the agreement is very good (the relative difference between the simulations and our analytic model is less than 1%).

## QUADRUPOLE MODE

The quadrupole mode frequency $\omega_2$ can be evaluated using the dispersion relation of the axisymmetric modes in a cold fluid spheroid [4, 6], i.e.,

$$1 - \frac{\omega_p^2}{\omega_2^2} = \frac{k_2}{k_1} \frac{P_2(k_1) Q_2^{0\prime}(k_2)}{P_2'(k_1) Q_2^0(k_2)} \tag{8}$$

126

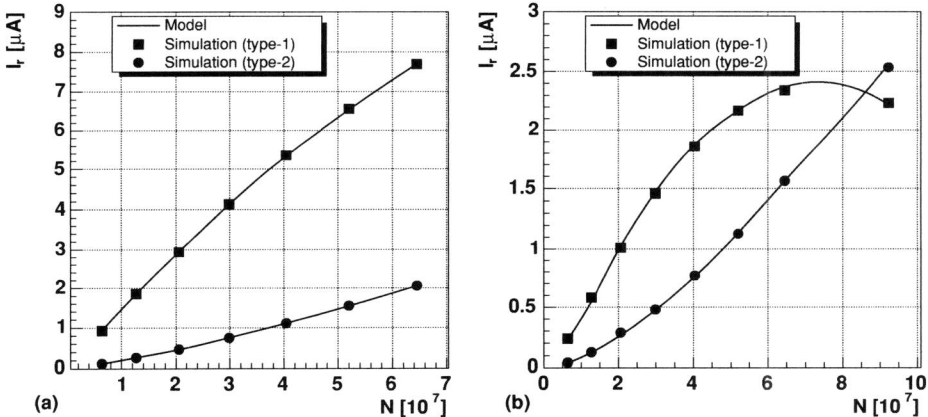

**FIGURE 4.** (a) Simulated induced current on the trap electrodes due to the oscillation of the center-of-mass for different plasmas ($z_{\text{offset}} \simeq 0.3$ mm). The solid lines represent the values of the induced currents evaluated by means of Eq. (7). (b) Simulated induced current on the trap electrodes due to the quadrupole mode for different plasmas ($z_{\text{stretch}} \simeq 0.3$ mm). The solid lines represent the values of the induced currents evaluated by means of Eq. (10).

where $P_2$ is a Legendre polynomial, $Q_2^0$ is its singular partner, $\omega_p = \sqrt{nq^2/\varepsilon_0 m}$ is the plasma frequency, $k_1 = \alpha/\sqrt{\alpha^2 - 1 + \omega_p^2/\omega_2^2}$, and $k_2 = \alpha/\sqrt{\alpha^2 - 1}$.

The model developed for the evaluation of the center-of-mass excitation and detection can be extended in order to describe the quadrupole mode oscillations. The extension is based on the introduction of coupling factors $h_i(\alpha, z_p)$ similar to the functions $g_i(\alpha, z_p)$ described above. In this case the coupling functions reflect the "stretching nature" of the quadrupole mode,

$$h_i(\alpha, z_p) = \frac{3\alpha^2 r_w}{z_p^3} \int_{-z_p}^{z_p} dz \frac{z}{z_p} \int_0^{z_p \alpha^{-1} \sqrt{1 - z^2/z_p^2}} r\, dr F_{zi}(r, z). \tag{9}$$

Figure 2 shows a comparison between $g_i(\alpha, z_p)$ and $h_i(\alpha, z_p)$ values.

Using the factors $h_r(\alpha, z_p)$ evaluated for the receiving electrode, a relationship similar to that of Eq. (6) can be written:

$$I_r = h_r(\alpha, z_p) \frac{qN}{2r_w} \omega_2 z_{\text{stretch}}, \tag{10}$$

where $\omega_2 z_{\text{stretch}}$ can be identified as the maximum velocity of the particles in the plasma due to the oscillation of the mode.

An antisymmetric linear stretch of the plasma along the $z$ axis is used to excite the quadrupole mode in the simulations. Despite the simplicity of Eq. (10) the agreement between the simulation results and the model predictions is good (see Fig. 4).

# EXPERIMENTAL RESULTS

The experimental implementation of this diagnostic is extensively discussed in Ref. [3]. In this section an example of monitoring the plasma parameters during a long period of evolution is shown. In these measurements a plasma of about $4 \times 10^7$ positrons was confined in the nested trap for about 1000 s without the injection of antiprotons.

The number density and aspect ratio were extracted using the measured frequencies ($\omega_1$, $\omega_2$) of the dipole and quadrupole modes and combining equations (1) and (8). The evolution of these two parameters during the plasma expansion is shown in Fig. 5.

For a complete non-destructive diagnostic it is necessary to measure another parameter: the total number of particles $N$, or either $r_p$ or $z_p$ ($N = 4\pi n z_p^3/3\alpha^2 = 4\pi n \alpha r_p^3/3$). None of these quantities can be determined by using the frequency data even if higher order longitudinal modes are detected. But the missing parameter can be extracted by a complete study of the way the plasma mediates the transmission of a driving signal on one electrode to a different receiving electrode. The signal transmitted through the plasma for a driving frequency near $\omega_1$ is characterized by a resonance shape. Using the model discussed in the previous sections it is possible to approximate the plasma response to that of a resonant RLC circuit to an external drive. The parameters of the resulting circuit depend on all of the plasma properties, so that the shape of the resonant response makes it possible to evaluate these parameters, and consequently the remaining plasma properties. Figure 6 shows the time evolution of the plasma length and radius, while the reconstructed number of positrons obtained from the resonance curve is shown in Fig. 7(a). The fluctuations in the data are mainly due to the experimental signal-to-noise ratio of the quadrupole mode signal. Improving the signal-to-noise ratio by increasing the mode driving amplitude would, however, result in an unfavorable plasma heating effect.

An example of the application of this diagnostic to positrons during antihydrogen

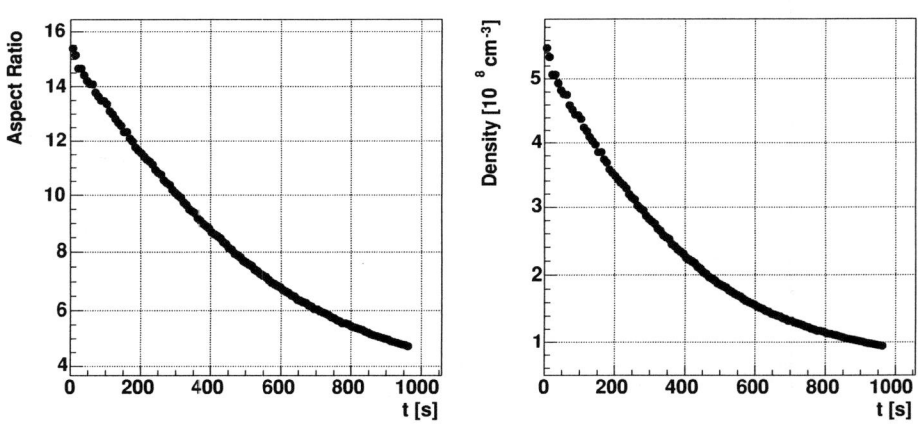

**FIGURE 5.** The evolution of the aspect ratio and the number density for a plasma of about $4 \times 10^7$ positrons confined in the nested trap for about 1000 s without the injection of antiprotons.

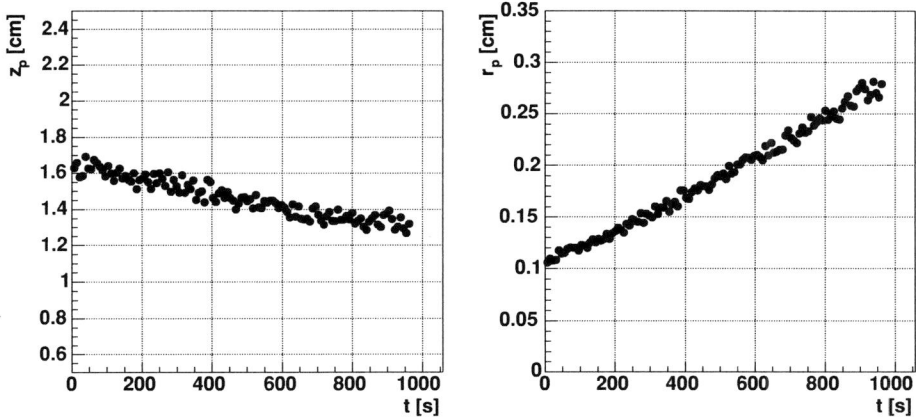

**FIGURE 6.** The evolution of the plasma semi-length and radius for the same plasma as Fig. 5.

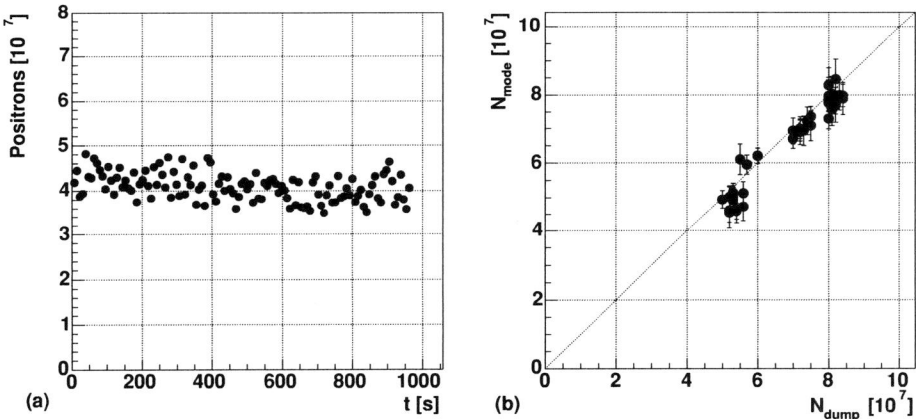

**FIGURE 7.** (a) Total number of positrons estimated by means of the transmission diagnostic for the plasma evolution of Fig. 5. (b) The total number of positrons obtained by using the transmission diagnostic is plotted against the number obtained by extracting the positrons to a Faraday cup.

production is shown in Fig. 7(b). Here the total number obtained by using the model is plotted against the number found by extraction to a Faraday cup. In this regime the linearity and good correspondence in the absolute number show that both the model and its implementation constitute a complete non-destructive plasma diagnostic. Typical measured properties of the ATHENA positron plasma for antihydrogen production were $7 \times 10^7$ positrons at a density of about $1.7 \times 10^8$ cm$^{-3}$ in a plasma approximately 3.2 cm long ($2z_p$) with a radius of about 0.25 cm and a storage time of many hundreds of seconds.

# CONCLUSION

We have extended the plasma mode diagnostic method to provide comprehensive characterization of the cold, dense positron plasma employed in the ATHENA antihydrogen experiment. The method has already been utilized to great advantage in ATHENA, and promises to be an essential element of future experiments. The technique, while particularly useful for non-destructive measurements on difficult-to-produce species such as positrons, has immediate applicability to other Penning trap plasmas.

# ACKNOWLEDGMENTS

This work was supported by INFN (Italy), FAPERJ (Brasil), MEXT (Japan), SNF (Switzerland), NSRC (Demark), EPSRC (UK), and the European Union.

# REFERENCES

1.  M. Amoretti, *et al.*, Nature (London) **419**, 456 (2002).
2.  M. Amoretti, *et al.*, Phys. Rev. Lett. **91**, 055001 (2003).
3.  M. Amoretti, *et al.*, Phys. Plasmas. **10**, 3056 (2003).
4.  D. H. E. Dubin, Phys. Rev. Lett. **66**, 2076 (1991).
5.  D. H. E. Dubin, Phys. Fluids B **5**, 295 (1993).
6.  M. D. Tinkle, R. G. Greaves, C. M. Surko, R. L. Spencer, and G. W. Mason, Phys. Rev. Lett. **72**, 352 (1994).
7.  D. J. Heinzen, J. J. Bollinger, F. L. Moore, W. M. Itano, and D. J. Wineland, Phys. Rev. Lett. **66**, 2080 (1991).
8.  J. J. Bollinger, D. J. Heinzen, F. L. Moore, W. M. Itano, D. J. Wineland, and D. H. Dubin, Phys. Rev. A **48**, 525 (1993).
9.  C. S. Weimer, J. J. Bollinger, F. L. Moore, and D. J. Wineland, Phys. Rev. A **49**, 3842 (1994).
10. M. D. Tinkle, R. G. Greaves, and C. M. Surko, Phys. Plasmas **2**, 2880 (1995).
11. H. Higaki and A. Mohri, Jpn. J. Appl. Phys. Part 1 **36**, 5300 (1997).
12. T. B. Mitchell, J. J. Bollinger, X.-P. Moore, and W. M. Itano, Optics Express **2**, 314 (1998).
13. H. Higaki, N. Kuroda, T. Ichioka, K. Yoshiki Franzen, Z. Wang, K. Komaki, Y. Yamazaki, M. Hori, N. Oshima, and A. Mohri, Phys. Rev. E **65**, 046410 (2002).
14. R. L. Spencer, S. N. Rasband, and R. R. Vanfleet, Phys. Fluids B **5**, 4267 (1993).
15. S. A. Prasad and T. M. O'Neil, Phys. Fluids **22**, 278 (1979).
16. H. Raimbault-Hartmann, D. Beck, G. Bollen, M.König, H.-J. Kluge, E. Schark, J. Stein, S. Schwarz, J. Szerypo, Nucl. Instr. and Meth. **B 126**, 378 (1997).
17. M. D. Sirkis and N. Holonyak, Am. J. Phys. **34**, 943 (1966).
18. C. A. Kapetanakos and A. W. Trivelpiece, J. Appl. Phys **41**, 4841 (1971).
19. D. J. Wineland and H. G. Dehmelt, J. Appl. Phys **46**, 919 (1975).

# Three Dimensional Annihilation Imaging of Antiprotons in a Penning Trap

M.C. Fujiwara[ab1], M. Amoretti[c], G. Bonomi[d], A.Bouchta[d], P.D. Bowe[e], C. Carraro[cf] C.L. Cesar[g], M. Charlton[h], M. Doser[d], V. Filippini [i], A. Fontana[j], R. Funakoshi[a], P. Genova [j], J.S. Hangst[f], R.S. Hayano[a], L.V. Jørgensen[h], V. Lagomarsino[cf], R. Landua[d], D. Lindelöf [k], E. Lodi Rizzini[l], M. Macri[c], N. Madsen[e], M. Marchesotti [i], P. Montagna [ij], H. Pruys [k], C. Regenfus[k], P. Rielder[d], A. Rotondi [ij], G. Testera[c], A. Variola[c], D.P. van der Werf [h]

(ATHENA Collaboration)

[a] *Department of Physics, University of Tokyo, Tokyo 113-0033 Japan*
[b] *Atomic Physics Laboratory, RIKEN, Saitama, 351-0189 Japan*
[c] *Istituto Nazionale di Fisica Nucleare, Sezione di Genova, 16146 Genova, Italy*
[d] *EP Division, CERN, Geneva 23 Switzerland*
[e] *Department of Physics and Astronomy, University of Aarhus, DK-8000 Aarhus, Denmark*
[f] *Dipartimento di Fisica, Università di Genova, 16146 Genova, Italy*
[g] *Instituto de Fisica, Universidade Federal do Rio de Janeiro, Rio de Janeiro 21945-970, Brazil*
[h] *Department of Physics, University of Wales Swansea, Swansea SA2 8PP, UK*
[i] *Istituto Nazionale di Fisica Nucleare, Sezione di Pavia, 27100 Pavia, Italy*
[j] *Dipartimento di Fisica Nucleare e Teorica, Università di Pavia, 27100 Pavia, Italy*
[k] *Physik-Institut, Zürich University, CH-8057 Zürich, Switzerland*
[l] *Dipartimento di Chimica e Fisica per l'Ingegneria e per i Materiali, Universit`a di Brescia, 25123 Brescia, Italy*

**Abstract.** We demonstrate three-dimensional annihilation imaging of antiprotons trapped in a Penning trap. Exploiting unusual feature of antiparticles, we investigate a previously unexplored regime in particle transport; the proximity of the trap wall. Particle loss on the wall, the final step of radial transport, is observed to be highly non-uniform, both radially and azimuthally. These observations have considerable implications for the production and detection of antihydrogen atoms.

## INTRODUCTION

Imaging techniques have played an important role in trapped particle studies. Dumping particles onto a collimated Faraday cup, or on a screen viewed by a CCD camera, is now a standard technique which gives a $z$-integrated plasma shape [1-3] ($z$ is the direction along the magnetic axis). Detection of laser fluorescence is another

---

[1] Corresponding author: e-mail: Makoto.Fujiwara@cern.ch

CP692, *Non-Neutral Plasma Physics V*, edited by M. Schauer et al.
© 2003 American Institute of Physics 0-7354-0165-9/03/$20.00

common technique for trapped atomic ions where convenient transition lines exist [4,5]. This method can also provide particle velocity from the laser Doppler shift [6].

Antiparticle annihilation imaging can give information complementary to the above more conventional methods. Depending on the density of the residual gas in the system (see below), antiprotons can annihilate either on gas, or on the trap wall as a result of radial transport. If the vacuum is sufficiently high and the annihilation on the gas is negligible, antiproton imaging is uniquely sensitive to particle losses at the trap wall. It thus allows investigation of particle transport processes in the yet un-explored regime, the proximity of the trap wall.

O'Neil's confinement theorem [7] states that, for an axially symmetric system in a uniform magnetic field, due to the conservation of canonical angular momentum, the mean-square radius of trapped particles is approximately constant, ensuring confinement of non-neutral plasmas. In actual experiments, however, plasmas expand at a finite rate, eventually leading to deconfinement. Starting from the pioneering works of the 1980s [8], radial particle transport across magnetic field lines has been the subject of extensive studies. As such, there is now a large body of evidence which suggests that the radial transport is driven by mechanical and field asymmetries, inherent in all trap constructions [9]. However, while there are notable recent developments [10], the exact mechanism is not yet completely understood. With our imaging technique, we show in this report that the particle loss at the trap wall, the final step of the radial transport, occurs in a manner that is highly non-uniform, both axially and azimuthally.

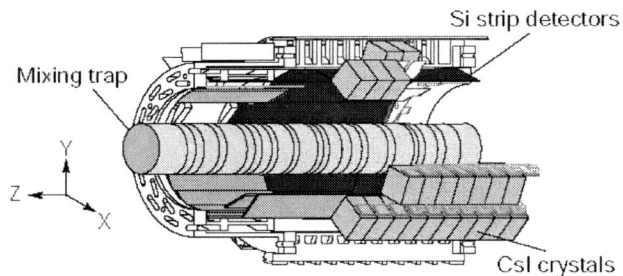

Figure 1. A schematic view of the apparatus. Two layers of double-sided silicon strip detectors and 192 CsI crystals surround the mixing trap. The signal from total of 8192 detector channels are read out via flash analogue-to-digital converters (ADCs) and written onto a disc at a rate of up to 40 Hz.

## EXPERIMENT

The present measurements are performed using the ATHENA apparatus, which recently achieved the first production of cold antihydrogen atoms [11]. Detailed descriptions of ATHENA are given elsewhere [11-15]. The Antiproton Decelerator (AD), located at the European Organization for Nuclear Research (CERN) in Geneva, provides a pulse of 5 MeV antiprotons every 100 s. This is presently the world's only source of low energy antiprotons. The antiprotons are dynamically captured by briefly opening the potential wall at the entrance side. Cold electrons, preloaded in the trap, cool the antiprotons via Coulomb collisions [16]. The antiproton capture and cooling can take place in either of two separate traps; the catching trap, or the mixing trap. In the former case, the cooled antiprotons are adiabatically transferred to the mixing trap after the catching and cooling [17]. The traps (inner radius 1.25 cm) are held at a temperature between 15 and 40 K. The imaging detector is kept at 140 K and housed in a separate vacuum (Fig. 1). Typically, $10^3$ to $10^4$ antiprotons together with $10^7$ to $10^8$ electrons are stored in the mixing trap for these measurements. Antiprotons annihilate either on the residual gas, or if they reach the trap wall, on the surface of gold-plated electrodes.

We observe that keeping the electrons together with the antiprotons shortens the storage time of the latter. This effect, illustrated by the data presented in Fig. 2, is possibly due to an enhancement of the radial transport of the antiprotons due to electron collective effects, and merits further dedicated study. For the purpose of the work here, however, we used this effect simply to accelerate the radial loss. It should be noted, though, that the number of the electrons, or lack thereof, does not affect the main conclusions reported here.

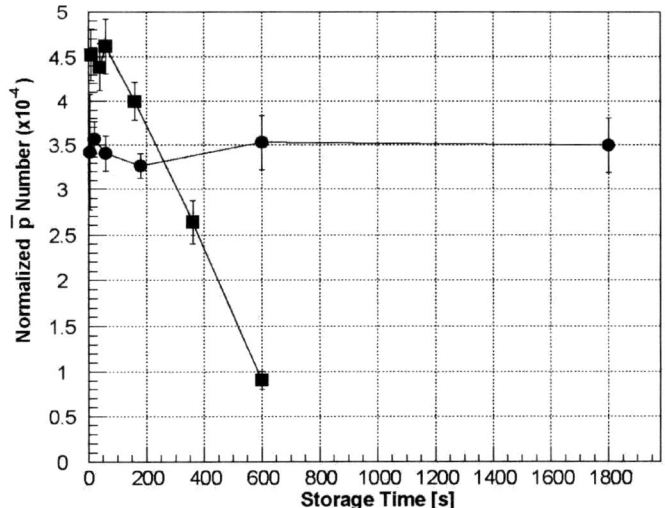

**FIGURE 2.** Number of antiprotons as a function of storage time. The squares refer to measurements with $10^8$ electrons kept in the trap, whilst the circles are with less than $10^5$ electrons remaining. The trapped antiprotons are counted with external annihilation detectors [17], and are normalized to the incoming antiproton number, which in turn is measured by a calibrated beam detector [18].

## ANTIPROTON IMAGES

Antiproton annihilations produce several charged and neutral mesons (mostly pions). The average charge multiplicity depends on the target nucleus; 2.6 for gold and 3.0 for a proton [19]. The charged particles are detected by two layers of double-sided silicon micro-strip detectors. A signal in one of the layers is a "hit", and two hits from two layers are fitted to a straight line to determine a "track". From the intersection of two or more charged tracks, an annihilation "vertex" is determined. A collection of the vertices thus represents the three-dimensional distribution of antiproton positions at the time of their annihilations. Unmeasured curvature of the pion tracks in the 3 T magnetic field is the dominant source of the 4 mm (1$\sigma$) vertex reconstruction uncertainty.

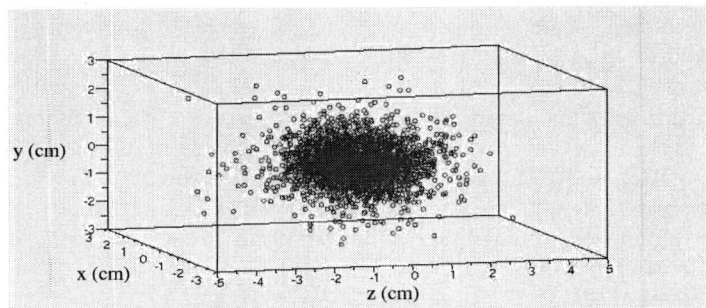

**Figure 3.** Three dimensional imaging of trapped antiprotons

Figures 3 and 4-I (a) show measurements in a harmonic trap with a depth of 30V and length about 5 cm with a relatively high gas pressure in the system, of the order of $10^{-11}$ mbar as estimated from the antiproton lifetime. In these conditions, annihilations on the residual gas (or ions) dominate. Thus, the image obtained corresponds to the distribution of antiprotons in a quadratic potential trap, and is azimuthally symmetric, as expected.

A striking pattern emerges if the residual gas pressure is reduced to below $10^{-13}$ mbar, as shown in Fig. 4-I (b). Annihilations are then observed to be highly non-uniform both azimuthally and axially, and are localized in a few "hot spots". As illustrated in Figs. 4-II and III, the existence of hot spots is a universal feature of charged particle loss in our trap.

Figure 4-II shows images from a series of measurements using only one electrode to create a trap well. A potential of –140 V, with respect to the rest of the grounded electrodes, was applied. The antiproton annihilations take place in the potential well regions, as expected, but they are localized in both $z$ and $\phi$, where $\phi$ is the azimuthal angle of the vertices seen from the trap axis. Figure 4-III illustrates results of measurements with different numbers of electrodes used to form wells. Either –140 V or –50 V was applied, the value of which did not change the features of the images. Again, annihilations are localized. We observe that the annihilation hot spots are clustered near the edge of the electrodes, and their number grows with the number of electrodes used for the well. Azimuthally segmented four-sectored electrodes are seen to enhance annihilations in some (but not all) cases. For example, see the top panel of Fig. 4-III, where the segmented electrodes are depicted as "SE".

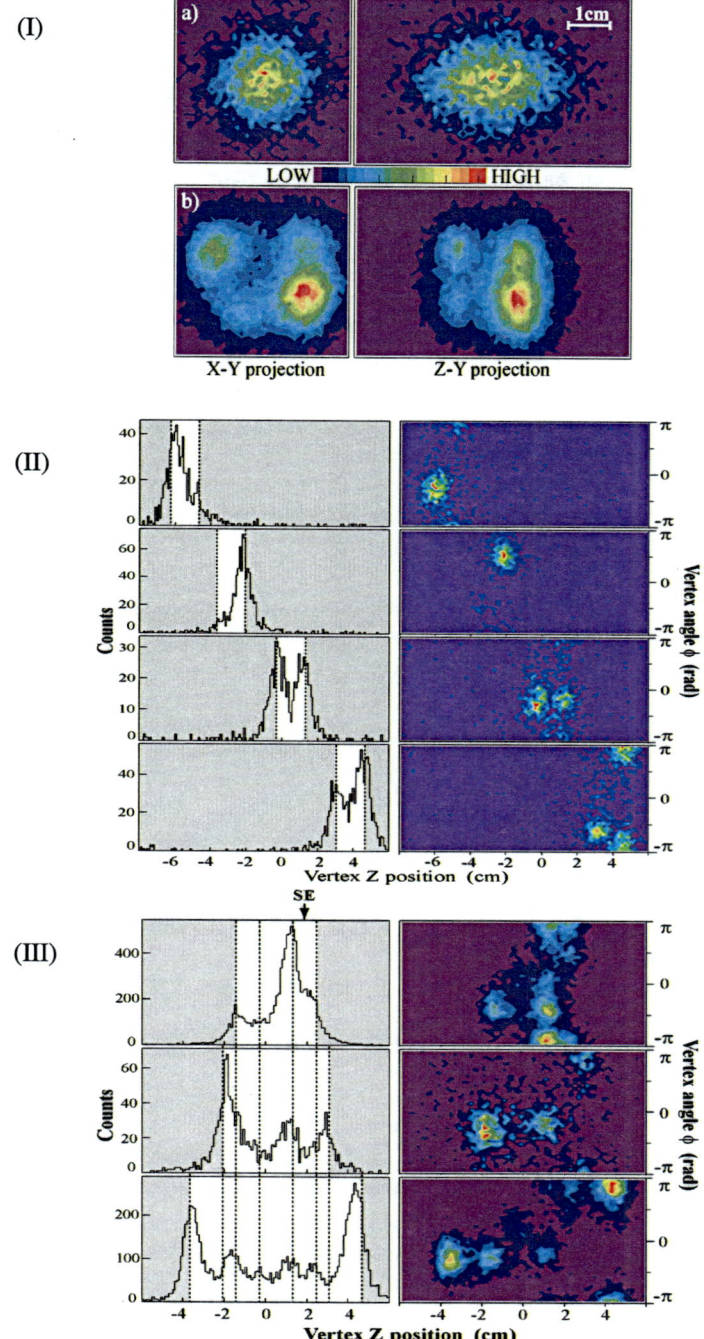

**Figure 4.** Antiproton annihilation images. See text for details.

135

# MONTE CARLO SIMULATIONS

In order to quantitatively understand the observed images, we have performed detailed Monte Carlo (MC) simulations, based on the GEANT simulation package. In our simulations, antiproton annihilation on protons is assumed, and a tabulated branching ratio is used to generate annihilation products, both charged and neutral. Interactions of these particles with our detector and the apparatus are simulated, including electromagnetic and hadronic cascades in the magnet materials. The apparatus geometry is directly imported from a CAD program, and the measured module-by-module detector efficiencies are included in the calculations. The code generates simulated data in the same structure as the experiment, and the same analysis program is applied to both.

The results of the simulation are compared with the data in Fig. 5. A radial distribution of antiproton annihilations for the high vacuum case (from Fig. 4-I (b)), and a simulated distribution assuming a point annihilation source on the trap wall are plotted in Fig. 5 (a). The good agreement between the experiment and the simulation establishes that most of the annihilations occur on the wall (r=1.25 cm). The structure near the peak of the simulated distribution is due to reconstruction errors caused by the curvature of the charged track. It disappears in simulations with the magnetic field set to zero.

The radial distribution of the measured data for the high density case from Fig. 4-I (a) are also plotted in Fig. 5 (a). As is evident, our imaging can clearly distinguish between the distributions for predominantly gas annihilations and those resulting from wall annihilations.

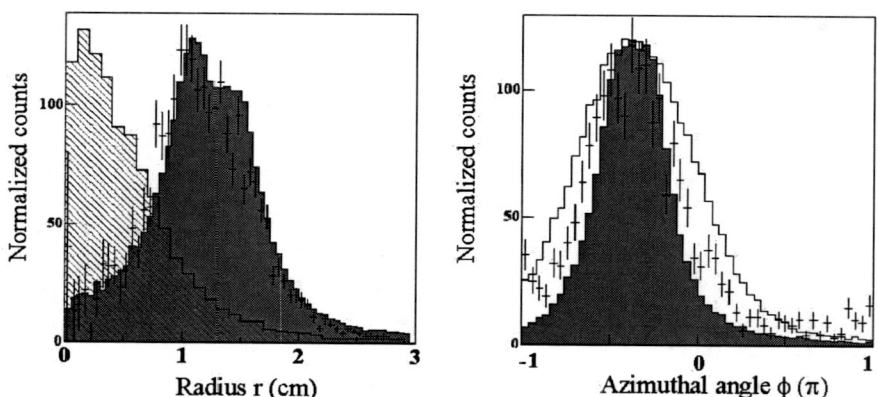

**FIGURE 5.** (a) Comparison of radial (r) annihilation distributions (dN/rdr) for the data from the measurement at low background gas pressures (error bars) and the MC simulations assuming annihilations on the trap wall (dark histogram). Also shown are the data for the high pressure measurement (diagonally filled histogram). (b) The azimuthal ($\phi$) angular distribution of the annihilation (error bars) and its comparison with the MC assuming point source annihilation (dark histogram). Also shown is a MC assuming an extended annihilation source (±4 mm).

We now focus on one of the hot spots from Fig. 4-I (b), in order to study the extent to which it is localized. Figure 5 (b) is a comparison of the annihilation azimuthal angle distributions for the experimental data and the simulations. Also included are histograms for Monte Carlos assuming both a point source annihilation on the wall, and assuming an extended source spot of ±4 mm. From the comparison, we can exclude a source extent of this size.

# DISCUSSION

While it is clear that the presence of hot spots is a result of asymmetries in the system, the underlying mechanism resulting in loss localization is not completely understood at the present time. The effects due to image charge and surface field irregularities complicate the dynamics of particle transport in the proximity of the trap wall.

Various measurements were performed to establish the universality of the loss localization. We observed the hot spots in all cases when antiprotons annihilate on the wall. This was regardless of the details of the antiproton and electron (re)loading procedure and the values of the potentials. Hot spots are present, even though electrons were removed the electrons from the trap, indicating that the loss localization mechanism is not due to collective plasma effects of the electrons, but is dominated by the single particle transport properties of the antiprotons. Note that the antiproton density of is low ($<10^3$ cm$^{-3}$) in these measurements.

The present observations have major implications for the detection of antihydrogen annihilations. Our initial observation of antihydrogen was based on simultaneous detection of antiproton and positron annihilations at the same place. While antiproton annihilation detection is efficient (the vertices can be reconstructed with about 50% efficiency), positron detection is more difficult due to the low intrinsic efficiency of the CsI crystals, and the presence of background. Thus, the overall efficiency was about 0.2% for fully reconstructed antihydrogen events, where both the charged vertex and back-to-back gamma rays were detected. Our finding that neutral antihydrogen atoms annihilate on the wall in a radially uniform manner [11], whereas charged antiprotons produce hot spots, can provide a new and effective signature of antihydrogen annihilations. An antihydrogen detector without the need to register gamma-rays can be envisioned, a considerable simplification when compared to the present system.

Imaging profiles of antiprotons obtained in high density cases (Fig. 4-I (a)) can provide useful information relevant in our quest to understand the antihydrogen production processes. The image obtained represents the spatial distribution of antiprotons, convoluted with the gas (or ion) distribution on which they annihilate. The observed distributions of the antiproton cloud in Figs. 3 and 4-I (a) have an aspect ratio of approximately 2. This is in contrast to the positron plasma aspect ratio (4 −7), as determined by modes analysis [15] based on Dubin's cold fluid model [20]. If we assume that the effects of residual gas on the particle dynamics is negligible on the time scale of the measurements (a few minutes), the observed difference in the cloud radii may explain the apparent partial mismatch in the radial overlap between antiprotons and positrons, indicated by our measurements of positron cooling of antiprotons [21].

Another application of antiproton imaging is illustrated in Fig. 6. By moving the trap well, and measuring the annihilation positions, we can determine our trap electrode positions, relative to the detector, over a wide range of axial positions as shown in Fig. 6 [22]. Imaging of antiproton annihilations thus permitted detector position calibration at 1 mm precision, a task otherwise nontrivial in the present setup. The precise calibration shown here is an important input to the physics analyses using the ATHENA antihydrogen annihilation detector.

As we have previously stated, our imaging position resolution is, at present, limited by the unmeasured curvature of the charged tracks. This could be improved in a future apparatus by using three or more layers of Si strips. Our detector readout rate (~40 Hz) limits the physical processes that can be imaged to relatively slow ones (such as the radial loss reported here), and much faster processes, e.g. bursts of annihilations due to diocotron instability, cannot be readily imaged. The implementation of the signal level discrimination at the ADC level (so called zero suppression) is in progress and can improve the readout rate by up to a factor of 10.

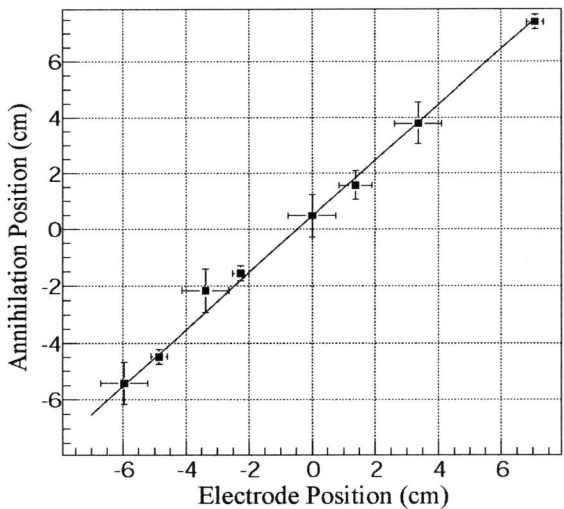

**Figure 6**. The correlation between the trap well positions and the measured annihilation positions.

## CONCLUSIONS

In this paper, we reported imaging of antiproton distributions via reconstruction of annihilation vertices. With this new technique, we probed the previously unexplored final step in the radial transport of trapped charged particles; a regime in the proximity of the trap wall. We observed that antiproton annihilations on the trap wall are localized in all cases, an effect which may be applicable to other Penning (and related) systems. Several implications for antihydrogen production and detection were discussed.

The main disadvantage of the antiproton imaging technique for trapped particle studies is the scarcity of antiprotons. In the near future, we will extend our antiparticle imaging studies by using positron annihilations, much like positron emission tomography in medical applications.

# ACKNOWLEDGMENTS

We gratefully thank CERN's AD crew and J. Rochet for providing essential support, Professors J. Fajans, H. Higaki, A. Mohri, and Y. Yamazaki for valuable discussions, and A. Cavanagh for a critical reading of the manuscript. This work was supported in part by MEXT and RIKEN (Japan), CNPq (Brazil), SNF (Denmark), INFN (Italy), SNF (Switzerland), and the EPSRC (UK).

# REFERENCES

1. Dubin, D. H., and O'Neil T. M., Rev. Mod. Phys. **71**, 87 (1999).
2. Davidson, R. C., *Physics of Nonneutral Plasmas*, (Imperial College Press, London, 2001)
3. *Non-Neutral Plasma Physics*, edited by Anderegg, F., et al. Vol. 4 (AIP Conference Proceedings , New York, 2002).
4. Larson, D. J., et al., Phys. Rev. Lett. **57**, 70 (1986).
5. Brewer, L. R., et al., Phys. Rev. A **38**, 859 (1988).
6. Mitchell, T. B., et al., Opt. Express **2**, 314 (1998).
7. O'Neil, T. M., Phys. Fluids **23**, 2216 (1980).
8. Malmberg, J. H., and Driscoll, C. F., Phys. Rev. Lett. **44**, 654 (1980); Driscoll, C. F. and Malmberg, J. H., ibid. **50**, 167 (1983); Driscoll, C. F., et al., Phys. Fluids **29**, 2015 (1986).
9. Notte, J., and Fajans, J., Phys. Plasmas **1**, 1123 (1994); Eggleston, D. L., and O'Neil, T. M., ibid. **6**, 2699 (1999); Chao, E. H., et al., ibid. **7**, 831 (2000); Kriesel, J. M., and Driscoll, C. F., Phys. Rev. Lett. **85**, 2510 (2000); Kabantsev, A. A., et al., ibid. **87**, 225002 (2001); Sarid, E., et al., ibid. **89**, 105002 (2002).
10. Kabantsev, A. A., and Driscoll, C. F., Phys. Rev. Lett. **89**, 245001 (2002); Kabantsev, A. A., et al., Phys. Plasmas **10**, 1628 (2003).
11. Amoretti, M., et al., Nature 419, 456 (2002).
12. Amoretti, M., et al. submitted to Nucl. Instr. Meth. A., 2003
13. Regenfus, C., Nucl. Instrum. Methods A **501**, 65 (2003).
14. Fujiwara, M. C., et al., Nucl. Instrum. Methods B. in press (e-print archive: hep-ex/0306023).
15. Amoretti, M., et al., Phys. Plasmas **10**, 3056 (2003); Phys. Rev. Lett. **91**, 055001 (2003).
16. Gabrielse, G., et al., Phys. Rev. Lett. **57**, 2504 (1986); **63**, 1360 (1989).
17. Fujiwara, M. C., et al., Hyperfine Interact. **138**, 153 (2001).
18. Fujiwara, M. C., and Marchesotti, M., Nucl. Instrum. Methods A **484**, 162 (2002).
19. Bendiscioli, G., and Kharzeev, D., Rivista Nuovo. Cim. 17(6), 1 (1994).
20. Dubin, D. H., Phys. Rev. Lett. **66**, 2076 (1991).
21. Amoretti, M., et al., to be published.
22. Bouchta, A., and Fujiwara, M. C., ATHENA Technical report (2002).

# Design and Performance of a Trap-Based Positron Beam Source

R. G. Greaves and J. Moxom

*First Point Scientific, Inc., 5330 Derry Avenue, Suite J, Agoura Hills, CA 91301.*

**Abstract.** Positron traps have been demonstrated to have the capability for producing high-quality positron beams and intense positron pulses for a variety of scientific and technological applications including atomic physics experiments and probes for materials science. Techniques for accumulating positrons in traps have now advanced to the point where the development of commercial trap-based positron beam systems is feasible. Trap-based beams offer advantages over conventional positron beam systems, including the ability to produce ultra-cold positron beams, giant positron pulses, and ultra-short pulses. They also enable high-efficiency brightness enhancement schemes. The design and performance of a new trap-based positron beam system is described.

## INTRODUCTION

Low-energy positron beams are extensively employed for a variety of scientific and technological applications including atomic physics [1], analytical probes for materials science [2] and mass spectrometry [3]. Furthermore, intense positron pulses have a wide range of potential applications such as positronium physics experiments, production of Bose-Einstein condensation of positronium atoms, demonstration of the positron annihilation gamma-ray laser [4], and as a diagnostic of anomalous electron transport in Tokamaks [5]. Techniques to accumulate positrons in Penning traps facilitate the production of high quality positron beams and intense pulses for these and other applications. These traps provide significant advantages over currently available techniques for the production of such beams.

Although a variety of techniques has been demonstrated for accumulating positrons in traps [6], the only method that has a high enough efficiency for practical beam systems is the buffer gas method developed by Surko *et al.* [7]. The principle behind this technique is illustrated in Fig. 1. Positrons from a conventional solid neon-moderated radioactive source [8] are injected into a specially modified Penning-Malmberg trap. As shown in Fig. 1, this trap has a pressure gradient created by differential pumping of nitrogen that is continuously introduced into the trap. It also has three stages of successively lower electrostatic potential created by suitably biasing the confining electrodes. Positrons are trapped by inelastic collisions with the nitrogen gas atoms, and accumulate in the lowest potential (and pressure) region of the trap by making multiple inelastic scattering collisions with the gas, where they cool to room temperature. Using this technique, up to $3\times10^8$ positrons have been accumulated in a few minutes. A trap of similar design was recently employed by the Athena collaboration to create the first low-energy antihydrogen atoms [9].

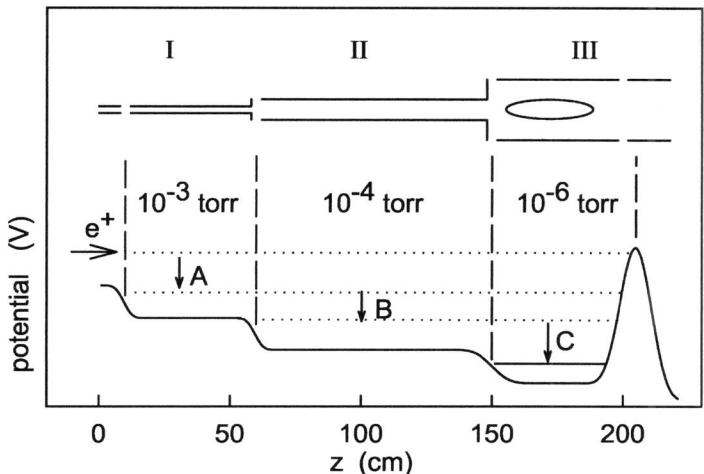

**FIGURE 1.** Layout of the Surko-type buffer gas trap showing the three regions of successively lower pressure and electrostatic potential. An axial magnetic field in the *z*-direction provides radial confinement.

## TRAP CAPABILITIES

Once positrons have been trapped and cooled, they can be released from the trap in a controlled manner to form an ultra-cold positron beam [10]. These beams have been recently exploited to investigate a variety of topics in positron-molecule interactions including the first systematic measurements of the vibrational excitations of molecules by positrons [11], the first measurement of the state-resolved absolute electronic excitation cross sections of molecules by positrons [12], and the first experimental evidence of the role of vibrational excitations in producing anomalously large annihilation rates in large molecules [13]. The potential also exists to apply this technique to create low-energy positron beams for materials science applications. Two powerful capabilities of trap-based positron beams are of specific interest for materials science applications, namely, beam bunching, and high efficiency brightness enhancement.

## Bunching

The positron is a very sensitive probe of vacancy-type defects [2]. This property has a wide range of applications in research and industry and has been used to study such diverse issues as ion-implantation-induced damage in semiconductors [14], the properties of low-*k* dielectrics used as insulators in the IC manufacturing industry [15], accelerated aging in polymers [16], metal fatigue [17], and hydrogen embrittlement in structural metals [18]. Since the positron lifetime can vary from a few hundred picoseconds to several hundred nanoseconds, positron beams with pulse widths of the order of ~200 ps and interpulse periods >100 ns (i.e., frequencies < 10 MHz) are desirable. Conventional schemes for bunching positron beams are based on rf techniques [19]. Such techniques can produce the required short pulses, but are restricted in their frequency range to >50MHz, and are, thus, not suitable for measuring long lifetime components. Positron traps, in contrast, have the potential for meeting both requirements.

Two techniques that are applicable for bunching positrons using traps are shown in Fig. 2. Figure 2(a) shows the harmonic potential method [20] where a potential of the form $V(z) = V_0(z_0-z)^2$ is applied along the flight path of the positrons, leading to a time focusing effect at the minimum of the potential. Figure 2(b) illustrates the timed potential method [20], where the trapped positrons are injected into a bunching electrode and a potential of the form

$$V(t) = V_0\left[1 - \frac{t_0^{\,2}}{(t-t_0)^2}\right] \qquad (1)$$

is applied to this electrode as the positions exit, again leading to a time focusing effect. Here, $t_0$ is the time-of-flight from the end of the bunching electrode to the time focus at $z_0$ in the absence of a bunching waveform. Both of these techniques benefit greatly from the ultra-cold nature of the positrons in traps, as well as the ability of traps to produce positron pulses with arbitrary interpulse periods.

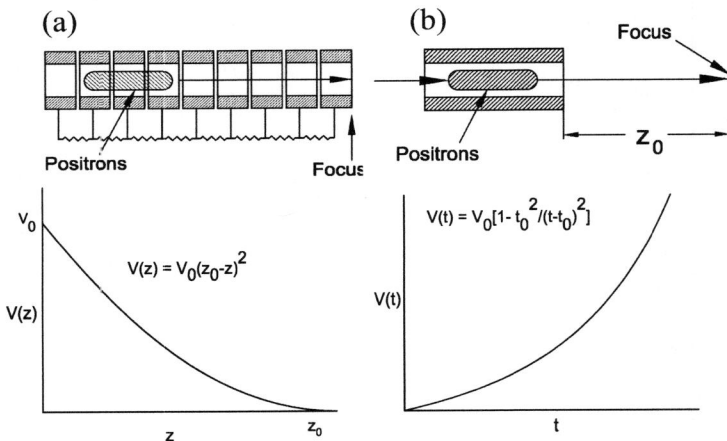

**FIGURE 2.** Two methods for bunching positrons in a trap: (a) harmonic potential method and (b) timed potential method.

# Brightness enhancement

For some applications, positron beams with small beam diameter are desirable for imaging small features in a sample such as microstructures in microelectronic circuits. For these applications, positrons beams down to one micron in diameter or less are desirable. Currently available techniques for producing such beams result in an attenuation of the beam flux by about two orders of magnitude, leading to very long imaging times [21]. Positron traps have the potential to produce small diameter beams with much higher efficiency. The concept of trap-based brightness enhancement exploits two plasma properties of trapped positrons through two distinct effects, as illustrated in Fig. 3.

The first of these effects is the radial compression of positron plasmas by applying a rotating electric field [21]. This is accomplished by applying suitably phased sine waves to an azimuthally segmented electrode confining the positrons. This couples angular momentum into the plasma by exciting rotating Trivelpiece-Gould waves, leading to a reduction in the plasma diameter. The necessary cooling can be provided by a buffer gas such as $CF_4$ [22].

The second effect is the preferential release of positrons from the center of the trap that occurs when only part of the plasma is released. This effect results because the space potential is highest at the center of the trap due to the positron space charge. Based on electron plasma experiments, beams as narrow as $4\lambda_D$ in diameter can be created [23], where the Debye length, $\lambda_D = (k_B T/4\pi n e^2)^{1/2}$, $k_B$ is Boltzmann's constant, $n$ is the positron number density, and $T$ is the positron temperature. If positrons are released in this manner while the plasma is continuously replenished from the source as the rotating electric field is applied, the density cavity that results from positron release will be continuously filled by inward radial transport.

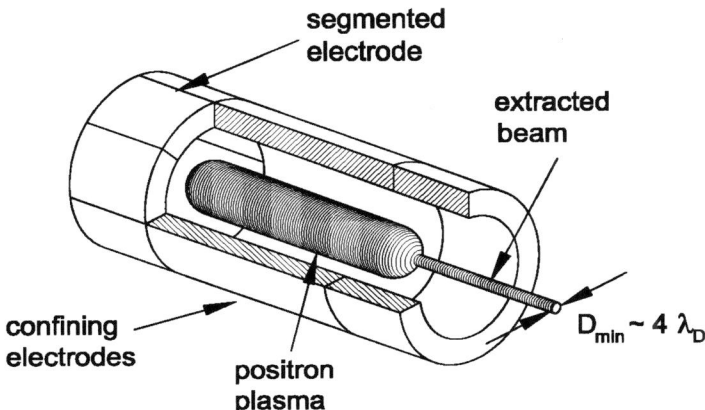

**FIGURE 3.** Cutaway view of a positron trap illustrating the concepts for trap-based brightness enhancement.

# TRAP DESIGN

Figure 4(a) shows the hardware configuration of the Surko trap in more detail, including the pumping arrangement and magnetic field coil. First Point Scientific, Inc., has re-examined the design with a view to optimization for use as a commercial positron beam source, and has modified the design accordingly. The principle considerations for this design were to minimize cost and optimize the performance of the trap for pulsed beam and micro beam production. These considerations led to the separation of the device into two components, as illustrate in Fig. 4(b), namely: (1) a two-stage trap, designed to produce pulsed beams with a wide range of repetition rates from 10 Hz upward; and (2) an accumulator, designed to capture positron pulses from the two-stage trap for producing giant positron pulses or for implementing advanced plasma-based brightness enhancement schemes that require large numbers of positrons. This new design permits the size and cost of the magnet and associated power supplies (the largest capital cost in the system) to be drastically reduced. It also makes available, at reduced cost, a compact two-stage trap for users who do not require large numbers of positrons or micro beams. M. Charlton *et al.* [24] are also developing a two-stage trap based on similar concepts for studies of positronium physics.

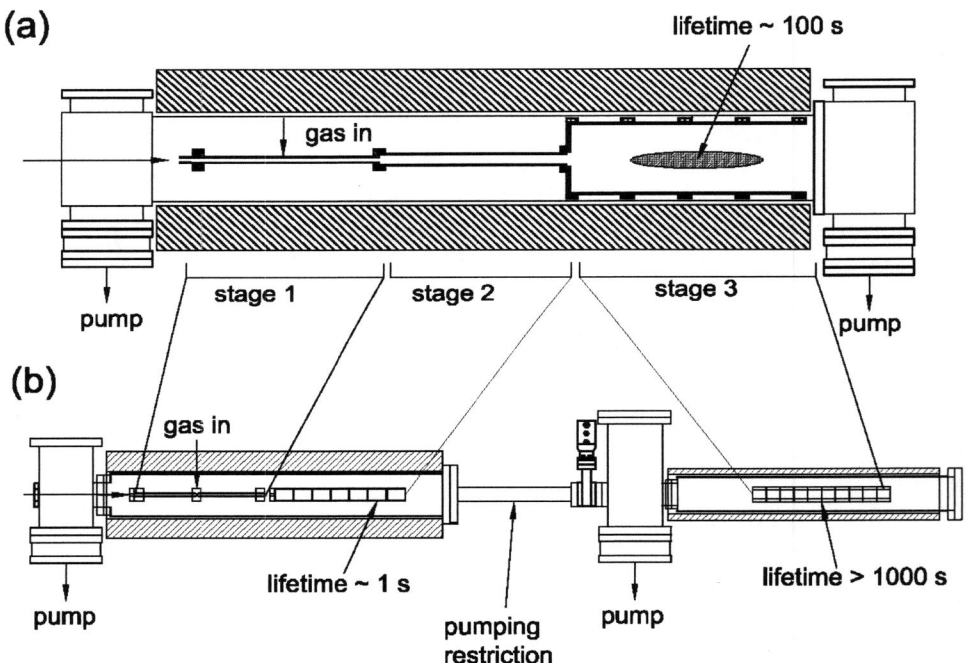

**FIGURE 4.** Comparison between (a) existing positron trap and (b) new design optimized for use as a commercial positron beam source showing the two-stage trap to the left of the pumping restriction and the positron accumulator to the right.

# PERFORMANCE CHARACTERISTICS

## Two-Stage Trap Operation

Measurements of positron accumulation in the two-stage trap show that the positron lifetime in the second stage is about 1 s, which permits efficient operation when the system is operated at a repetition rate of 10 Hz or greater. Trapping efficiencies of up to 26% have been obtained and further improvements are possible by fine-tuning of the operating parameters and small modifications to the trap electrode geometry. The plasma in the second stage has a diameter of about 6 mm. This will be reduced by incorporation of a rotating electric field electrode (described below), which is also expected to improve trapping efficiency.

The time structure of pulses released from the two-stage trap is shown in Fig. 5(a). The pulse width in this case was 18 ns, which is too long for lifetime measurements. This was reduced by timed potential bunching using a simple linear ramp. A typical result is shown in Fig. 5(b), indicating a pulse width of ~4 ns. This will be further reduced by applying the ideal waveform specified by Eq. (1) using an arbitrary waveform generator. This is expected to reduce the pulse width to a few hundred picoseconds, which will be suitable for lifetime measurements in solids.

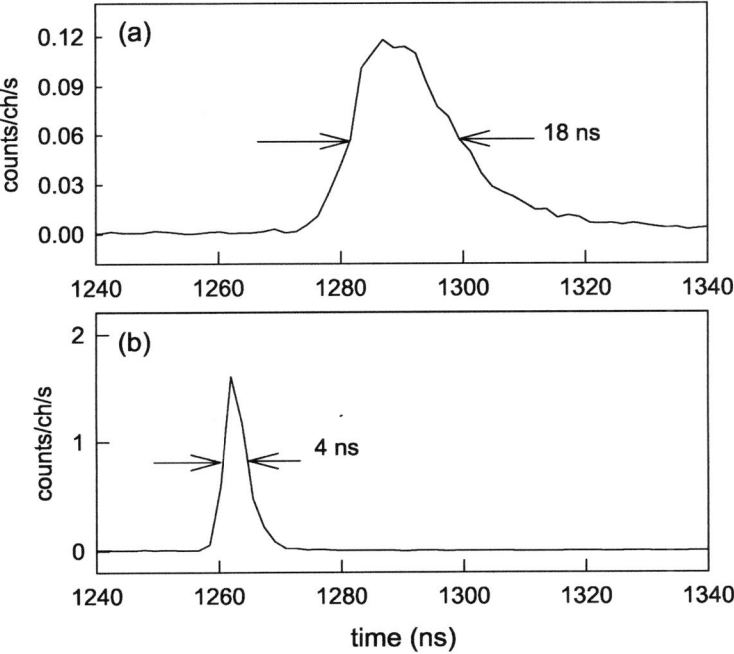

**FIGURE 5.** (a) Time structure of pulses released from the two-stage trap without any bunching signal applied. (b) Pulse obtained using a linear ramp for bunching.

# Accumulator operation

The accumulator is operated by re-trapping and stacking pulses from the two-stage trap at a rate of 5 to 10 Hz. The re-trapping is accomplished using the "ballistic" method employed by the Athena group [25]: positrons are released from the two-stage trap, injected into the accumulator, and trapped by raising the potential on the accumulator entrance gate electrode. As shown in Fig. 6(a), 100% capture efficiency can be obtained for a single pulse by adjusting the timing on the gate electrode. For accumulating multiple pulses ("stacking"), a secondary electrostatic well is created in the accumulator by suitably biasing the electrodes. Most of the trapped positrons diffuse into this well before the entrance gate is opened to admit the next pulse, thus permitting high efficiency stacking. Figure 6(b) presents the measured stacking efficiency for various values of the confining magnetic field in the accumulator. These data show that stacking efficiencies of almost 90% can be obtained.

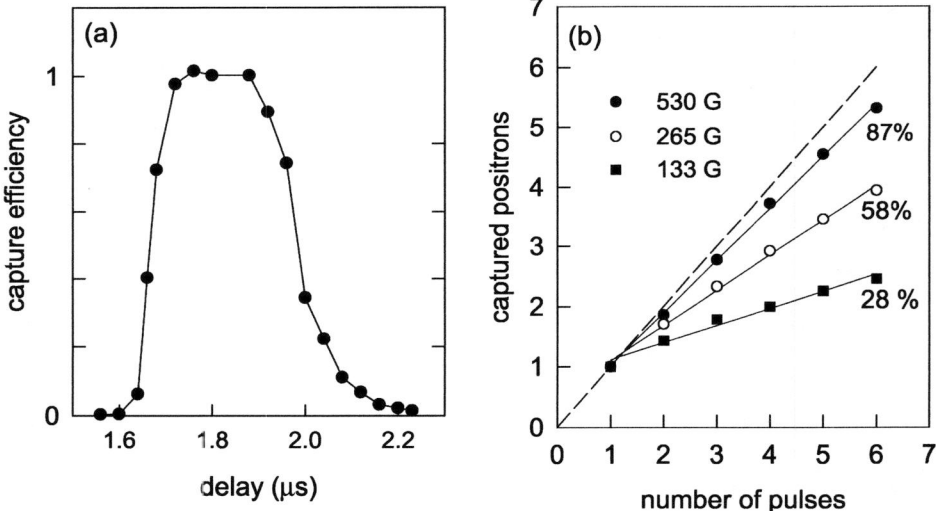

**FIGURE 6.** (a) Capture efficiency for "ballistic" retrapping of single positron pulses in the positron accumulator shown in Fig 4(b). (b) Stacking efficiency in the accumulator for multiple positron pulses using various values of magnetic field strength.

Positrons stacked in the accumulator can be compressed using the rotating electric field technique described above. Figure 7(a) presents the central density of a positron plasma as a function of time following the application of a rotating electric field, showing the very high compression rates that can be obtained. The inset to this figure presents radial profiles before and after compression, showing a reduction in the plasma diameter, $D$, (FWHM) from 7.4 mm to 0.57 mm and an increase in central density of more than two orders of magnitude. Since the beam brightness in a magnetic field scales as $D^4$ [26], this represents a brightness enhancement by a factor of $8 \times 10^8$.

146

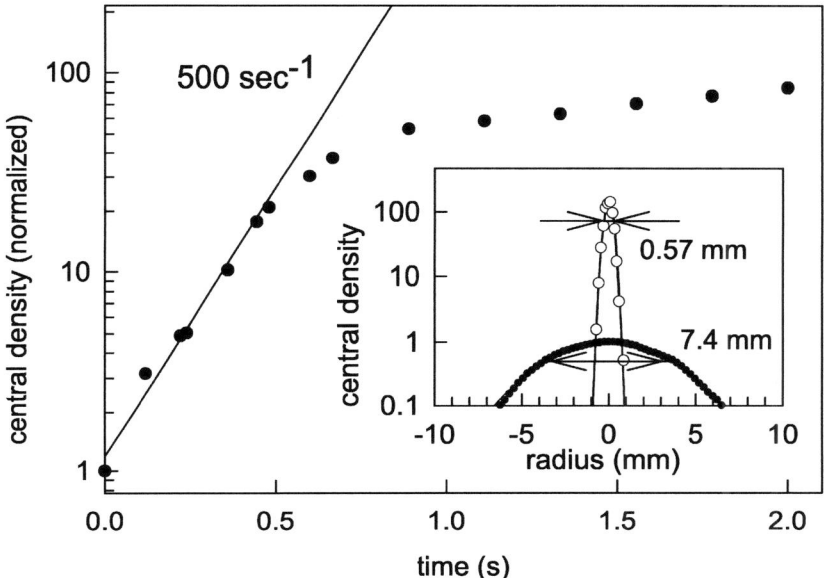

**FIGURE 7**. Central density of a positron plasma as a function of time following application of a rotating electric field. Inset: plasma profiles at $t=0$ and $t=4$ s.

## DISCUSSION AND SUMMARY

Although still in its infancy, the ability to accumulate and cool positrons in Penning traps has led to advances in the fields of atomic physics, plasma physics and antihydrogen production. Commercial positron beam systems for materials science applications are now being developed. The ability to accumulate large numbers of positrons is being exploited for studies of various topics in positronium physics such as the creation of the first positronium molecules, the demonstration of Bose-Einstein condensation of positronium atoms, and the development of the positron annihilation gamma-ray laser. A new trap-based positron beam system for these and other applications has been designed and preliminary test have been conducted. These tests show the viability of a two-stage trap for producing short-pulsed positron beams suitable for materials science and atomic physics experiments. They also show the viability of a positron accumulator for implementing advanced trap-based brightness enhancement schemes, as well as accumulating large numbers of positrons for various purposes. Future work will focus of optimizing the performance of the system.

## ACKNOWLEDGMENTS

This work was supported by the Office of Naval Research and the National Science Foundation.

# REFERENCES

1.  Mills, Jr., A. P., *Science* **218,** 335-40 (1982).
2.  Schultz, P. J., and K. G. Lynn, *Reviews of Modern Physics* **60**, 701–779 (1988).
3.  Hulett, L.D., Jr., *et al.*, *Chemical Physics Letters.* **216**, 236-40 (1993)
4.  Mills, A. P., Jr., *Nucl. Instrum. Methods.* **B192**, 107 (2002).
5.  Surko, C. M., *et al.*, *Rev. Sci. Instrum.* **57**, 1862-7 (1986).
6.  Greaves, R. G., and C.M. Surko, *Phys. Plasmas* **4,** 1528 (1997).
7.  Surko, C. M., *et al., Phys. Rev. Lett.* **62,** 901 (1989).
8.  Mills, Jr., A. P., and E. M. Gullikson, *Applied Physics Letters* 49, 1121-3 (1986)
9.  Amoretti, M., *et al.*, *Nature* **419,** 456 (2002).
10. Gilbert, S. J., *et al., Appl. Phys. Lett.* **70,** 1944 (1997).
11. Sullivan, J. P., *et al., Phys. Rev. Lett.* **86,** 1494 (2001).
12. Sullivan, J. P., *et al., Phys. Rev. Lett.* **87,** 073201-1 (2001).
13. Gilbert, S. J., *Phys. Rev. Lett.* **88,** 043201-1 (2002).
14. Knights, A. P., F. Malik and P. G. Coleman, *Applied Physics Letters* **75,** 466-8 (1999).
15. Petkov, M. P., *et al., Applied Physics Letters* **74,** 2146-8 (1999).
16. Chen, H., *et al.*, *Applied Surface Science* **194**, 168-175 (2002).
17. W. Egger *et al., Applied Surface Science* **194**, 214-217 (2002).
18. H. Schut *et al., Applied Surface Science* **194**, 239-244 (2002).
19. Sperr, P., G. Kogel, P. Willutzki, and W. Trifthauser, *Applied Surface Science* **116**, 78-81, (1997).
20. Mills, A. P., Jr., *Applied Surface Science* **22,** 273-276 (1980).
21. Anderegg, F., E. M. Hollmann, and C. F. Driscoll, Phys. Rev. Lett. 81, 4875 (1998).
22. Greaves, R. G., and C.M. Surko, *Phys. Rev. Lett.* **85,** 1883 (2000).
23. Danielson, J. R., *et al.*, these proceedings.
24. Clarke, J., *et al.*, these proceedings.
25. van der Werf, D. P., *et al.*, these proceedings
26. Greaves, R. G., S. J. Gilbert, and C. M. Surko, *Applied Surface Science* **194,** 56 (2002).

# A Cryogenic, High-field Trap for Large Positron Plasmas and Cold Beams

J. R. Danielson, P. Schmidt, J.P. Sullivan[1], and C. M. Surko

*Department of Physics, University of California, San Diego, La Jolla, CA 92093-0319*
*1. Photon Factory, High Energy Accelerator Research Organization, Japan*

A new Penning-Malmberg trap using a 5 tesla magnetic field and a cryogenic electrode structure (T~10K) has been constructed with the goal of producing large (N ≥ $10^{10}$), high-density positron plasmas and cold positron beams ($\Delta\varepsilon$ ~ 1 meV). With background pressures ≤ $10^{-11}$ torr and rotating electric fields to counteract plasma expansion due to background asymmetries, this trap is designed to be a nearly ideal reservoir of positrons with very long confinement and annihilation times. This paper describes recent experiments using electron plasmas to optimize confinement and plasma compression, and minimizing the diameters of extracted beams. Further, it is shown that this trap will be an excellent device in which to study the physics issues associated with a recently proposed multi-cell trap.

## I.    Introduction

There are a number of motivations to develop the capability to accumulate large numbers of positrons and create cold, high-density positron plasmas. Applications include positron storage for antihydrogen production [1, 2], electron-positron plasma studies [3], Bose-condensed positronium [4], high resolution measurement of positron scattering and annihilation cross sections with atoms and molecules [5], and the development of portable antimatter traps [6]. The excellent confinement properties of Penning-Malmberg traps make them ideal for the storage of non-neutral plasmas. This paper describes a new 5T trap designed and built to address the physics of cold and high-density positron plasmas. This device is also well suited for studying the issues involved in the accumulation and storage of very large numbers of positrons.

The main factors that inhibit the confinement of high-density single component plasmas are expansion from inherent trap asymmetries and plasma heating from this expansion. A phenomenological model of the background transport combined with limitations from the plasma space charge suggests that the two goals of low temperatures and large total number favor different operating regimes. Consequently, it is of interest to determine the dominant transport mechanisms for high-density non-neutral plasmas (n~$10^{10}$ cm$^{-3}$) for temperatures ranging from 1 meV to a few eV, a parameter regime that has yet to be fully explored.

CP692, *Non-Neutral Plasma Physics V*, edited by M. Schauer et al.
© 2003 American Institute of Physics 0-7354-0165-9/03/$20.00

Cold, low-energy positron beams can be extracted from the trap by decreasing the confinement potential. Cyclotron cooling in large magnetic fields is an effective way to reduce the plasma temperature [7]. Since the energy resolution of a beam formed in this way is determined by the plasma temperature, beams with energy spreads $\sim 1$ meV should be possible with an electrode structure maintained at 10K. Current positron beams are limited to about 20 meV with room temperature electrodes, so this could improve the achievable energy resolution of these beams by more than an order of magnitude. Also, for plasmas with appreciable space charge, beams with very small diameters ($D \sim 4\lambda_D$) are possible. For example, it is possible to produce a 10$\mu$m diameter beam from a positron plasma with $n \sim 10^{10}$ cm$^{-3}$ and $T \sim 10^{-2}$ eV.

To test these ideas, a high-field Penning-Malmberg trap has been constructed. This trap, shown schematically in Fig. 1, uses a warm bore superconducting magnet with a maximum field of 5 tesla, and a two stage pulse-tube cryogenic refrigerator that is designed to cool the electrodes to 10 K inside a 30 K thermal shield. The electrodes (radius $R_w$=1.27 cm) include two rings that are segmented azimuthally (one 4 sector, and one 8 sector electrode) for the application of rotating electric field perturbations. The density is measured by dumping the plasma out the end onto a phosphor screen and imaged with a CCD camera. For testing purposes, we have installed a thermionic electron gun to fill the trap with electrons.

This paper is organized as follows: In Section II, we describe some fundamental transport issues and basic trap limitations. We combine a simple model of the transport and heating to place limits on the expected operation for various plasma parameters. We also describe measurements of the background transport rate using trapped electrons and compare these measurements to the results of previous experiments on similar traps and to an empirical transport model. Section III describes our initial electron experiments using a rotating electric field to counter the outward transport and produce plasma compression. The observed compression is limited by the outward transport. Data demonstrating a dynamic balance of outward transport and plasma compression is compared to a simple model. Two compression regimes are identified, and differences between the two are highlighted. The technique of extracting small diameter beams is discussed in Section IV. Also discussed is an extension of this technique that could be useful as a diagnostic for the plasma temperature. The design of a multi-cell trap for the storage of more than $10^{12}$ positrons is discussed in Section V. Concluding remarks are made in Section VI, as well as a discussion of future plans to create specially tailored positron plasmas and beams using the techniques discussed here.

## II.    Confinement and Cooling

A non-neutral plasma in a Penning-Malmberg trap is confined radially by an axial magnetic field and axially by applied electrostatic potentials. There are several fundamental limits to particle storage in these traps. The Brillouin density limit ($n_B \sim 2 \times 10^{13}$ cm$^{-3}$ for B=5T) [8], is about 100 times greater than the plasma density considered in this paper, so this limit is not a problem. A second limit is the maximum space charge that can be confined by a given potential, $V_e$, applied to the end electrodes. For a total

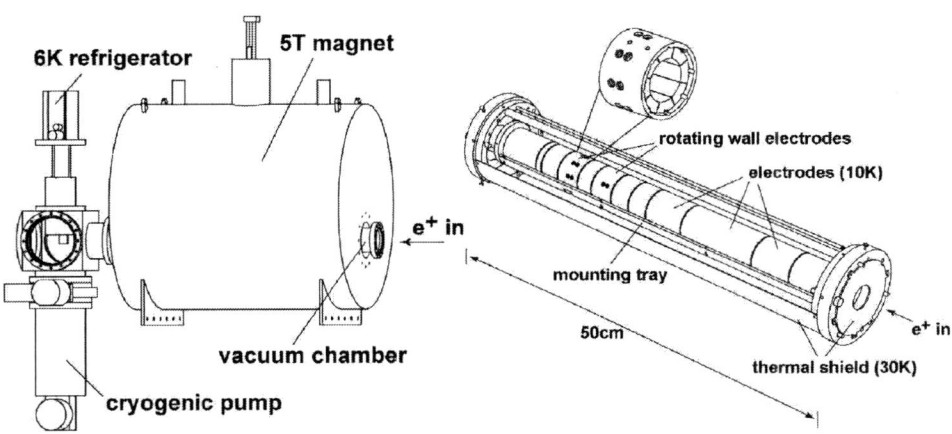

FIGURE 1. Schematic diagram of the high-field trap.

charge $Q_{tot} = eN_{tot}$, plasma length $L_p$, and radius $R_p$, in a grounded tube of radius $R_w$, the space charge potential for a centered rod of positrons is given by:

$$\Phi_{sp} = \frac{eN_{tot}\eta}{L_p},\tag{1}$$

where $\eta = 1 + 2\ln(R_w/R_p)$. For example, a plasma with $N_{tot}=10^9$, L=10 cm, $R_p/R_w=0.1$, corresponds to $\Phi_{sp} \sim 80$ V. Thus confining $10^{10}$ positrons in the same geometry ($\Phi_{sp} \sim 800$ V) would require $V_e \sim 1$ kV on the end electrodes. For the electron plasmas considered in this paper, we use $V_e \leq -100$ V; however, kilovolt potentials will be used in the future to test the physics of the multi-cell trap described in Section V.

The third limit to storing large numbers of positrons is transport from inherent trap asymmetries. For low background gas pressures, the plasma lifetime in Penning-Malmberg traps is limited by small trap asymmetries (eg. electrostatic misalignment of confinement electrodes or inherent magnetic field variations) [9, 10]. The most recent work by Kabantsev et al. demonstrated a strong relationship between the observed transport and the damping of asymmetric trapped particle modes [11, 12]. The detailed link between these modes and particle transport has yet to be elucidated. However, over an extended range of density and temperature, measurements of the plasma transport rate, $\Gamma_O$, are observed to be (roughly) characterized by the inverse of the square of the plasma rigidity, i.e. $\Gamma_O \sim R^{-2}$; where $R = f_B/f_R$ is the rigidity, $f_B$ is the axial bounce frequency, and $f_R$ is the ExB rotation frequency [9, 13], as motivated by the theory of bounce-resonant transport [14]. This is observed to be valid in a regime where the plasma is relatively collisionless, and indeed, the $R^{-2}$ dependence is observed to fail in regimes where the plasma is highly collisional [15, 16].

For the purpose of estimating the expected transport rate as a function of plasma parameters, we assume:

$$\Gamma_O = A \times \left(\frac{f_B}{f_R}\right)^{-2} = \frac{A}{R^2}, \tag{2}$$

where A is some trap dependent constant. We choose a curve with A=0.016 sec$^{-1}$, as shown by the solid curve in Fig. 2; this is approximately the best confinement that has been achieved in electron experiments in high B fields (1-5 Tesla), over a broad range of density and temperature [13].

When the plasma expands, it becomes closer to its image charge in the wall resulting in a decrease in electrostatic energy. This decrease in energy results in an increase in the plasma temperature. The heating rate from the background expansion can be written as:

$$\Gamma_{heat} = \left(\frac{1}{\eta} \frac{e\phi_{sp}}{T_e}\right)\Gamma_O. \tag{3}$$

This formula shows that, for large plasmas, with correspondingly large $\Phi_{sp}$, small background expansion rates are required in order to achieve low plasma temperatures. For the trap described here, this heating is balanced by cyclotron cooling, with a rate given approximately by [7]:

$$\Gamma_{cool} = \frac{1}{4}\left(\frac{B}{1T}\right)^2, \tag{4}$$

For B=5T, Eq. 4 gives $\Gamma_{cool} \sim 6.3$ sec$^{-1}$.

The plasma transport rate, $\Gamma_O$, for electron plasmas in the high-field trap was measured for several different confinement lengths. Typical plasma parameters were $n_0 \sim 2 \times 10^8$ cm$^{-3}$, $T_e \sim 300$ K, and $5 < Lp < 30$ cm. These transport measurements are compared with measurements from similar traps in Fig. 2. This transport rate is obtained from the rate of change of the central density, which is approximately exponential with a time constant $\tau = 1/\Gamma_O$. The measured rates are comparable to those predicted by the model, and are close to the best performance observed to date in Penning-Malmberg traps. This gives us confidence in using the model to predict transport rates for trap design purposes, as discussed in Section V.

## III.    Rotating Wall Experiments

The use of a rotating electric field to counteract inherent trap asymmetries is now common. Electrons [16, 17], ions [18], and positrons [19], have all been successfully confined and compressed using the rotating wall technique. However, a simple model for the compression of the plasma from the applied torque has yet to be developed, particularly in the case of large transport rates. Thus, although it is known experimentally that plasmas can be controlled and compressed, at least to some degree, it

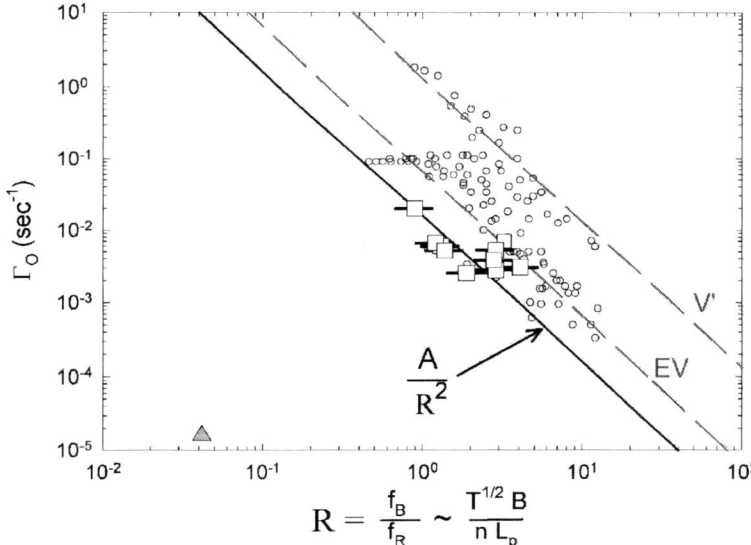

$$R = \frac{f_B}{f_R} \sim \frac{T^{1/2} B}{n L_p}$$

**FIGURE 2.** Measured outward transport rate versus plasma rigidity. The hollow squares are data from the 5T trap. The hollow circles are data from an older electron trap [14]. The two dashed lines are the older from the V' and EV machines [9]. The solid line corresponds to Eq. 2 with $A=0.016$ s$^{-1}$. The triangle is data from Reference [16], and represents the high-collisionality regime.

is currently only through empirical data that the limits of this technique can be determined. A general model has been difficult to develop because the empirical data from different traps and with different mass particles gives different experimental scalings. For example, work with electron plasmas observed only narrow regimes of frequency compression linked to resonances with asymmetric plasma modes. These experiments observed compression rates that scaled linearly with the applied amplitude (i.e. $\Gamma_c \sim V_a^1$) [17]. In contrast, an experiment on positrons, in a different range of plasma parameters, found a broad-frequency non-resonant regime, in which the rate scaled with the square of the applied amplitude (i.e. $\Gamma_c \sim V_a^2$), consistent with the predictions of linear theory [19]. Further, experiments with trapped, laser-cooled ion crystals found a totally different regime in which the plasma rotation phase locked to the frequency of the applied field [20].

Here, we present the results of experiments using a rotating electric field to balance the plasma expansion, including measurement of compression rates for different applied wall amplitudes and drive frequencies. The experimental data indicates two distinct regimes: a broad, non-resonant frequency regime for large drive amplitudes, and a narrow-spectrum, resonant regime at smaller drive amplitudes. Lastly, a simple compression model, utilizing Eq. 2 to describe a dynamic balance between compression and plasma expansion, is compared to experiments in the non-resonant regime.

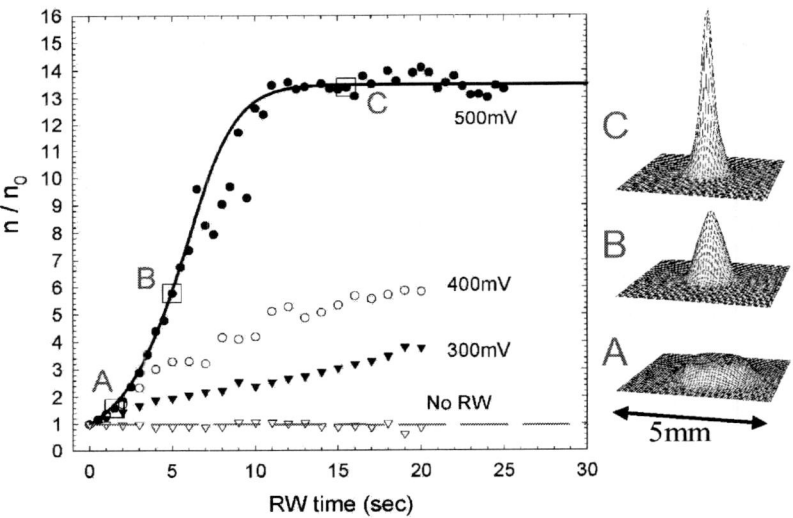

**FIGURE 3.** Rotating wall compression as a function of time for $f_{RW}$=8.0 MHz, $V_{RW}$=300 mV, 400 mV, and 500 mV. For comparison, the decay of the plasma with no applied RW is also shown (dashed line) corresponding to an expansion rate $\Gamma_0$=0.0045 s$^{-1}$. Two-dimensional density profiles are shown on the right, for the points A, B, C labeled on the plot. Fitting to Eq. 6 yields $\Gamma_{in}$=0.38 s$^{-1}$, $\Gamma_0$=0.0021 s$^{-1}$, and peak compression $n_f/n_0$=$(\Gamma_{in}/\Gamma_0)^{1/2}$=13.5, as shown by the solid curve.

These experiments used a 10 cm long electron plasma confined with the sectored electrode placed near one end, with initial parameters n ~ 5x10$^8$ cm$^{-3}$, T ~ 0.1 eV, and $R_p$ ~ 2mm. The applied wave is phased 0°-90°-180°-270° in the same direction as the plasma rotation (i.e. a dipole perturbation with $m_\theta$=1). The plasma is injected and allowed to equilibrate for 1 s. The RF is then turned on at a given applied amplitude ($V_{RW}$) and frequency ($f_{RW}$) for a specified amount of time. Then the electrons are accelerated to 6 kV and dumped on a phosphor screen, and the radial profile is measured with a high resolution CCD camera. The compression is measured by taking the ratio of the compressed density n(t) at the center to the initial central density $n_0$. The compression rate, $\Gamma_c$, is given by the ratio of the change in density to the initial density, divided by the compression time, i.e. $\Gamma_c$ = (n(t)-$n_0$)/($n_0$ $\Delta$t).

Figure 3 plots the compressed plasma density as a function of compression time for an experiment with $f_{RW}$=8 MHz and $V_{RW}$=300, 400, and 500 mV. For comparison the density evolution with no applied RW field is also shown. For the lower amplitudes, we observe a continual rise in the density with time. However, for $V_{RW}$=500 mV, over the same time period, a quick rise in central density is observed which then plateaus at a compression factor of ~13.

A simple extension of the transport model from Section II can be used to predict the plasma evolution under some very restrictive assumptions. Assuming that the inward compression rate ($\Gamma_{in}$) is independent of time, density, and temperature, and utilizing the empirical outward transport rate dependence on density ($\Gamma_O \propto n^2$), we can write an evolution equation for the plasma density of the form:

$$\frac{1}{n}\frac{dn}{dt} = \Gamma_{in} - \Gamma_O = \Gamma_{in} - \Gamma_0\left(\frac{n}{n_0}\right)^2. \qquad (5)$$

This has the analytic solution:

$$\frac{n(t)}{n_0} = \sqrt{\frac{\Gamma_{in}/\Gamma_0}{1+(\Gamma_{in}/\Gamma_0 - 1)e^{(-2\Gamma_{in}t)}}}. \qquad (6)$$

When this model for the time evolution of the density is fit to the data in Fig. 3, we find $\Gamma_{in} = 0.38$ s$^{-1}$, $\Gamma_0 = 0.0021$ s$^{-1}$ (solid line), with a steady-state compression ratio of $n_f/n_0=13.5$. It is interesting to note that the fit to $\Gamma_0$ is within a factor of two of the independently measured outward rate of $\Gamma_O = 0.0045$ s$^{-1}$. This agreement is somewhat surprising since we neglected the evolution of the plasma temperature. This may be due to the fact that the cyclotron cooling rate is rapid compared to the other time scales in the problem.

A similar equation for the plasma temperature can be written using Eqs. 3 and 4. This could then be solved self consistently with the assumed temperature dependence of the outward transport rate. These coupled equations can be solved numerically to obtain a more refined comparison with the experimental measurements. A more complete test of the model will require measuring the time dependence of the plasma temperature, which we plan to do.

Figure 4 shows the compression rate as a function of applied amplitude for three different applied frequencies. Over a broad range in amplitude, the compression rates scale like $\Gamma_C \propto V_{RW}^2$. This is similar to the scaling observed previously for warm positron plasmas [19].

Figure 5 shows the measured compression rate as a function of applied frequency for three different applied amplitudes. Over a broad range in frequency, the compression rates scale closely to $\Gamma_C \propto f_{RW}^2$. At present, we have no explanation for this scaling. However, since we have no data on the heating associated with the compression, this frequency scaling could be an indication of higher plasma temperatures caused by an increase in heating at higher frequencies. The model suggests that as the temperature increases, the background transport rate decreases. Thus, for a given input torque (fixed drive amplitude), we would observe a higher compression rate at higher temperatures, which is qualitatively consistent with the data. These scalings must be compared to an experimentally measured plasma temperature before definite conclusions can be made.

All of the data presented thus far have been at relatively high drive amplitudes (V > 200mV). Frequency scans were also performed at low amplitudes (e.g. 50mV and 100mV) and are shown in Fig. 6. In this regime, we do not observe a broad frequency dependence to the compression rate, but rather the compression is now sharply peaked at several discrete frequencies. These frequencies are close to the calculated Trivelpiece-

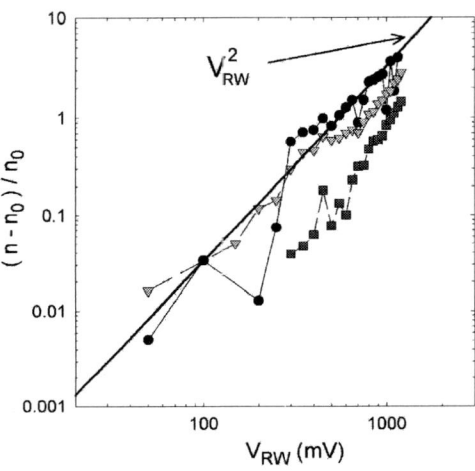

**FIGURE 4.** Rotating wall compression rate (after 2 s) as a function of applied drive amplitude for three different frequencies. Circles, triangles, squares are 8 MHz, 6 MHz, 4 MHz, respectively. The solid line is $\Gamma_C \propto V_{RW}^2$.

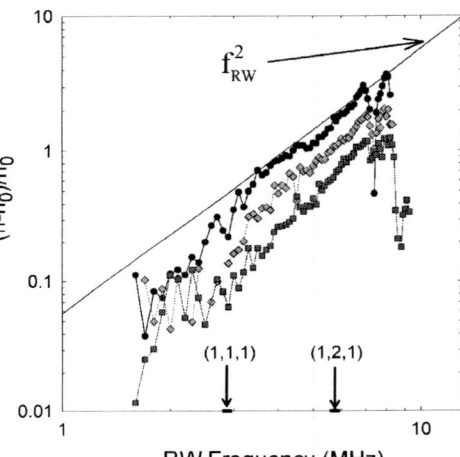

**FIGURE 5.** Rotating wall compression rate (after 2 s) as a function of applied frequency for three different applied amplitudes. The circles, diamonds, squares, are 1.0 V, 0.8 V, 0.6 V, respectively. The lowest calculated TG mode frequencies ($m_\theta$, $m_z$, $m_r$) are shown by the arrows. The solid line is $\Gamma_C \propto f_{RW}^2$.

**FIGURE 6.** Plasma compression after 30 s versus frequency for low amplitude drive. The solid circles are $V_{RW}=50$ mV, the grey triangles are $V_{RW}=100$ mV. The calculated values of the lowest Trivelpiece-Gould modes ($m_\theta$, $m_z$, $m_r$) are shown by the arrows.

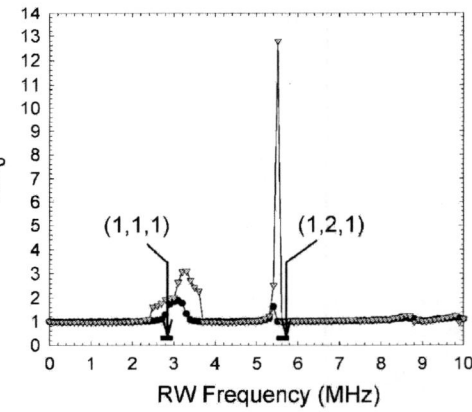

Gould mode frequencies, as indicated by the arrows [21]. This resonant regime is similar to that observed previously in electron plasmas [17]. The exact crossover between the two different regimes has yet to be identified, but it appears that there is a relatively distinct transition between the two.

A closer look at the compressed profiles shows a qualitative difference between the two regimes. Figure 7(a) shows the initial flat-top profile before compression. Figure 7(b), shows the steady-state profile after 15 s with a large drive amplitude (500 mV).

156

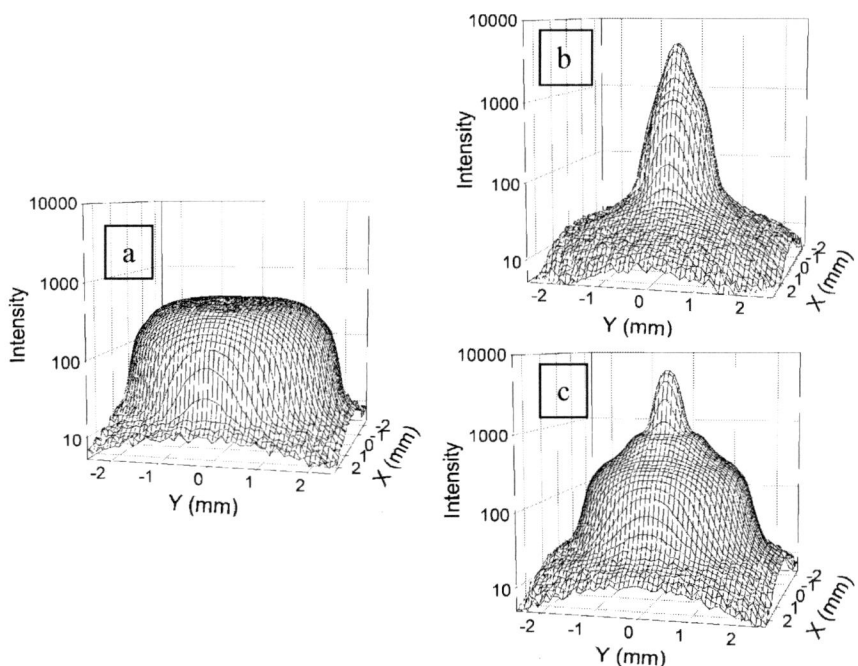

**FIGURE 7.** Comparison of steady-state profiles for non-resonant compression and resonant compression of an initial flat-top plasma: (a) initial plasma. (b) after 15 s with $f_{RW} = 8$ MHz and $V_{RW} = 500$ mV. (c) after 30 s of resonant compression with $f_{RW} = 5.5$ MHz and $V_{RW} = 100$ mV.

Profile 7(b) is peaked with a low level, diffuse background just above the noise floor. Figure 7(c) shows an example of one of the resonant peaks (5.5 MHz) at low drive amplitude (100 mV). This profile has a narrow peak at the center and a rather broad halo that has a much higher density as compared with the large-amplitude case. These different profiles are likely indications that different mechanisms are responsible for the compression in the two regimes. For example, the low amplitude case could reflect coupling to weakly damped TG modes [17], whereas the large amplitude case could be due to particle bounce-resonance coupling [22].

## IV.    Extraction of Small Diameter Beams

If the end confinement electrode is slowly lowered to a potential close to the plasma space charge potential, particles with the highest energy (i.e. the high energy tail of the Maxwellian velocity distribution) are able to leave the trap. Since the confinement potential provided by the electrodes is a minimum on the trap axis, the first particles

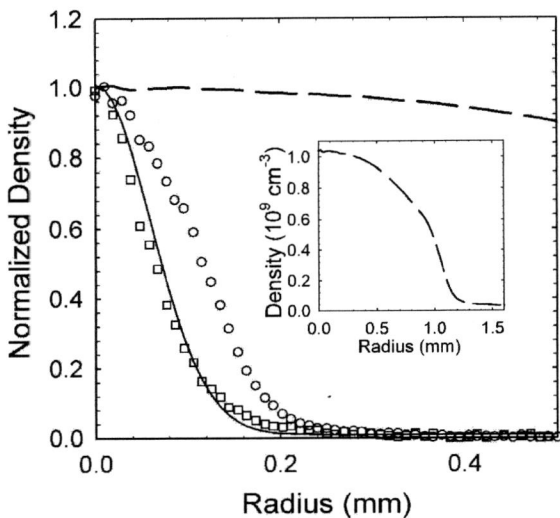

**FIGURE 8.** Radial profile of small beams ejected from a broad, dense electron plasma. The dashed curve is the initial plasma. The hollow circles and squares are for two different levels of ejection, $n_b/n_0=0.51$ and $n_b/n_0=0.32$ respectively. The solid curve is a fit of Eq. 7 to the hollow squares, with a beam radius of $\sim 60$ μm. The inset is an expanded view of the initial plasma profile.

exiting the trap will be near the plasma center. For small particle loss (i.e. $\Delta n_0 < n_0/2$), the radial profile of the ejected plasma will be approximately Gaussian, with a beam density, $n_b(r)$, of the form [23]:

$$n_b(r) = Cn_0(r)\exp\left(-\left(\frac{r}{2\lambda_D}\right)^2\right),$$ (7)

where $\lambda_D$ is the plasma Debye length and the constant C depends on the ratio of the electrode potential to the plasma space charge potential. For plasmas with $R_p \gg \lambda_D$, the ejected beam will have a diameter, $D \sim 4\lambda_D$ (which for cold, $T\sim1$ meV, and dense, $n \sim 10^{10}$ cm$^{-3}$, plasmas can be less than 10 μm).

The radial profiles of two ejected beams are shown in Fig. 8 along with the profile of the initial plasma. For this case, the central plasma density is $1.2\times10^9$ cm$^{-3}$; assuming the plasma is at a temperature $T \sim 300$ K (i.e. 25 meV ), we estimate a Debye length of 34 μm. Fitting a Gaussian to the smallest beam, gives a measured half-width of 60 μm. From Eq. 7, the beam half-width is $2\lambda_D \sim 68$ μm, which is within about 10% of measured beam radius.

Measurement of the profile of an ejected beam can also be used to estimate the plasma Debye length, $\lambda_D$. Thus, if the density profile is known, an estimate of the plasma temperature can be made. This technique is limited by the assumption of uniform plasma temperature as a function of radius, and that the particle ejection is fast compared to the transport and equilibration times, but slow compared to the particle bounce frequency, $f_b$, and any plasma mode frequency. Further, only the initial pulse from the unperturbed plasma can be used. After such a pulse, the plasma will be slightly hollow, and instabilities can develop which would alter the ejected plasma profile.

## V.    Multi-cell Trap

Recently a novel, multi-cell Penning-Malmberg trap has been proposed that is designed to confine more than $10^{12}$ positrons in plasmas with lifetimes of days or longer [24]. A 10 tesla magnetic field and 10 kV electrode potentials are assumed to confine the plasma in a cryogenically cooled electrode structure (T~10 K). One set of constraints on the operating parameters arises from the previously described outward, asymmetry-driven transport and the associated expansion heating. The transport rate was assumed to be given by the empirical transport scaling from Sect. II (Eq. 2). This inherent expansion will be balanced by radial compression from a rotating electric field. It was also assumed that the heating associated with this transport will be balanced by cyclotron cooling in the 10 T field ($\Gamma_{cool} \sim 25$ s$^{-1}$). End potentials of $V_e = 10$ kV, provide the capability to confine plasmas up to a maximum plasma space charge potential of 10 kV. The transport and cooling considerations favor high plasma temperatures; for positrons, this is limited by positronium atom formation on the residual background gas (p~$10^{-12}$ torr). These considerations lead to a multi-cell design in which separate plasmas are confined in cylindrical electrodes ~ 1 cm in diameter and 1 cm in extent along the magnetic field. Typical plasma parameters are proposed to be $T_e \sim 2$ eV, n~$10^{11}$ cm$^{-3}$ and a plasma radius ~ 1.5 mm. A trap for $10^{13}$ positrons could be confined in $10^3$ cells housed in an electrode structure 15 cm in diameter and 30 cm long (immersed in the common 10 T field).

Although the proposed trap is ambitious, the basic physics issues involved with the operation of such a trap can be investigated using the high field trap described here. A 1 kV confining potential and 5 T magnetic field, will allow for the confinement of $10^{10}$ positrons, with $L_p$=10 cm, $R_p$=1 mm, $R_w$=1.27 cm, and n~$3\times10^{10}$ cm$^{-3}$. Assuming a plasma temperature of 2 eV, Eq. 2 predicts an outward rate of $\Gamma_O$~0.033 s$^{-1}$, and Eq. 3 predicts a heating rate of $\Gamma_{heat}$~2.7 s$^{-1}$, which is less than the cyclotron cooling rate at 5T of $\Gamma_{cool}$~6.3 s$^{-1}$. Further, the trap will accommodate an electrode stack with up to 10 radial cells, and 10 axial cells. Assuming $10^{10}$ positrons per cell, this would give a total of $10^{12}$ stored positrons. Such a device could form the basis for a portable positron trap. For example, a 25 mCi $^{22}$Na source and neon moderator can produce about $10^6$ moderated positrons per second. For a trap with $10^{12}$ stored positrons, this is equivalent to $10^6$ seconds, or about 11 days of continuous operation.

Several challenges need to be addressed before this can be accomplished. First, the transport physics of high-density non-neutral plasmas (n ~$10^{10}$ cm$^{-3}$) needs to be characterized. Further, a robust model of plasma confinement using a rotating electric field needs to be validated and verified in the high-density regime. A technique for the stacking and manipulation of large numbers of positrons, specifically into different radial cells, needs to be developed, as well as an efficient method of extraction from these cells.

These challenges can be addressed with a small number of cells in the current high-field trap.

An additional challenge is operating with a suitable positron source. The current buffer gas trap can provide about 500 million positrons every 5 minutes. Consequently, it would take about 7 days of continuous operation to fill to $10^{12}$ positrons. Thus, to obtain the full potential of the multi-cell trap described here (e.g. $N_{e+} > 10^{12}$), will require either a reactor-based or LINAC positron source or a stronger radioactive source (e.g. $\geq 1$ Ci $^{22}$Na).

## VI.   Concluding Remarks

We report first results using a specially designed high-field, cryogenic Penning-Malmberg trap that is designed to confine large numbers of positrons and produce cold plasmas and beams. Electron plasmas were used to determine the optimum operating regime with the electrode structure at 300 K. The experiments indicate that a simple model of rotating wall compression, balancing expansion heating and cyclotron cooling, qualitatively describes the plasma compression and resulting dynamic steady state. Density compression factors $\geq 10$ were obtained, resulting in plasma densities $\geq 5 \times 10^9$ $cm^{-3}$. First attempts to extract small-diameter beams from the plasma center were successful with beam diameters, D, approaching the expected theoretical limit $D \sim 4\lambda_D$.

Experiments with positrons are planned for the near future. Bursts of positrons (N $\sim 5 \times 10^8$) will be transferred from the existing buffer-gas trap to the high-field trap on a 5 minute cycle. The positrons will be transferred through a pulsed valve to minimize contamination of the cryogenic, UHV trap. Based upon previous experience [1] and using a set of specially designed electrodes in the region of rapidly varying magnetic field, transfer efficiencies $\geq 70\%$ are expected. Positron plasmas with $N \sim 10^{10}$ can be accumulated in $\sim 2$ hours. The trap described here is also expected to be well suited to testing the design of a multicell Penning-Malmberg trap capable of confining orders of magnitude more positrons, since the optimum parameters of the multicell trap are not far from the operating range of the present device.

## VII.   Acknowledgements

This work was supported by the Office of Naval Research, grant #N00014-02-1-0123. The authors acknowledge the expert technical assistance of E. A. Jerzewski and R.G. Greaves for helpful conversations and collaboration on the multi-cell trap.

Finally, we take this opportunity to acknowledge the long-term support of ONR. The positron plasma research that led to the development of the efficient buffer-gas trap [25-27], solid neon positron moderator [28], cold positron beam [29], rotating wall compression of positron plasmas [19, 30], a wide variety of plasma and atomic physics results [30-33], and the high-field cryogenic trap described here would not have been possible without the ONR support and the support and encouragement of ONR program officer C. W. Roberson.

# References:

1.    Amoretti, M., Amsler, C., Bonomi, G., *et al.*, *Nature* **419**, 456 (2002).
2.    Gabrielse, G., Bowden, N., Oxley, P., *et al.*, *Phys. Rev. Lett.* **89**, 213401 (2002).
3.    Greaves, R. G. and Surko, C. M., *Non-Neutral Plasma Physics IV, edited by F. Anderegg, et al. (American Institute of Physics)* , 10-23 (2002).
4.    Mills, A. P., Jr., *Nucl. Instrum. and Meth.* **192**, 107-116 (2002).
5.    Sullivan, J. P., Gilbert, S. J., Marler, J. P., *et al.*, *Phys. Rev. A* **66**, 042708-1-12 (2002).
6.    Greaves, R. G., Gilbert, S. J., and Surko, C. M., *Appl. Surf. Sci.* **194**, 56 (2002).
7.    O'Neil, T., *Phys. of Fluids* **23**, 725 (1980).
8.    Davidson, R. C., *"Physics of Nonneutral Plasmas" (Addison-Wesley, Reading, MA, 1990)* .
9.    Driscoll, C. F. and Malmberg, J. H., *Physical Review Letters* **50**, 167-70 (1983).
10.   Kriesel, J. M. and Driscoll, C. F., *Phys. Rev. Lett.* **85**, 2510 (2000).
11.   Kabantsev, A. A. and Driscoll, C. F., *Phys. Rev. Lett.* **89**, 245001 (2002).
12.   Hilsabeck, T. J., Kabantsev, A. A., Driscoll, C. F., *et al.*, *Phys. Rev. Lett.* **90**, 245002 (2003).
13.   Cluggish, B. P., *University of California, San Diego Thesis* (1995).
14.   Eggleston, D. L. and O'Neil, T., *Phys. of Plasmas* **6**, 2699 (1999).
15.   Malmberg, J., O'Neil, T. M., Hyatt, A. W., *et al.*, *Proc. of the Sendai Symposium on Plasma Nonlinear Electron Phenomena* , 31-35 (1984).
16.   Hollmann, E. M., Anderegg, F., and Driscoll, C. F., *Phys. of Plasmas* **7**, 2776-89 (2000).
17.   Anderegg, F., Hollmann, E. M., and Driscoll, C. F., *Phys. Rev. Lett.* **81**, 4875-4878 (1998).
18.   Huang, X. P., Anderegg, F., Hollmann, E. M., *et al.*, *Phys. Rev. Lett.* **78**, 875 (1997).
19.   Greaves, R. G. and Surko, C. M., *Phys. Rev. Lett.* **85**, 1883-1886 (2000).
20.   Huang, X. P., Bollinger, J. J., Mitchel, T. B., *et al.*, *Phys. Rev. Lett.* **80**, 730 (1998).
21.   Trivelpiece, A. W. and Gould, R. W., *J. of Appl. Phys.* **30**, 1784-1793 (1959).
22.   Crooks, S. M., *University of California, San Diego Thesis* (1995).
23.   Eggleston, D. L., Driscoll, C. F., and Beck, B. R., *Physics of Fluids B* **4**, 3432-9 (1992).
24.   Surko, C. M. and Greaves, R., *J. Rad. Chem.* **in press** (2003).
25.   Murphy, T. J. and Surko, C. M., *Phys. Rev. A* **46**, 5696-705 (1992).
26.   Greaves, R. G. and Surko, C. M., *Non-Neutral Plasma Phys. III, edited by John J. Bollinger, et al., (American Institute of Physics, Princeton, NJ)* **498**, 19-28 (1999).
27.   Greaves, R. G. and Surko, C. M., *Canadian Journal of Physics* **51**, 445-8 (1996).
28.   Gilbert, S. J., Kurz, C., Greaves, R. G., *et al.*, *Appl. Phys. Lett.* **70**, 1944-1946 (1997).
29.   Greaves, R. G. and Surko, C. M., *Phys. Plasmas* **8**, 1879-1885 (2001).
30.   Greaves, R. G. and Surko, C. M., *Phys. Plasmas* **4**, 1528-1543 (1997).
31.   Gilbert, S. J., Dubin, D. H. E., Greaves, R. G., *et al.*, *Phys. Plasmas* **8**, 4982-94 (2001).
32.   Iwata, K., Greaves, R. G., Murphy, T. J., *et al.*, *Phys. Rev. A* **51**, 473 (1995).
33.   Barnes, L. D., Gilbert, S. J., and Surko, C. M., *Phys. Rev. A* **67**, 032706 (2003).

161

# Kinetic Theory for
# Antihydrogen Recombination Schemes[1]

Ronald Stowell and Ronald C. Davidson

*Plasma Physics Laboratory, Princeton University, Princeton, NJ 08543*

**Abstract.** Guiding-center kinetic theory has been developed for antihydrogen recombination experiments, which are conducted with magnetic fields of $3$-$5\,\mathrm{T}$; temperatures of $4$-$10\,\mathrm{K}$; positron densities of $10^7$-$10^8\,\mathrm{cm}^{-3}$; and antiproton densities of $10^4$-$2\times10^7\,\mathrm{cm}^{-3}$. Collision operators provide the leading-order correction to weak-coupling theory as the coupling parameter increases. Six collision operators – three Landau analogs and three Balescu[1]-Guernsey[2]-Lenard[3] analogs – are found for particles of unlike charges. One operator is the multiple-species generalization of Dubin's and O'Neil's operator [4]. A stability analysis is performed for counter-streaming positrons and antiprotons occupying a cylindrical region coaxial with an outer conducting cylinder in a constant, axial magnetic field. The finite transverse geometry of the system is included, leading to a three-dimensional Penrose criterion, which is applied to drifting Maxwellian distributions to obtain the regime of stability as a function of the species' temperature ratio, density ratio and relative mean velocity. Collisional corrections are considered. Terms resulting from collisions between particles of the same species cancel under general assumptions satisfied by both O'Neil's operator [5] and Dubin's and O'Neil's operator [4]. The multiple-species generalization of Dubin's and O'Neil's operator is used for unlike-species collisions to find a collisionally corrected dispersion relation, which is applied to a detailed study of stability properties.

## INTRODUCTION AND BACKGROUND

Since the last Workshop on Nonneutral Plasmas, the first antihydrogen atoms were detected [6, 7]. With the achievement of this major milestone, increases in the rate of antihydrogen production, and therefore increases in plasma density, have now become important objectives. As antimatter is successfully accumulated, a statistical description of the particles will become more useful. Since the coupling parameter[2]

$$\Gamma_s \equiv \frac{e^2 \big/ (4\pi n_s/3)^{-1/3}}{k_B T_s} = 0.005 \text{ to } 0.14 \tag{1}$$

in this regime is small, kinetic theory may be employed. Here $e$ is the fundamental unit of electric charge, $s \in \{e^+, \bar{p}\}$ is some species, and $n$ is density and $T$ is temperature. Since the recombination times are long ($\approx 10^{-4}\,\mathrm{s}$), collective phenomena, which are familiar in other contexts, may appear before recombination occurs.

---

[1] Research supported by the United States Department of Energy, and in part by the Office of Naval Research.

[2] Also, $\Gamma_s = 3^{-1/3}\left(\dfrac{\lambda_{Ds}}{e^2/k_B T_s}\right)^{-2/3}$ and $\Lambda_s \equiv n_s\lambda_{Ds}^3 = \dfrac{1}{4\pi}\dfrac{\lambda_{Ds}}{e^2/k_B T_s}$, where $\lambda_D$ is the Debye length.

CP692, *Non-Neutral Plasma Physics V*, edited by M. Schauer et al.
© 2003 American Institute of Physics 0-7354-0165-9/03/$20.00

In particular, two-stream interactions, manifest as the electron-proton (e-p) instability in proton storage rings [8], could occur unless the constituent species are mixed at sufficiently low velocities. With this in mind, a linear stability analysis is performed. Since the coupling parameter does not have the minute size characteristic of typical plasmas, discrete-particle effects are substantial. In kinetic equations, one accounts for discrete-particle effects by means of collision operators.

The fundamental kinetic theory for this regime has not been fully developed, so it is necessary to advance the basic theory before proceeding with a stability analysis. Toward this end, two supporting results are presented. The first is a collection of collision operators valid in this general region of parameter space. The second is a Penrose criterion valid for systems with a cylindrical bounding wall.

The fundamental parameters for antihydrogen traps are temperatures $T_s$ of $4 - 15 \, \text{K}$; Positron densities $n_{e+}$ of $8 \times 10^6 - 3 \times 10^8 \, \text{cm}^{-3}$; antiproton densities $n_{\bar{p}}$ of $10^4 - 4 \times 10^6 \, \text{cm}^{-3}$; and magnetic fields $\mathbf{B}$ of $3 - 5 \, \text{T}$. Thus the frequencies and lengths satisfy

$$\omega_{cs} \gg \omega_{ps} \gg \omega_{Ds} \gg 2\pi v_{\text{rec}} \gg 2\pi v_{\text{cool}}$$

and

$$r_{\text{plasma}} \gg \lambda_{De+} \gg \frac{e^2}{k_B T_s} \gtrsim \rho_{L\bar{p}} \gg \frac{h}{m_{e+} v_{te+}} \gtrsim \rho_{Le+},$$

respectively, where $s$ is a species with cyclotron frequency $\omega_{cs}$, plasma frequency $\omega_{ps}$, diocotron frequency $\omega_{Ds}$, thermal distance of closest approach $e^2/k_B T_s$, gyroradius $\rho_{Ls}$ and thermal DeBroglie wavelength $h/m_s v_{ts}$; the recombination frequency is $v_{\text{rec}}$; the synchrotron cooling frequency of the positrons is $v_{\text{cool}}$; and the radius of the plasma is $r_{\text{plasma}}$. The ratio (2) of the positron-positron collision frequency $v_{e+e+}$ to the positron plasma frequency $\omega_{pe+}$ will be a small expansion parameter:

$$\frac{2\pi v_{e+e+}}{\omega_{pe+}} = \frac{\ln 4\pi \Lambda_{e+}}{4\pi \Lambda_{e+}} = 0.21. \tag{2}$$

# NEW COLLISION OPERATORS FOR $\text{sgn}(q_1 q_2) = -1$

Collision operators provide the leading-order correction to kinetic theory as the coupling parameter $\Gamma_s$ increases. To begin, it has been found, using multiple-scale analysis, that when $\rho_{Ls}/\rho \ll 1$, to leading order, the particles are effectively constrained to move parallel to the magnetic field over the time and space of a collision [9]. Thus, particles with similarly signed charges can either reflect (that is, scatter by $180°$) or, for sufficiently high relative kinetic energies, pass by each other.

Presently, there is no operator which completely describes collisions between particles with similarly signed charges, because one must simultaneously treat reflection and Debye shielding. If we treat reflection, as in O'Neil's Boltzmann analog [5], the shape of the Debye cloud is not included. If we include the shape of the Debye cloud, as in Dubin's and O'Neil's Balescu[1]-Guernsey[2]-Lenard[3] (henceforth BGL) analog [4], reflection is not included, because reflecting trajectories do not change by only a small amount due to the collision. Both operators are of the same order. However, particles with oppositely signed charges never reflect. This removes the obvious barrier to application of the Klimontovich formalism to derive BGL analogs for particles with oppositely signed charges.

When $\rho_{Ls} \ll q_1 q_2 / k_B T$, which is true for the positrons and often true for the antiprotons, typically the colliding particles are, at the closest, many gyroradii apart. Since, to leading order, the particles are effectively constrained to move parallel to the magnetic field, to leading order, the guiding-center Klimontovich equation [10] for species $s$ is

$$\frac{\partial N_s}{\partial t} + v_z \frac{\partial N_s}{\partial z} + \frac{q_s}{m_s} E_z \frac{\partial N_s}{\partial v_z} = 0, \tag{3}$$

where $t$ is time; $z$ is the coordinate parallel to the magnetic field; $v_z$ and $E_z$ are velocity and electric field, respectively, in the $z$ direction; $N_s \equiv \sum_{n=1}^{N_s} \delta(\mathbf{x} - \mathbf{x_n}) \delta(v_z - v_{zn})$; and n indexes each of the $N_s$ particles of species $s$, which have masses $m_s$ and charges $q_s$.

To derive a collision operator valid for small deviations in the particles' trajectories, one could begin with Eqs. (3) and work through a long exercise in Klimontovich analysis. But, alternatively, one could note that Eq. (3) is identical in form to the familiar Klimontovich equation for no magnetic field and particles constrained to move in the $z$ direction. Since the collision operators are determined entirely from the Klimontovich equations, identical Klimontovich equations lead to identical operators.

Thus we may simply insert the delta-function distributions $f_{sn}(\mathbf{v}) = \delta(\mathbf{v}_\perp) g_{sn}(v_z)$ into the familiar BGL operator. With very little effort, we can recover the spatially local version of Dubin's and O'Neil's BGL analog [4] when attention is restricted to one species. When the two species are different, the spatially local, multiple-species version Dubin's and O'Neil's BGL analog is found:[3]

$$\left( \frac{\partial g_{s_1}(v_{1z})}{\partial t} \right)_{coll} = \pi \frac{q_{s_1}^2}{m_{s_1}} \frac{\partial}{\partial v_{1z}} \int \frac{d^3 k}{(2\pi)^3} \sum_{s_2} n_{s_2} q_{s_2}^2 \int_{-\infty}^{\infty} dv_{2z} \frac{(\hat{z} \cdot \boldsymbol{\varepsilon}_\mathbf{k})(\hat{z} \cdot \boldsymbol{\varepsilon}_\mathbf{k}^*)}{\left| D_{\mathbf{k} k_z v_{2z}} \right|^2}$$

$$\times \delta [k_z(v_{1z} - v_{2z})] \left[ \frac{1}{m_{s_1}} \frac{\partial}{\partial v_{1z}} - \frac{1}{m_{s_2}} \frac{\partial}{\partial v_{2z}} \right] g_{s_1}(v_{1z}) g_{s_2}(v_{2z}), \tag{4}$$

where

$$D_{\mathbf{k}\omega} \equiv 1 + \sum_s \frac{\omega_{ps}^2}{k^2} \int_{-\infty}^{\infty} dv_z \frac{k_z}{\omega - k_z v_z} \frac{\partial g_s}{\partial v_z}$$

and $\boldsymbol{\varepsilon}_\mathbf{k} \equiv -4\pi i \mathbf{k}/k^2$. As $k\lambda_D \to \infty$, the leading-order contribution due to collisions is given by the Landau analog

$$\left( \frac{\partial g_{s_1}(v_{1z})}{\partial t} \right)_{coll}$$

$$= 4\pi \frac{q_{s_1}^2}{m_{s_1}} \ln \Lambda \sum_{s_2} n_{s_2} q_{s_2}^2 \frac{\partial}{\partial v_{1z}} \left[ \frac{1}{m_{s_1}} \frac{\partial g_{s_1}(v_{1z})}{\partial v_{1z}} g_{s_2}(v_{1z}) - \frac{1}{m_{s_2}} g_{s_1}(v_{1z}) \frac{\partial g_{s_2}(v_{1z})}{\partial v_{1z}} \right], \tag{5}$$

---

[3] All the following operators are limits of Rostoker's operator [11]. The advantage of the new operators is their simplicity.

where $\Lambda \equiv \dfrac{\lambda_D}{q_1 q_2 / k_B T}$. The operator (5) will be applied to a linear stability analysis in the next section. This operator has no velocity integral, so it happens that the operator is also local in velocity. This is because particles with very similar velocities, or small relative kinetic energies, influence each other much more than other particles.

In deriving a collision operator, one normally integrates over impact parameter from $q_1 q_2 / k_B T$ up to $\lambda_D$. To obtain an operator with a simple form, it has so far been necessary to exclude the gyroradii $\rho_{Ls}$ from the range of integration. In the Landau and BGL operators, $\lambda_D \ll \rho_{Ls}$. In O'Neil's operator [5], in Dubin's and O'Neil's operator [4], and in the preceding two operators (4) and (5), $\rho_{Ls} \ll q_1 q_2 / k_B T$.

However, in the present case, the argument $\Lambda$ of the Coulomb logarithm does not have its usual enormous size. It may happen in this general region of parameter space that the entire range of integration fits *between* the two gyroradii. More operators may be derived by capitalizing on the method of inserting the delta-function distributions into the BGL operator. This time we use the method on $s_1$ only: $f_{s_1}(\mathbf{v}) = \delta(\mathbf{v}_\perp) g_{s_1}(v_z)$. Thus, the particles with large gyroradii are influencing the particles with small gyroradii. The resulting BGL analog is

$$
\left( \frac{\partial g_{s_1}(v_{1z})}{\partial t} \right)_{coll} = \pi \frac{q_{s_1}^2}{m_{s_1}} \frac{\partial}{\partial v_{1z}} \int \frac{d^3\mathbf{k}}{(2\pi)^3} \sum_{s_2} n_{s_2} q_{s_2}^2 \int d^3\mathbf{v}_2 \frac{(\hat{z} \cdot \boldsymbol{\varepsilon}_\mathbf{k}) \boldsymbol{\varepsilon}_\mathbf{k}^*}{\left| D_{\mathbf{k} k_z v_{1z}} \right|^2}
$$

$$
\times \, \delta \left[ \mathbf{k} \cdot (v_{1z}\hat{z} - \mathbf{v}_2) \right] \left[ \frac{1}{m_{s_1}} \hat{z} \frac{\partial}{\partial v_{1z}} - \frac{1}{m_{s_2}} \frac{\partial}{\partial \mathbf{v}_2} \right] g_{s_1}(v_{1z}) f_{s_2}(\mathbf{v}_2), \quad (6)
$$

where

$$
D_{\mathbf{k}\omega} \equiv 1 + \frac{\omega_{ps_1}^2}{k^2} \int_{-\infty}^\infty dv_{1z} \frac{k_z}{\omega - k_z v_{1z}} \frac{\partial g_{s_1}}{\partial v_{1z}} + \frac{\omega_{ps_2}^2}{k^2} \int d^3\mathbf{v}_2 \frac{\mathbf{k}}{\omega - \mathbf{k} \cdot \mathbf{v}_2} \cdot \frac{\partial f_{s_2}}{\partial \mathbf{v}_2}.
$$

In the limit as $k\lambda_D \to \infty$, we obtain the corresponding Landau analog

$$
\left( \frac{\partial g_{s_1}(v_{1z})}{\partial t} \right)_{coll} = 2\pi \frac{q_{s_1}^2}{m_{s_1}} \ln \Lambda \sum_{s_2} n_{s_2} q_{s_2}^2 \frac{\partial}{\partial v_{1z}} \int d^3\mathbf{v}_2 \frac{u^2 \hat{z} - u_z \mathbf{u}}{u^3}
$$

$$
\cdot \left[ \frac{1}{m_{s_1}} \hat{z} \frac{\partial}{\partial v_{1z}} - \frac{1}{m_{s_2}} \frac{\partial}{\partial \mathbf{v}_2} \right] g_{s_1}(v_{1z}) f_{s_2}(\mathbf{v}_2), \quad (7)
$$

where $\mathbf{u} = v_{1z}\hat{z} - \mathbf{v}_2$.

Finally, we use the method on $s_2$ only: $f_{s_2}(\mathbf{v}) = \delta(\mathbf{v}_\perp) g_{s_2}(v_z)$. Now the particles with small gyroradii are influencing the particles with large gyroradii. The resulting BGL analog is

$$
\left( \frac{\partial f_{s_1}(\mathbf{v}_1)}{\partial t} \right)_{coll} = \pi \frac{q_{s_1}^2}{m_{s_1}} \frac{\partial}{\partial \mathbf{v}_1} \cdot \int \frac{d^3\mathbf{k}}{(2\pi)^3} \sum_{s_2} n_{s_2} q_{s_2}^2 \int_{-\infty}^\infty dv_{2z} \frac{\boldsymbol{\varepsilon}_\mathbf{k} \boldsymbol{\varepsilon}_\mathbf{k}^*}{\left| D_{\mathbf{k} k_z v_{2z}} \right|^2}
$$

$$
\times \, \delta \left[ \mathbf{k} \cdot (\mathbf{v}_1 - v_{2z}\hat{z}) \right] \left[ \frac{1}{m_{s_1}} \frac{\partial}{\partial \mathbf{v}_1} - \frac{1}{m_{s_2}} \hat{z} \frac{\partial}{\partial v_{2z}} \right] f_{s_1}(\mathbf{v}_1) g_{s_2}(v_{2z}), \quad (8)
$$

where

$$D_{\mathbf{k}\omega} \equiv 1 + \frac{\omega_{ps_1}^2}{k^2}\int d^3\mathbf{v}_1 \frac{\mathbf{k}}{\omega - \mathbf{k}\cdot\mathbf{v}_1} \cdot \frac{\partial f_{s_1}}{\partial\mathbf{v}_1} + \frac{\omega_{ps_2}^2}{k^2}\int_{-\infty}^{\infty}dv_{2z}\frac{k_z}{\omega - k_z v_{2z}}\frac{\partial g_{s_2}}{\partial v_{2z}}.$$

As before, for $k\lambda_D \to \infty$, we obtain a Landau analog, which in this case is

$$\left(\frac{\partial f_{s_1}(\mathbf{v}_1)}{\partial t}\right)_{coll} = 2\pi\frac{q_{s_1}^2}{m_{s_1}}\ln\Lambda\sum_{s_2}n_{s_2}q_{s_2}^2\frac{\partial}{\partial\mathbf{v}_1}\cdot\int_{-\infty}^{\infty}dv_{2z}\frac{u^2\mathbf{I}-\mathbf{u}\mathbf{u}}{u^3}$$

$$\cdot\left[\frac{1}{m_{s_1}}\frac{\partial}{\partial\mathbf{v}_1} - \frac{1}{m_{s_2}}\hat{z}\frac{\partial}{\partial v_{2z}}\right]f_{s_1}(\mathbf{v}_1)g_{s_2}(v_{2z}), \quad (9)$$

where $\mathbf{u} = \mathbf{v}_1 - v_{2z}\hat{z}$ and I is the $3\times3$ identity tensor. The four preceding operators are asymmetrical because the scattered and scattering species do not behave the same. This last pair of operators has the same regime of applicability as the preceding pair.

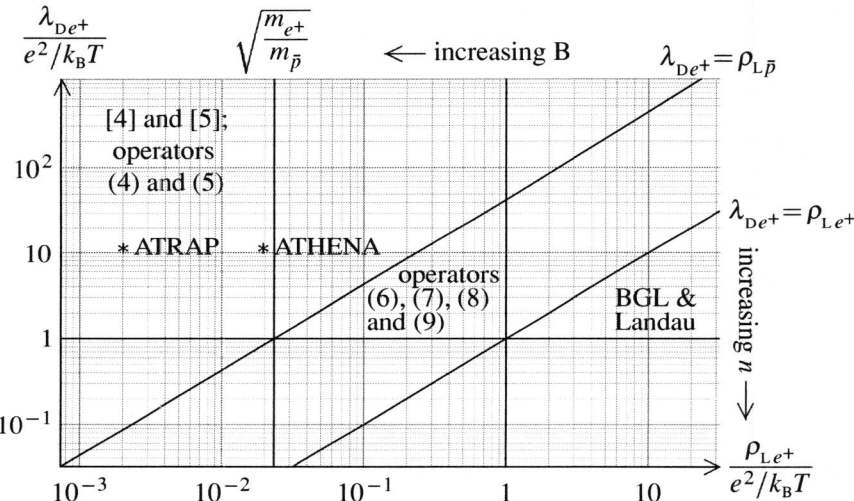

FIGURE 1. Regimes of applicability of the operators. In this plot, $T \equiv T_{e^+} = T_{\bar{p}}$ and $n_{e^+} \geq n_{\bar{p}}$. Typical parameters for ATHENA and ATRAP are marked with asterisks. O'Neil's [5] and Dubin's and O'Neil's [4] operators are at the left. If the ratio $T_{\bar{p}\perp}/T_{e^+\perp}$ of the temperatures perpendicular to the magnetic field B is increased, the regime of applicability of operators (6), (7), (8) and (9) will expand.

## LINEAR STABILITY ANALYSIS

When the positrons and antiprotons are mixed, they are not in thermal equilibrium. A two-stream instability could result in the conversion of the directed energy of the antiprotons into random kinetic energy, lowering the recombination rate. An estimate of the final effective temperature in a worst-case scenario, found by equating the final and initial thermal and directed kinetic energies yields temperature increases by a factor of between two and $10^3$ for current [12, 13] values of mixing parameters.

Growth rates for two-stream instabilities in cold plasmas scale like $\omega_{pe+}^{1/3}\omega_{p\bar{p}}^{2/3}$ [14]. Since, in practice, the growth rates can be of the same size in hot plasmas, a two-stream instability would provide a collective mechanism for very rapid randomization of directed kinetic energy. Since $\omega_{pe+}^{1/3}\omega_{p\bar{p}}^{2/3}$ is about the same size as the collision frequency $2\pi\nu_{e+\bar{p}}$ for the influence of antiprotons on positrons, collisions must be considered when computing the growth rate. The system of equations

$$\frac{\partial f_s}{\partial t} + \left[\frac{c\mathbf{E}\times\mathbf{B}}{B^2}\right]\cdot\nabla_{\perp}f_s + v_{\parallel}\nabla_{\parallel}f_s + \frac{q_s}{m_s}E_{\parallel}\frac{\partial f_s}{\partial v_{\parallel}} = \frac{\ln\Lambda_{e+}}{\Lambda_{e+}}\sum_{s'}C_{ss'}\left[f_s,f_{s'}\right] \quad \text{and} \quad (10)$$

$$\nabla^2\phi^{\text{self}} = -4\pi\sum_{s}n_s q_s\int d^3v f_s, \quad (11)$$

used to describe the antimatter consists of the collisional guiding center kinetic equations (10), which require that $\omega/\omega_{cs}\ll 1$, $k\rho_{Ls}\ll 1$ and $v_{ss'}/\omega_{ps}\ll 1$ for frequencies $\omega$ and wavenumbers $\mathbf{k}$ of interest; and Poisson's equation (11), which is valid if the longitudinal and transverse modes are decoupled. Thus the model is *electrostatic*. The right-hand side of Eq. (10) represents the collision operator.

In the present analysis, we perturb about a collisionless equilibrium. Collisions change both the unperturbed quantities $f_{s0}$ and $\phi_0$ and the perturbations $\delta f_s$ and $\delta\phi$. The effect of collisions on the perturbations is studied in the present analysis.

The species are studied during their initial mixing [15]. The geometry considered is shown in Figure 2. Two long, cylindrical, coaxial plasma columns, composed of

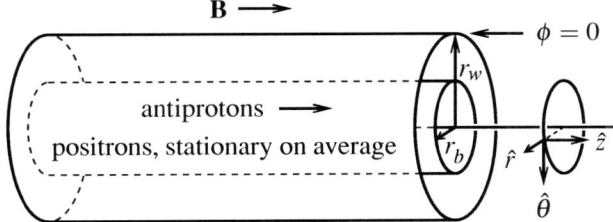

FIGURE 2. Geometry considered.

positrons and antiprotons, are coaxial with an outer, cylindrical, conducting wall, making the transverse geometry finite. The antiprotons move through the positrons, which are stationary on average. Finally, attention is restricted to radial profiles of constant density and equal radius $r_b$. The system is studied on an axial scale much smaller than the bunch lengths of the particles. Since the plasma is not electrically neutral, there is a DC radial electric field. The externally applied electrostatic potential $\phi^{\text{ext}}$ is constant, and the magnetic field $\mathbf{B}$ is constant and axial.

For the known collision operators $C_{s_1 s_2}\left[f_{s_1},f_{s_2}\right]$ [4, 5] for particles with small gyroradii and like charges, it is true that

$$m_{s_1}n_{s_1}\int d^2v_{\perp}C_{s_1 s_2}\left[f_{s_1},f_{s_2}\right] + m_{s_2}n_{s_2}\int d^2v_{\perp}C_{s_2 s_1}\left[f_{s_2},f_{s_1}\right] = 0, \quad (12)$$

so it is plausible (but not proven) that the operator which describes both reflection and Debye shielding in collisions between particles of the same species obeys (12). The symmetry relation (12) is equivalent to mass conservation at a given $v_z$. In the linear stability analysis, it happens that all the like-species collisional terms occur in pairs, as in the right side of (12), so they cancel. The remaining terms are for species with $\text{sgn}(q_1 q_2) = -1$, so in these terms we use the operator (5).

Eqs. (10) and (11) are linearized about collisionless equilibrium solutions $f_{s0}$ and $\phi_0$ which are independent of $\theta$ and $z$. The linearized equations are transformed using $\int_0^{2\pi} d\theta \int_{-\infty}^{\infty} dz \int_0^{\infty} dt\, e^{-i(\ell\theta + k_z z - \omega t)}$. The resulting dispersion relation accurate to $O\left((\ln \Lambda_{e+}/\Lambda_{e+})^2\right)$ is $D = 0$, where

$$D \equiv 1 - \frac{1}{\left(\dfrac{\lambda_{De+z}}{r_b}\right)^2 (k_r r_b)^2 + \left(k_z \lambda_{De+z}\right)^2} \int_{-\infty}^{\infty} dv_z \frac{v_{te+z}}{v_z - \dfrac{\omega - \ell\omega_E}{k_z}} \frac{\partial f^{coll}}{\partial v_z}, \tag{13}$$

$$f^{coll} \equiv f_0 + \frac{\ln\Lambda_{e+}}{\Lambda_{e+}} \frac{i}{k_z \lambda_{De+z}} \Delta f^{coll}, \qquad f_0 \equiv 2\pi v_{te+z} \sum_s \frac{\omega_{ps}^2}{\omega_{pe+}^2} \int_0^{\infty} dv_\perp v_\perp f_{s0},$$

$$\Delta f^{coll} \equiv \sum_s \frac{n_{f0}}{n_{e+0}} \frac{\omega_{ps}^2}{\omega_{pe+}^2} \frac{\dfrac{q_s}{m_s} - \dfrac{q_f}{m_f}}{\dfrac{q_s}{m_s}} \frac{m_{e+}}{m_s} v_{te+z}^3 m_{e+}$$

$$\times \left[ \frac{1}{2}\left(\frac{1}{m_s} + \frac{1}{m_f}\right) \frac{\partial}{\partial v_z}\left(\frac{\partial g_s}{\partial v_z}\frac{\partial g_f}{\partial v_z}\right) - \frac{1}{3}\frac{1}{m_s}\frac{\partial^2}{\partial v_z^2}\left(\frac{\partial g_s}{\partial v_z} g_f\right) \right],$$

and $g_s \equiv 2\pi v_{te+z} \int_0^{\infty} dv_\perp v_\perp f_{s0}$. Here $\omega_E \equiv -cE_0(r)/rB_0$ is the $\mathbf{E} \times \mathbf{B}$ rotation frequency, $v_{tsz} \equiv \sqrt{k_B T_{sz}/m_s}$ is the thermal velocity of species $s$ in the $z$ direction, $\lambda_{De+z}$ is the positron Debye length computed using $v_{e+z}$ and, in the sum, $f$ is the species which is not $s$. The finite transverse geometry of the system quantizes the values of $k_r(k_z)$, which are solutions to

$$|k_z| r_b \frac{K_\ell(|k_z|r_w)I_\ell'(|k_z|r_b) - I_\ell(|k_z|r_w)K_\ell'(|k_z|r_b)}{K_\ell(|k_z|r_w)I_\ell(|k_z|r_b) - I_\ell(|k_z|r_w)K_\ell(|k_z|r_b)} - k_r r_b \frac{J_\ell'(k_r r_b)}{J_\ell(k_r r_b)} = O\left(\frac{\omega_{ps}}{\omega_{cs}}\right), \tag{14}$$

where $I_\ell$, $J_\ell$ and $K_\ell$ are Bessel functions. The wavenumber $k_r$ is of order $1/r_b$, but $k_z$ is of order $1/\lambda_{De+z}$. The ratio $\lambda_{De+z}/r_b = 0.004 - 0.01$ is small, but its size need not be used.

## COLLISIONLESS PENROSE STABILITY CRITERION IN FINITE TRANSVERSE CYLINDRICAL GEOMETRY

When deriving the Penrose criterion [16] for the kinetic stability of a plasma, one normally encounters a stability condition for each mode. With finite transverse geometry, there is an additional term $k_r^2$ in the condition, which is

$$v_{te+z}\int_{-\infty}^{\infty}dv_z\frac{f_0(v_z)-f_0(u_0)}{(v_z-u_0)^2} < \lambda_{De+z}^2(k_r^2+k_z^2), \tag{15}$$

where $u_0$ is the central minimum of the total distribution function $f_0$. The necessary and sufficient criterion for the stability of the system is the strictest of the conditions (15), which is the one with the smallest right-hand side. Accordingly, we seek to minimize $k_r^2+k_z^2$. To do this, it is first proven that the solutions $k_r$ to (14) are monotonically

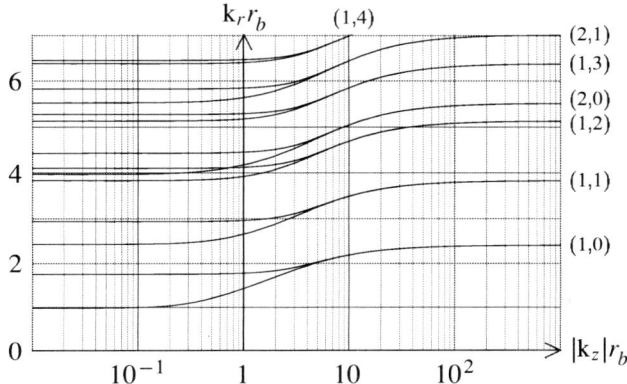

**FIGURE 3. Solutions $k_r(k_z)$ to Eq. (14).** For a given radial mode labeled $(n,\ell)$, the lower and upper branches are for $r_b/r_w = 0.16$ (ATHENA's value) and 0.7 (ATRAP's value), respectively. The index $n \in \{1,2,3,\cdots\}$ is defined so that $k_r$ increases with $n$.

increasing in $|k_z|$. Second, it is proven that $k_r$ increases with $\ell$. The index $n$ is defined so that $k_r$ increases with $n$. Thus, when $n=1$, $\ell=0$ and $k_z=0$, the right-hand side of (15) is minimized, and the radial equation (14) becomes

$$k_r r_b\frac{J_0'(k_r r_b)}{J_0(k_r r_b)} = \left(\ln\frac{r_b}{r_w}\right)^{-1}. \tag{16}$$

Accordingly, the *necessary and sufficient* condition for the stability of the system is

$$v_{te+z}\int_{-\infty}^{\infty}dv_z\frac{f_0(v_z)-f_0(u_0)}{(v_z-u_0)^2} < \left(k_r\lambda_{De+z}\right)^2, \tag{17}$$

where $k_r$ is the least solution ($n=1$) to Eq. (16). This result is applicable to problems outside of antihydrogen recombination.

## STABILITY OF DRIFTING MAXWELLIAN DISTRIBUTIONS

We now specialize this analysis to drifting Maxwellian unperturbed distribution functions $f_{s0} = \left(h_s(v_\perp)/v_{tsz}\sqrt{2\pi}\right)\exp\left[-(v_z-u_{sz})^2/2v_{tsz}^2\right]$, where $u_{sz}$ is the mean velocity of species $s$ in the $z$ direction, and $h_s(v_\perp)$ are arbitrary but normalized according to $2\pi\int_0^{\infty}dv_\perp v_\perp h_s(v_\perp) = 1$. The resulting collisionless stability boundary is shown in Figure 4.

169

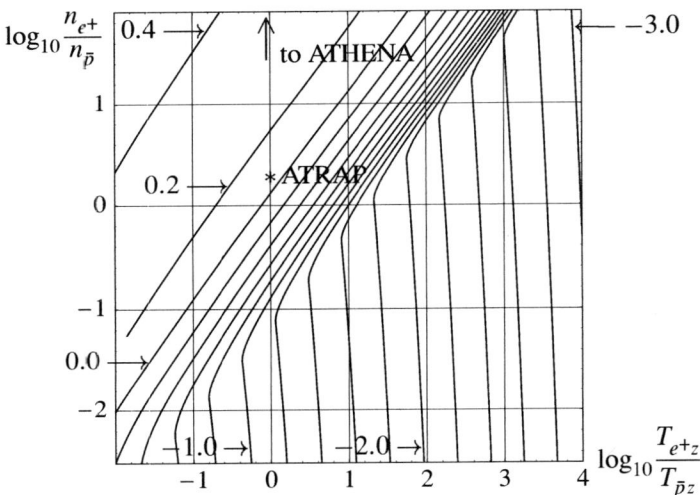

**FIGURE 4. The collisionless stability boundary.** This is a contour plot of $\log_{10}\left(u_c/v_{te+z}\right)$ for $\lambda_{De+z}/r_b \ll 1$, where $u_c$ is the critical value of the mixing velocity below which the system is stable. To compute the critical antiproton release voltage for stability, susbtitute the value of $u_c/v_{te+z}$ obtained from this plot into the formula $\Delta V = (0.08\,\text{V})\left(T_{e+z}/1\,\text{K}\right)\left(u_c/v_{te+z}\right)^2$. For typical mixing parameters, the critical voltages are of the same size as those used by the experimentalists [12, 13].

The effect of mixing velocity on collisionless growth rate is shown in Figure 5. As

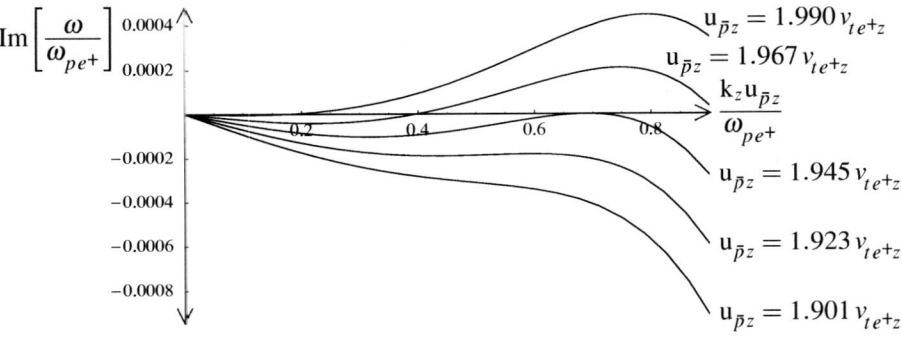

**FIGURE 5. The effect of mixing velocity $u_{\bar{p}z}$ on collisionless stability.** This is a plot of the growth rate of the instability for $T_{e+z}/T_{\bar{p}z} = 1$, $n_{e+}/n_{\bar{p}} = 2$ and $\lambda_{De+z}/r_b \ll 1$. The critical mixing velocity $u_c$ for stability is $1.945\,v_{te+z}$.

the mixing velocity $u_{\bar{p}z}$ is decreased, the system becomes stable. It has similarly been found that the stability of the system increases moderately with increasing $n_{e+}/n_{\bar{p}}$ and with increasing $T_{\bar{p}z}/T_{e+z}$.

The effect of mixing velocity on collisional growth rate is shown in Figure 6. Here we

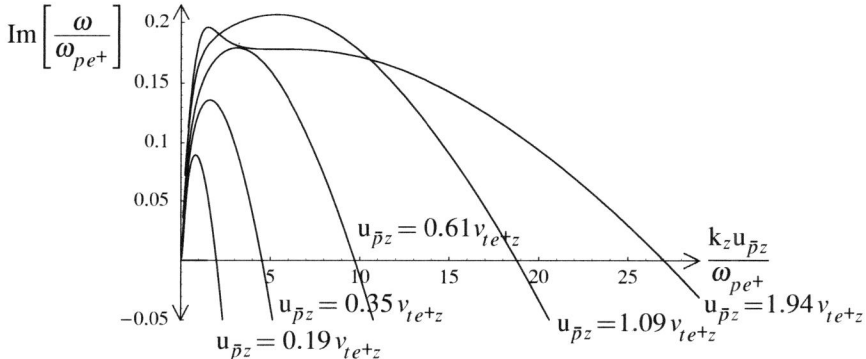

**FIGURE 6. The effect of mixing velocity $u_{\bar{p}z}$ on collisional stability.** This is a plot of the growth rate of the instability for $T_{e+z}/T_{\bar{p}z} = 1$, $n_{e+}/n_{\bar{p}} = 2$ and $\lambda_{De+z}/r_b \ll 1$. The size of the change in the frequency $\omega \sim \omega_{pe+}$ is consistent with the collisional correction given by Eq. (2).

are scanning in the mixing velocity over an order of magnitude bounded above by the mixing velocity that would have been critical without collisions. While the growth rate curve never fully dips below the real axis, the maximum growth rate still goes to zero as the mixing velocity goes to zero.

For the purpose of making antihydrogen, the most important conclusions are that the stability of the system improves with increasing $n_{e+}/n_{\bar{p}}$, with decreasing $T_{e+}/T_{\bar{p}}$, with decreasing coupling parameter $\Gamma_s$, and with decreasing mixing velocity $u_{\bar{p}z}$. Other important results are the operators (4) - (9) and the stability criterion (17).

# REFERENCES

1. Balescu, R., *Phys. Fluids*, **3**, 52 (1960).
2. Guernsey, R. L., *Phys. Fluids*, **5**, 322 (1962).
3. Lenard, A., *Ann. Phys.*, **10**, 390 (1960).
4. Dubin, D. H. E., and O'Neil, T. M., *Phys. Rev. Lett.*, **78**, 3868 (1997), the operator is implicit in Eq. (8).
5. O'Neil, T. M., *Phys. Fluids*, **26**, 2128 (1983), see Eq. (34).
6. Amoretti, M. *et al.*, *Nature*, **419**, 456 (2002).
7. Gabrielse, G. *et al.*, *Phys. Rev. Lett.*, **89**, 2134011 (2002).
8. Neuffer, D. *et al.*, *Nucl. Instr. and Meth. in Phys. Res.*, **A321**, 1 (1992).
9. O'Neil, T. M., *Phys. Fluids*, **26**, 2128 (1983), see § II for like-species results.
10. Dubin, D. H. E., and O'Neil, T. M., *Phys. Rev. Lett.*, **78**, 3868 (1997), see Eq. (2).
11. Rostoker, N., *Phys. Fluids*, **3**, 922 (1960).
12. Jørgensen, L. V., Private communication (2003).
13. Oxley, P. K., Private communication (2003).
14. Davidson, R. C., *Physics of Nonneutral Plasmas*, Addison-Wesley, 1990, see Eq. (5.13.8), p. 271.
15. Dolliver, D. D., and Ordonez, C. A., "Self-Consistent Static Analysis of Using Nested-Well Plasma Traps for Achieving Antihydrogen Recombination," in *Nonneutral Plasma Physics III* (AIP Conference Proceedings 498), edited by J. J. Bollinger, R. L. Spencer, and R. C. Davidson, AIP, 1999, p. 65, see Figure 2(b), p. 67.
16. Penrose, O., *Phys. Fluids*, **3**, 258 (1960).

# Transfer, stacking and compression of positron plasmas under UHV conditions

D. P. van der Werf[1,*], M. Amoretti[†], G. Bonomi[**], A. Bouchta[**], P. Bowe[*], C. Carraro[†‡], C. L. Cesar[§], M. Charlton[*], M. Doser[**], V. Filippini[¶], A. Fontana[¶‖], M. C. Fujiwara[††‡‡], R. Funakoshi[††], P. Genova[¶‖], J. S. Hangst[§§], R. S. Hayano[††], L. V. Jørgensen[*], V. Lagomarsino[†‡], R. Landua[**], E. Lodi Rizzini[¶], M. Macri[†], N. Madsen[§§], G. Manuzio[†‡], P. Montagna[¶‖], H. Pruys[¶¶], C. Regenfus[¶¶], A. Rotondi[¶‖], G. Testera[†] and A. Variola[†]

[*]Department of Physics, University of Wales Swansea, Swansea SA2 8PP, United Kingdom
[†]Istituto Nazionale di Fisica Nucleare, Sezione di Genova, I-16146 Genova, Italy
[**]EP Division, CERN, Geneva, Switzerland
[‡]Dipartimento di Fisica di Genova, I-16146 Genova, Italy
[§]Instituto de Fisica, Universidade do Brasil, Rio de Janeiro 21945-970, Brazil and CEFET-CE, Fortaleza 60040-531, Brazil
[¶]Istituto Nazionale di Fisica Nucleare Sezione di Pavia, I-27100 Pavia, Italy
[‖]Dipartimento di Fisica Nucleare e Teorica, Universita' di Pavia, I-27100 Pavia, Italy
[††]Department of Physics, University of Tokyo, Tokyo 153-8902, Japan
[‡‡]Atomic Physics Laboratory, RIKEN, Saitama 351-0198, Japan
[§§]Department of Physics and Astronomy, University of Aarhus, DK-8000 Aarhus C, Denmark
[¶¶]Physik-Institut, Zürich University, CH-8057 Zürich, Switzerland

**Abstract.** A ballistic method is presented for transferring positron plasmas emanating from a region with a low magnetic field and relatively high pressure into a 15 K Penning-Malmberg trap immersed in a 3 T magnetic field with a base pressure of the order of $10^{-13}$ mbar. Subsequent stacking resulted in a plasma containing $4.2 \times 10^8$ positrons. Using a rotating wall electric field a plasma containing 90 million positrons was compressed to a density of $3.6 \times 10^9$ cm$^{-3}$.

## INTRODUCTION

Recently, large amounts of cold antihydrogen atoms have been produced by the ATHENA collaboration [1] at the CERN Antiproton Decelerator. Subsequently, a similar result was reported by the ATRAP collaboration [2]. In both experiments anti-hydrogen atoms are formed by mixing antiprotons and positrons in a nested Penning trap [3]. The expected reaction mechanisms are radiative and 3-body combination [4], the reaction rates being proportional to the positron density, $n$, and $n^2$, respectively. In order to rapidly acquire large numbers of positrons to mix with antiprotons we have constructed a positron accumulator utilising nitrogen as a buffer gas [5, 6]. This type of accumulator currently has the highest reported trapping efficiency. The ATHENA

---

[1] Corresponding author (D.P.van.der.Werf@Swansea.ac.uk)

CP692, *Non-Neutral Plasma Physics V*, edited by M. Schauer et al.
© 2003 American Institute of Physics 0-7354-0165-9/03/$20.00

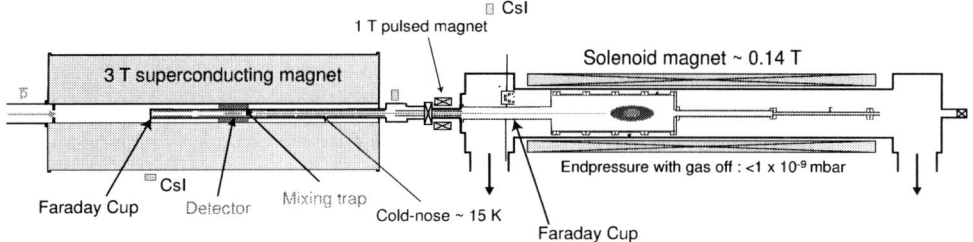

**FIGURE 1.** Schematic of the ATHENA experimental apparatus

antihydrogen apparatus [7] is designed using a modular approach. It consists of four main parts: a positron accumulator, an antiproton catching trap, a mixing trap and an antihydrogen annihilation detector as shown in Fig. 1. The accumulator, with a relatively high gas pressure, and the ultra high vacuum (lower than $10^{-13}$ mbar) mixing trap are connected by a transfer section consisting of a vacuum separation valve, a pumping restriction, a number of transfer electrodes and a pulsed transfer magnet with a field of 1 T. The transfer magnet is necessary because the magnetic field in the narrow-bore pumping restriction due to the accumulator solenoid and the superconducting magnet around the mixing region is not high enough to allow all the positrons to pass through. The number of particles caught in the mixing region can be detected destructively by dumping them onto a Faraday cup. Measures of both the total charge and the positron annihilation signal are recorded. As described below, positrons can be repeatedly stacked in the mixing region thus increasing the total number of positrons available for antihydrogen experiments. In order to increase the density of the plasma inside the 3 T solenoid it can be compressed by employing a rotating wall electric field [8]. A non-destructive diagnostic technique has recently been developed/improved using electrostatic mode analysis [9, 10] and we are now able to measure the compression in real time while using the rotating wall. In this paper we will describe the method used for the magnet-to-magnet transfer of positron plasmas and report the results of transfer, stacking and compression experiments.

# TRANSFER

## Experimental

Positrons are accumulated in a relatively low (0.14 T) magnetic field and at nitrogen buffer gas pressures of $10^{-6}$ mbar while applying a rotating wall electric field in order to compress the plasma [6]. After 200 seconds, obtaining a plasma consisting of about 150 million positrons with a diameter of 4-5 mm, the buffer gas is pumped out until a pressure of the order of $10^{-9}$ mbar has been reached. Subsequently, the vacuum separation valve is opened and the transfer magnet is energized for 1 second. The positrons in the accumulator are released by lowering the gate electrode of the accumulator (see Fig.

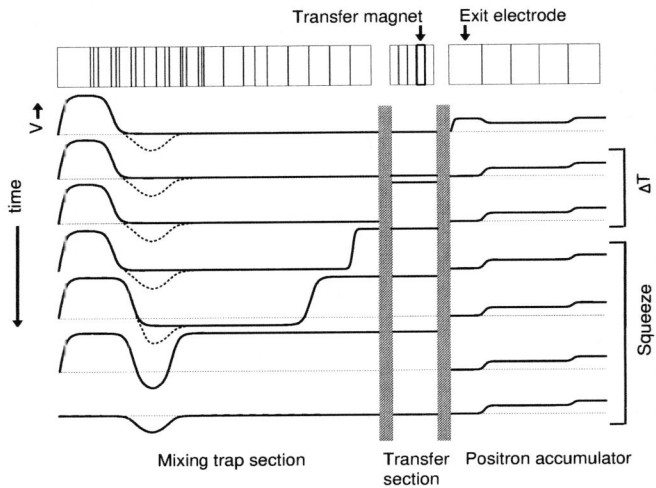

**FIGURE 2.** The potentials of the electrodes during positron transfers. Note that the time step between the subsequent potential lines is not uniform. The dashed lines represent the potentials when stacking subsequent plasmas.

2) from 140 V to 0 V with a fall time of 35 ns and retrapped in the mixing section by closing the electrodes in the transfer section a time ΔT later. The positrons are initially trapped in the entire length of the mixing trap and subsequently sqeeuzed into the central part. There they cool to the ambient temperature of 15 K by emission of synchrotron radiation. After cooling the high potential walls surrounding the positrons are lowered until the well becomes harmonic.

## Simulations

We simulated the transfer process using SIMION [11] assuming single particle trajectories confined to the axis of the instrument. The magnetic field was not taken into account. We presume the plasma to be intact just after the gate electrode has opened since the fall time of the voltage on this electrode is of the order of 1 ns over the plasma space charge of around 5 V. In order to mimick the space charge, we assign each particle a kinetic energy, randomly distributed between 0 and 5 eV, on top of the 25 eV orginating from the bottom of the well in the accumulator. We simulated trajectories for 36 values of ΔT starting at 0 with a spacing of 0.2 $\mu$s. For each point we used 2001 particle trajectories.

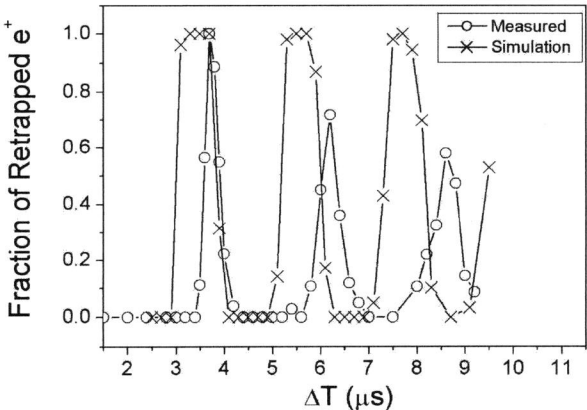

**FIGURE 3.** Fraction of retrapped positrons as a function of the time difference, ΔT, between opening the gate electrode and closing the transfer section. The curves are normalized on the first peak.

# Results

In Fig. 3 we plot the fraction of positrons transferred before squeezing as a function of ΔT for the experiment and the simulations. The experimental data were obtained by dumping the positrons on the Farady cup in the 3 T magnet 50 ms after retrapping. In both cases it is clear that the plasma can move back and forwards a number of times showing that ballistic transport is possible. The width of the recaptured peak increases due to the initial energy spread. The difference in bouncing time between the experiment and the simulation is attributed to the change of parallel (with respect to the magnetic axis) energy into perpendicular energy (not simulated) when the particles enter the 3 T field present in the mixing area. Based upon this we estimate that the particles lose about 3 eV of axial kinetic energy entering the mixing area, which corresponds to a perpendicular energy of about 0.14 eV in the 0.14 T field of the accumulator and about 20 meV in the regions of lowest magnetic field in the transfer section. The value of the perpendicular energy can most likely be attributed to imperfect alignment of the successive solenoids and/or electrodes giving the positrons an extra angular deviation. While there are no losses at the first peak in the simulation, the experimental efficiency of the transfer before squeezing is 55%. The squeeze itself gives rise to losses up to 38% giving a overall efficiency of about one third, *i.e.* we are able obtain about 50 million transferred positrons each time we transfer. These losses are not understood in detail but could also be a result of imperfect alignments.

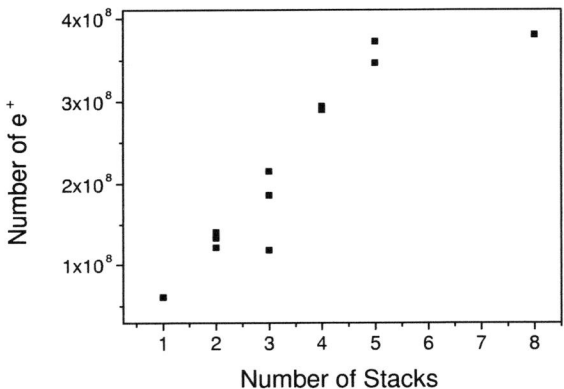

**FIGURE 4.** Number of positrons as a function of the number of stacks. Each shot contains 75 million positrons

## STACKING

Positrons have been stacked before [12] but within a homogenous magnetic field. Here the positrons transit different fields between about 0.02 and 1 T before stacked in a 3 T field. We performed a number of stacking experiments using the dashed potentials as depicted in Fig. 2 after the first shot. These potentials ensure that the positrons which have already been transferred cannot escape. The results are shown in Fig. 4 where each point has been measured by dumping the plasma onto the Faraday cup. The low point at 3 stacks is probably the result of a missed transfer in the stacking sequence. The stacking curve is linear up to 5 stacks after which it levels off. This is due to reaching the space charge limit given by the well depth. In a different set of measurements where the well depth was gradually increased we were able to obtain a plasma containing $4.2 \times 10^8$ positrons.

## COMPRESSION

After positron transfer a rotating wall electric field with a freqeuncy between 2.5 and 3.5 MHz was applied to the plasma for a duration of 200 seconds. During the process the plasma parameters were measured using mode analysis, in particular the (1.0) dipole and (2.0) quadrupole frequencies (as described in [10]), as a function of time. For a plasma containing 90 million positrons with a initial density of $3.0 \times 10^8$ cm$^{-3}$ we were able to compress the plasma by a factor of about 10 resulting in a density of about $3.6 \times 10^9$ cm$^{-3}$. This number is close to the previously reported maximum positron density of $4 \times 10^9$ cm$^{-3}$ [13] but there the plasma contained only a couple of thousand particles.

We are not yet able to reliably measure the (compressed) densities of plasmas with a larger number of positrons because of the limitation of the mode analysis detecting system and/or non-linearities in well potentials [10].

# CONCLUSIONS

We have been able to transfer positron plasmas between two solenoids with an overall efficiency of 34 %. Subsequently, stacking of a number of positron shots shows a linear behaviour until the space charge limit of the well has been reached obtaining a plasma containing 380 million positrons. Compression by a rotating wall electric field of 90 million positrons resulted in a positron plasma with a density of $3.6 \times 10^9$ cm$^{-3}$.

# ACKNOWLEDGMENTS

This work was supported in part by MEXT and RIKEN (Japan), CNPq (Brazil), SNF (Denmark), INFN (Italy), SNF (Switzerland), and the EPSRC (UK).

# REFERENCES

1.  Amoretti, M., *et al., Nature*, **419**, 456 (2002).
2.  Gabrielse, G., *et al., Phys. Rev. Lett.*, **89**, 213401 (2002).
3.  Gabrielse, G., Rolston, S. L., Haarsma, L., and Kells, W., *Phys. Lett.*, **A 129**, 38 (1988).
4.  Holzscheiter, M. H., and Charlton, M., *Rep. Prog. Phys.*, **62**, 1 (1999).
5.  Murphy, T. J., and Surko, C. M., *Phys. Rev. A*, **46**, 5696 (1992).
6.  Jørgensen, L. V., van der Werf, D. P., Watson, T. L., Charlton, M., and Collier, M. J. T., in *Nonneutral Plasma Physics IV*, edited by F. Anderegg, L. Schweikhard, and C. F. Driscoll, AIP Conference Proceedings 606, American Institute of Physics, New York, 2002, p. 35.
7.  Amoretti, *et al., submitted to NIM A (e-print archive: CERN-EP/2003-051)* (2003).
8.  Huang, X., Anderegg, F., Hollmann, E., Driscoll, C., and Neil, T. O., *Phys. Rev. Lett.*, **78**, 875 (1997).
9.  Amoretti, M., *et al., Phys. Rev. Lett.*, **91**, 55001 (2003).
10. Amoretti, M., *et al., Phys. Plasmas*, **10**, 3056 (2003).
11. Dahl, D., *Simion 3D Version 7.0*, Bechtel BWXT, Idaho, 2000.
12. Greaves, R. G., Tinkle, M. D., and Surko, C. M., *Phys. Plasmas*, **1**, 1439 (1994).
13. Jelenković, B. M., Newbury, A. S., Bollinger, J. J., Itano, W. M., and Mitchell, T. B., *Phys. Rev. A*, **67**, 063406 (2003).

# Developments in the Trapping and Accumulation of Slow Positrons using the Buffer Gas Technique

J. Clarke, D. P. van der Werf, M. Charlton, D. Beddows, B. Griffiths, and H. H. Telle

*Department of Physics, University of Wales Swansea, Singleton Park, Swansea, SA2 8PP. U.K.*

**Abstract.** A compact, two-stage positron accumulator has been designed, constructed and tested at Swansea, as part of a larger experiment to study Rydberg states of the positronium atom. It uses the well-understood nitrogen buffer gas technique to trap and accumulate positrons obtained from a Na-22 radioactive source, and slowed by a solid neon moderator. The main development this system embodies over previous accumulators is reduced physical dimension, made possible by the shorter storage times required.

## INTRODUCTION

Though positronium has been well studied since its discovery over fifty years ago, little work has been undertaken on the Rydberg states of the system in high magnetic fields. Studies to probe these states are planned at Swansea, using laser spectroscopy and other methods. To provide positrons on demand for the generation of Ps, a compact two-stage accumulator (based on the Surko three-stage technique [1]) using a buffer gas system has been constructed and preliminary tests conducted. It is designed to provide the density of trapped positrons necessary to produce $\sim 10^4$ positronium atoms when dumped on a silica powder target, at a 10 Hz frequency.

## POSITRON BEAM PRODUCTION

The beam-line at Swansea (see Fig. 1) is a direct descendant of the positron section of the ATHENA (AnTiHydrogEN Apparatus) [2,3] installed at CERN, Geneva. The beam delivered to the accumulator is generated via moderation of the particles produced during the radioactive decay of a 50 mCi Na-22 source. The moderator used is a solid neon film, held at $\sim 7$ K. It is produced by venting neon gas into the chamber, where it freezes onto the surface of a copper cone-shaped extension located on top of the source, at the end of a rod mounted on a cryogenic coldhead assembly. A particle flux of $\sim 5 \times 10^6$ s$^{-1}$, with a $\sim 3$ eV resolution FWHM, is then transported using magnetic fields to the accumulator, situated two metres away.

CP692, *Non-Neutral Plasma Physics V*, edited by M. Schauer et al.

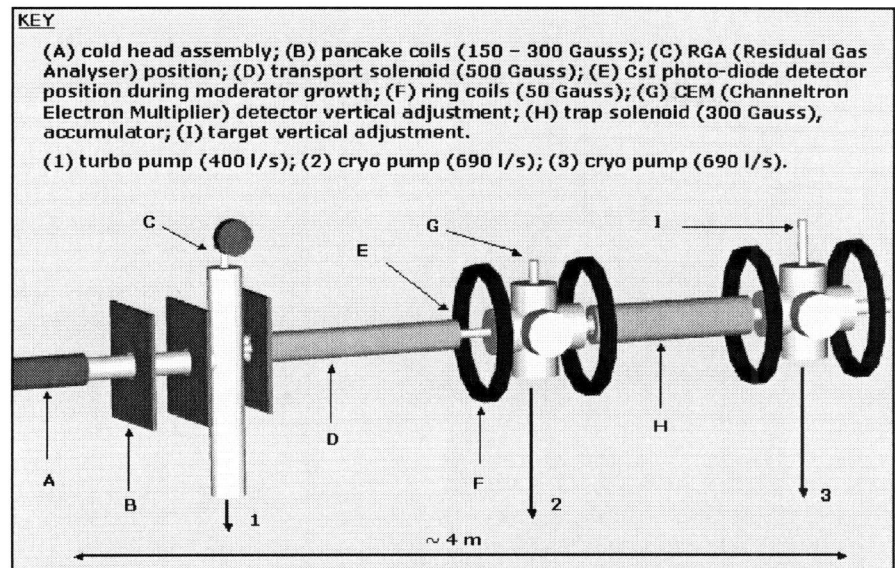

**FIGURE 1.** Simplified AutoCAD technical drawing of the positron beam-line at Swansea.

The physics involved in the energy loss mechanisms of positrons during moderation in rare gas solids has been well documented since the discovery of the method [4,5]. However, during the moderator growth cycle adopted at Swansea, unusual behaviour in the positron flux has occasionally been displayed as detected upon their annihilation by a CsI photo-diode crystal. An example is shown in Fig. 2.

Our 'recipe' for moderator growth is first to allow the cold-head assembly to warm to ~ 40 K to ensure that any solidified neon from a previous moderation cycle is removed, a process which can be monitored via a pressure gauge in the chamber. The cup is then cooled to a base temperature of ~ 7 K, and neon gas (99.999% pure at the bottle) is fed into the chamber through a computer-controlled piezo-electric valve. Early experimentation with the system showed that $5 \times 10^{-4}$ mbar was a good balance between maximizing the final slow positron flux, and the time taken for moderator growth. At this pressure, a yield of $5 \times 10^6$ s$^{-1}$ is typically reached after an hour. The piezo-electric valve is then closed, and a manual valve located between it and the chamber is also closed, to further ensure that the vacuum in the chamber is isolated. Any fast annealing cycles of the moderator to be performed are accomplished by switching off the compressor unit and allowing the conical cup to warm to the required temperature. The annealing temperature used is typically 10 K, at which point the compressor is restarted. Controlled annealing cycles can be performed if required by activating the heater elements found on the second stage of the cold-head assembly via a Lakeshore temperature controller. The latter was also used to measure all of the temperatures stated above.

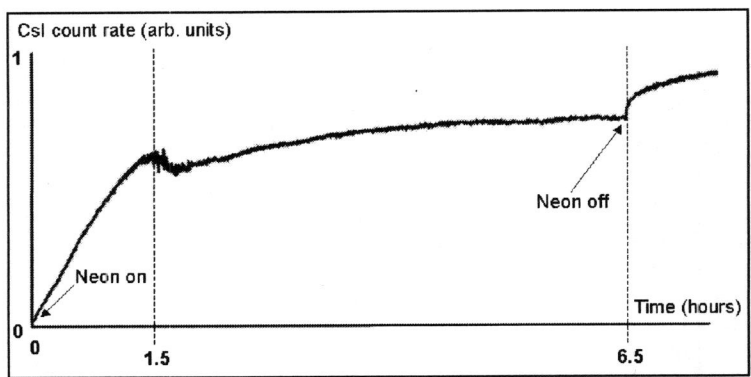

**FIGURE 2.** Evolution of positron count rate in time during moderator growth in January 2002 (neon input at $5 \times 10^{-5}$ mbar).

The unusual behaviour in Fig. 2 centred at 1.5 hours, in which the positron flux was observed to fluctuate well beyond expectations from counting statistics, was noted to be dependant on the input pressure of the neon gas, such that growing a moderator at $2.5 \times 10^{-5}$ mbar (i.e. half the pressure of that used in the example shown in Fig. 2) resulted in the same feature occurring at a time of ~ 3 hours. This suggested that the thickness of the neon film was an important factor. This behaviour was not seen however when the neon supply was replaced. Subsequent investigation revealed that the slow positron flux was highly dependant on the purity of the neon. Thus, measures were introduced to further purify the gas supply between bottle and experiment, leading to construction of a small in-line purification chamber. This chamber is filled with sorbent material, used to absorb most impurities, and cooled in a bath of liquid nitrogen to freeze out any remaining impurities. With this stage in place, anomalous features were seen again (see Fig. 3), though they do not appear to be of the same form as those observed previously. Further investigations into this phenomenon will be continued in the future, in the background to the main Ps experiment.

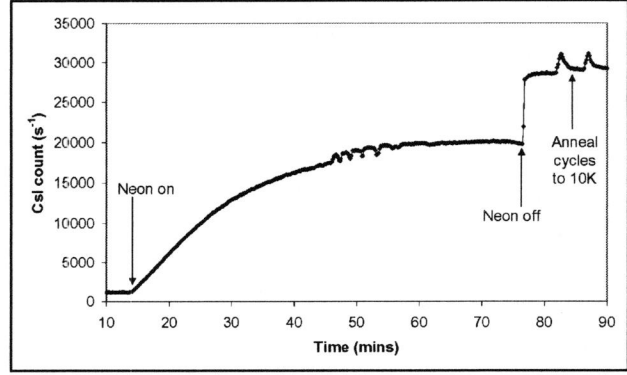

**FIGURE 3.** Evolution of positron count rate in time during moderator growth in July 2003 (neon input at $5 \times 10^{-4}$ mbar).

180

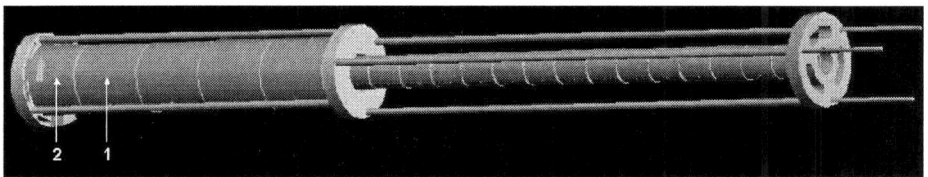

**FIGURE 4.** AutoCAD technical drawing of the electrode structure of the positron accumulator at Swansea, with electrode 1 being used to generate a potential well, and electrode 2 the restraining wall.

# ACCUMULATION

The accumulator (see Fig. 4) consists of two pressure regimes, created by differential pumping between two electrode systems with different radii. These regimes have been calculated to be ~ $10^{-3}$ mbar in the first stage, and $10^{-4} – 10^{-5}$ mbar in the second stage. The first stage consists of 15 gold-plated electrodes, of internal diameter 16 mm, and length 24 mm. These are biased to have a slight potential difference across them, of approximately 1 V, produced via a resistor chain located in an external bias box. The nitrogen buffer gas is fed into the accumulator through a bored hole in the eighth electrode in this first stage. The second stage consists of 6 electrodes, also gold-plated, of internal diameter 41 mm and length 49 mm. These are biased to create a potential well (see Fig. 5) into which the positrons may become trapped after losing energy via inelastic collisions with the nitrogen molecules. They are expelled by raising the base of the well whilst simultaneously lowering the wall at the far end of the trap. The biasing of the electrodes in the whole system is steered via LabVIEW computer control. The trapped positrons are detected by dumping them onto an annihilation target. The sharp bursts of emitted γ-rays are integrated by a CsI photo-diode detector located in proximity to the target, the signal of which, at present, is read out via an oscilloscope.

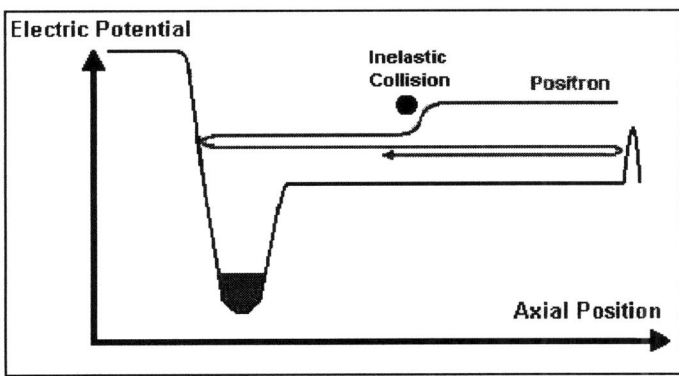

**FIGURE 5.** Schematic diagram showing the trapping process using a buffer gas approach.

# RESULTS

For our first series of measurements, only the bias of the restraining wall was switched between the trap/expel stages, from 80 V to 0 V, with the well bias being left constant. Experimentation to find the optimum bias value revealed that most positrons were trapped when the potential well was located in the second to last electrode (labelled 1 in Fig. 4), of ~ 7 V depth. The positron annihilation intensity was measured for various storage times at a number of different buffer gas pressures. An example of one such data plot is shown in Fig. 6. These plots were then fitted with the function $y = a(1 - e^{-bx})$, where the reciprocal of the variable b is the positron lifetime in the trap. The values of a and b for each pressure could be used to calculate a value for the positron accumulation rate, as shown in Fig. 7. It is important to note that the stated buffer gas pressures were not measured in the trap but instead in a neighbouring chamber.

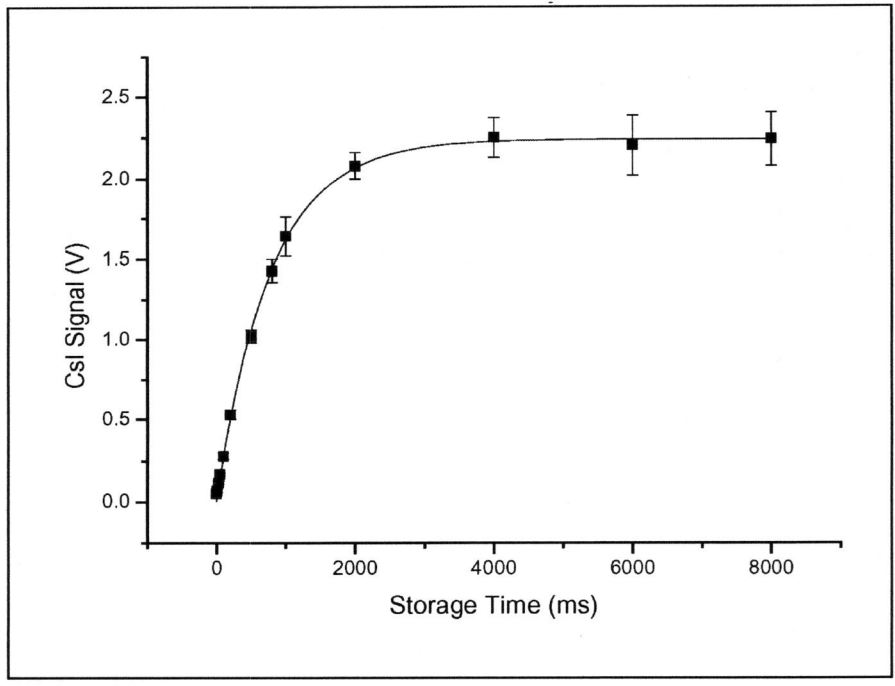

**FIGURE 6.** Relationship between storage time in the trap and the amplitude of the resulting annihilation signal, at a buffer gas pressure of $2 \times 10^{-6}$ mbar. (Fit values: a = 2.24 +/- 0.02; b = (1.29 +/- 0.03) x $10^{-3}$, see text).

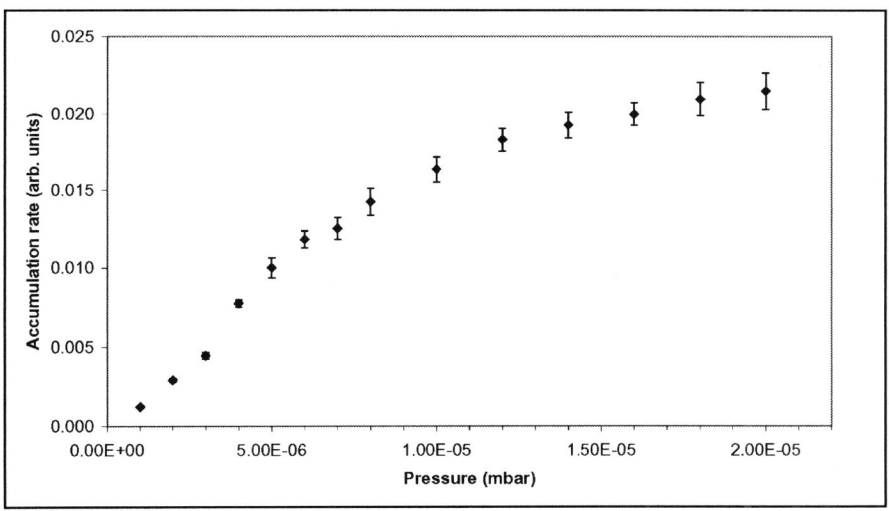

**FIGURE 7.** The variation of the positron accumulation rate in the trap with buffer gas pressure.

## CONCLUSION

A compact two-stage slow positron trap, using the buffer gas technique, has been successfully constructed, installed and is undergoing testing at Swansea. Experiments have shown that the system can accumulate positrons effectively, and operate in the 10 Hz range necessary for the Rydberg Ps experiments scheduled for late 2003. Unexpected fluctuations in moderation efficiency have been observed during the growth cycle of a solid neon film. The mechanism responsible for this phenomenon has been found to depend upon the purity of the neon used.

## REFERENCES

1.  T. J. Murphy and C. M. Surko, Phys. Rev. A 46, 5696 (1992); R. G. Greaves and C. M. Surko, Phys. Plasmas 4, 1528 (1997)
2.  M. J. T. Collier, L. V. Jørgensen, O. I. Meshkov, D. P. van der Werf and M. Charlton, in Nonneutral Plasma Physics III, (American Institute of Physics, New York, 1999), Vol. AIP 498, p. 13 (1999)
3.  D. P. van der Werf, L. V. Jørgensen, T. L. Watson, M. Charlton, M. J. T. Collier, M. Doser and R. Funakoshi, Appl. Surf. Sci. 194, 312 (2002)
4.  E. M. Gullikson and A. P. Mills, Jr., Phys. Rev. Lett. 57, 376 (1986)
5.  A. P. Mills, Jr. and E. M. Gullikson, Appl. Phys. Lett. 49, 1121 (1986)

# Long-Term Confinement of Dense Positron Plasma

Kirby J. Meyer*, Namdoo Moon*, Gerald A. Smith*, George Spalek[†], L. E. Thode*, Kevin J. VanderJack*, and Hoanh Vu*

*Positronics Research LLC, 4001 Office Court Dr. Ste. 303, Santa Fe, NM 87507
[†]Spalek Consulting, Santa Fe, NM 87501

**Abstract**. Efforts to increase the storage levels of positrons in Penning-Malmberg and other types of traps are important for many applications. Positronics Research LLC (PRLLC) is developing a storage trap that will host several positron, electron, and mixed plasma experiments intended to maximize storage conditions for positrons with acceptable lifetimes. A warm-bore 5 T cryomagnet with three independently controlled solenoids is used for experiment. The central solenoid is used for Penning-Malmberg trap studies. The end coils are used for magnetic mirror and other pinch-storage research experiments. The end potentials of the Penning-Malmberg trap will be raised to 100 kV, creating an effective well of length 100 cm. Optimization of the radio-frequency drive system, designed to control the plasma using Trivelpiece-Gould modes, and assessment of diocotron instabilities of the non-neutral plasma are being investigated using a 16-node, distributed Linux cluster.

## INCREASING POSITRON STORAGE

Positrons are the antimatter equivalent of electrons and have been successfully created and studied since the early 20th century. A positron annihilates with an electron counterpart to yield 1.02 MeV of gamma ray energy per event. Interest in confinement of unprecedented numbers of positrons for long periods of time has its origins in high-density energy storage with myriad long-term applications, including transportation, medicine and national defense. However, current trapping research is limited to perhaps $10^{10}$ positrons[1]. Various applications considered imply that the storage level be raised at least five orders of magnitude so that scaled applications experiments can take place.

The majority of positron traps are arranged in a coaxial electrode configuration, such as shown in Fig. 1(a). A set of a minimum of three electrodes is located within an axial magnetic field, and the outer electrodes are supplied a potential difference with respect to the central electrode that creates an effective 'well' for positrons to be confined axially. The magnetic field B confines the plasma radially. This configuration corresponds to a Penning-Malmberg trap, of which maximum storage number of the nonneutral plasma must obey two fundamental conditions. First, the Brillouin limit, or the ratio of the magnetic pressure to the kinetic energy of the plasma, must be much less than unity to restrict plasma diffusion in the radial direction. Second, the space charge limit, determined by the amount of stored charge over a cylindrical volume comprising the Penning-Malmberg trap, must be less than the potential difference of the electrodes to prevent positrons from leaving the system in the axial direction.

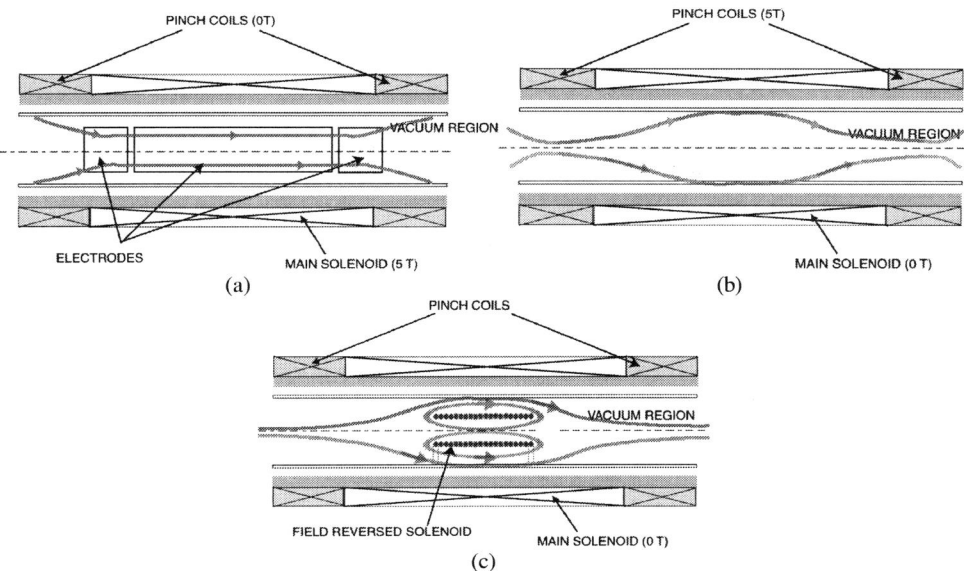

**FIGURE 1.** Positron storage configurations within a three-body magnet system: (a) Penning-Malmberg mode, with the main solenoid set to 5 T, (b) Magnetic mirror, with Pinch Coils at 5 T, (c) Field Reverse Configuration, with inclusion of FRC solenoid at lower field than pinch coils. The current in the FRC solenoid is introduced opposite to the pinch coils, creating a local region of closed field lines.

Typically, traps with axial magnetic fields of strength greater than 3 Tesla are not limited by the Brillouin limit. Indeed, it is usually the space-charge limit that restricts the total accumulation values within a Penning-Malmberg trap. Current trapping research is limited to roughly 10 kV on the end potentials. By increasing the electric potentials, as well as increasing the overall storage volume, it is possible to reach $10^{15}$ positrons within a laboratory environment.

Magnetic mirrors are the appropriate device for neutral plasmas and can defeat the space-charge limit. However, they suffer from several engineering problems and one critical physics problem. Positrons (and electrons) with large axial momentum versus gyrating momentum can leave the system in the infamous 'loss cone' [2]. The corresponding leak rate is quite high for most plasmas, and so the storage lifetimes are small. One means of bottling the loss cone for the positron species is to introduce end electrodes at an elevated potential. The less precious electrons can be resupplied from a semi-continuous cathode. In this assembly, positrons and electrons would be delivered from both ends of the mirror trap. An additional source used to neutralize the positron cloud presents unique challenges in both injection control and plasma diagnostics.

Finally, another neutralized system is the Field Reversed Configuration (FRC). Field Reversed Configurations have been used in the past to study fusion [3]. Two separate magnets create a magnetic mirror region. For fusion applications, a second set of coils within the mirror region is pulsed with a current that creates a central field that opposes the field of the mirror region. This intense current sufficiently drives a closed-loop magnetic field within the hot, neutral plasma. However, these self-induced magnetic fields reach a so-called 'tilt instability', which inhibits the plasma from reaching fusion conditions. Nevertheless, this has promising use in the positron-electron arena, for the plasmas are not hot, and the densities are lower than fusion plasmas; the required magnetic field may no longer be intense. This implies that the second set of coils or central solenoid in the FRC can be left in a constant state to force electrons and positrons to follow the field lines. Besides the magnetic pressure, the other concern for this system is the means of injecting the charged particles into a closed-loop region. Once the particles are attached to the field lines, they may be trapped indefinitely.

# EXPERIMENTAL APPARATUS

An assessment of the apparatus used to confine positrons was based primarily on the Penning-Malmberg system. Earlier systems [1] that describe a storage limit of 3 x $10^9$ positrons consisted of an axial length of approximately 60 cm with an electrode wall radius of 2.7 cm. The end potentials are set to 200 V. To increase the storage limit, the trap has a high-homogeneity magnetic field length of 1 m, followed with 100 kV potential barriers. The wall radii are increased to 3.9 cm. The space charge limit $N_b$ can be derived from the equation

$$V_s \cong 1.4 \times 10^{-7} \frac{N_b}{L} \left[ 1 + 2 \ln \left( \frac{R_w}{R_p} \right) \right],$$
(1)

where $V_s$ is the holdoff voltage, $L$ is the length of the trap, and $R_w/R_p$ is the ratio of the wall radius to the plasma radius. Letting $L = 100$ cm, $R_w/R_p = 2$, and $V_s = 100$ kV, the limit for this trap is 3 x $10^{13}$ positrons. The number density may reach 5 x $10^9$ cm$^{-3}$.

## 5 T Cryomagnet and Vacuum Equipment

To meet these geometrical conditions, Positronics Research LLC (PRLLC) has chosen a cryomagnet with a 125 cm central solenoid with a maximum field strength of 5 T shown in Fig. 2. The bore size is 7" (18 cm). Two 15.25 cm pinch coils reside adjacent to the central solenoid, creating a three-body magnet system that can be used for both magnetic mirror studies and for a pure Penning-Malmberg trap. The pinch coils have a maximum strength of 5 T, and the mirror ratio with central solenoid turned off is 10:1. While the pinch coils will not normally be active during this operation, the pinch coils can extend high field region into the injection system and extraction diagnostics should conditions warrant it. The overall length of this horizontal, warm-bore cryomagnet system is 188 cm, with a diameter of 76 cm. The cryomagnet is also sized to house additional FRC coils or solenoids in the central cavity with a strength of no larger than 1.0 T when the 5 T central solenoid is at maximum operation.

Storage lifetimes depend on the ultimate vacuum of the system. A warm-bore cryostat has the advantage of easy insertion and removal of vacuum components, but has no ability to cryo-pump gases like its cold-bore counterpart. To allow electron-electron or positron-positron collisions to dominate, the background density $n_b$ must be less than the density $n$ of the plasma. The background density is given by the formula

$$n_b = \frac{P}{kT} = 9.64 \times 10^{24} \frac{P(torr)}{T(K)}.$$
(2)

For a 298°K system, this implies a pressure less than $10^{-7}$ Torr. However, following sections show that electrons are being brought in pulses, perhaps 100 pulses in total. The first pulse implies n = 5 x $10^7$ cm$^3$. Therefore, a pressure much less than $10^{-9}$ Torr is necessary. To facilitate evacuation of the system down to these pressures, baking at 300°C and glow discharge cleaning using a 90/10 Ar/O$_2$ mixture reduces the outgassing characteristics by a factor of 14 [4]. The discharge gas is also used to strengthen the electrodes against Penning discharge when the electrodes are given large potential. The entrance of the gas feedthrough port is located opposite to the turbomolecular pump, as to provide flow through the electrode body inside the system.

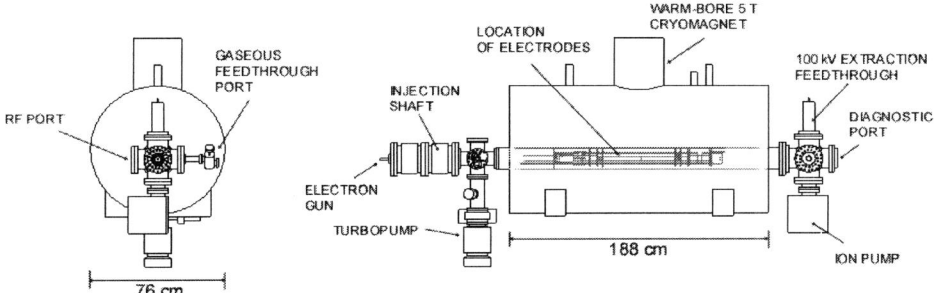

**FIGURE 2.** Vacuum and turbomachinery assembly of the PRLLC Plasma Trap (support structure of the cryomagnet and extraction diagnostics not shown).

## 100 kV Potential Well

The central electrode body shown in Fig. 3 is located at the center of the 5 T main solenoid. Three, 7.8 cm-diameter electrodes contact each other to form this body. For electron studies, these are shorted to ground ($V_0 = 0$) and have an effective length of 84 cm. Adjacent to the central body are the end electrodes. Under maximum storage conditions for electrons, these end electrodes ($V_1$ and $V_2$) are supplied –100 kV, creating an effective potential well that is 100 cm in length. This matches the high homogeneity magnetic field region of the central solenoid.

Near each end of the central body are two sets of radiofrequency (RF) ring electrodes. The RF octuplet used to drive Trivelpiece-Gould modes in the nonneutral plasma and provide safeguard against instabilities is located near the extraction end of the system. The RF quadruplet used to provide non-intrusive mode detection in the plasma (to verify plasma heating from the drive) is located near the injection end.

## INJECTION, STORAGE, AND EXTRACTION

All initial experiments will involve injection and extraction of electrons in the Penning-Malmberg configuration. Because of the large potential, an electron gun mounted near the entrance of the trap must rely on hybrid pulsing procedures to reach maximum storage limits in the system.

At the start of injection, $V_0 = 0$ V, $V_1 = -1.5$ kV, $V_2 = 0$ V. The pulser electrode described in Figure 3 is supplied –1.5 kV relative to $V_2$, the injection electrode potential.

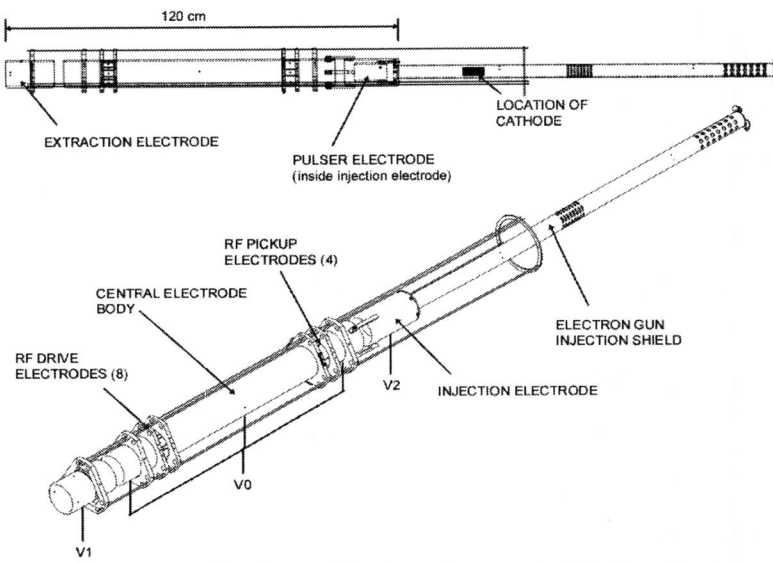

**FIGURE 3.** Electrode assembly for Electron injection and storage in Malmberg-Penning configuration. Electrons are injected from right. At ultimate storage, $V_0 = 0$ V, $V_1 = -100$ kV, $V_2 = -100$ kV.

The gun, which is optimized in the region of the diverging magnetic field lines at approximately 88 cm from the center of the trap, emits a pulse. The pulser electrode is lowered to 0 V relative, allowing the packet to enter the trap. Before the packet returns back to the cathode, the pulser is lowered once again to $-1.5$ kV relative, creating a potential well to confine the electrons. After a period of cooling, $V_1 = -3$ kV, $V_2 = -1.5$ kV, and the pulser remains $-1.5$ kV relative to $V_2$. The pulser is again brought to 0 V relative, allowing a second pulse to enter the trap before dropping to $-1.5$ kV relative. This process of pulsing and cooling continues until $V_1 = -100$ kV, where $V_2$ is lowered to $-100$ kV and the pulser electrode is switched off.

## Radiofrequency Drive and Feedback

Four of the RF drive electrodes are initially used to drive the (2,1,1) Trivelpiece-Gould mode in the plasma. This is based on the (l,m,n) conventionality, where l = 2 is the rotational mode number, m = 1 is the radial mode number, and n = 1 is the axial mode number. Because of the variation of number density during injection, the drive frequencies are transferred to a waveform generator via data file or other empirical means. Signals from the four pickup RF electrodes are interpreted through a spectrum analyzer, which will determine if the RF heating is broadcasting through the length of the plasma and also indirectly determine plasma characteristics such as its number density.

## Extraction Procedures and Diagnostics

The pickup RF electrodes provide a non-intrusive manner of detecting the plasma density within the system. Raw counts will be determined from downstream extraction diagnostics, which require destructive measurements. In the case of electrons, equipment such as Rogowski coils, Faraday cups, and Cerenkov counters are presently being considered. Such diagnostics must be accurate; secondary emission of electrons from the devices must be minimized.

The extraction electrode is ramped down sequentially upon activation of the diagnostics. Performed in steps, this provides counts at decreasing potentials, which can provide a thermal profile of the electrons within the system.

## COMPUTATIONAL VERIFICATION

Theory predicts that the long length and greater aspect ratio of this cylindrical Penning-Malmberg trap body will increase the onset of nonneutral plasma instabilities. PRLLC is able to investigate such behavior through the development of a new, 3-D Particle-In-Cell (PIC) code VPE3D operating on a multi-node Linux computer cluster. The PIC code uses cylindrical coordinate space to completely resolve the cylindrical electrodes of a Penning-Malmberg trap. Static magnetic fields and static potentials are computed separately and passed to the PIC code through external data file. This means that hybrid topics such as a nonneutral magnetic mirror can be investigated.

Verification problems are based on the diocotron instability. An example is an annular positron plasma column inside a cylindrical geometry. Results from VPE3D are compared to frequencies derived through several dispersion relationship codes. Trivelpiece-Gould mode frequencies are also currently being computed through these separate codes.

## SUMMARY

When improving positron storage limits within a laboratory system, both computational and experimental research must be performed. The 5 T Positronics Research LLC Plasma Trap employs a 1-meter, 100 kV Penning-Malmberg trap, where electrostatic breakdown and plasma instability studies are greatly emphasized. Plasma behavior is being investigated through a 3-D computational code, which may a play a greater role in trap design for future projects.

## ACKNOWLEDGMENTS

This project is sponsored through the USAF Air Force Research Laboratory, Eglin AFB, FL.

## REFERENCES

1. E. M. Hollmann, F. Anderegg, and C. F. Driscoll, *Phys of Plasmas* **7** (7), 2776 (2000).
2. H. Boehmer, M. Adams, and N. Rynn, Applied Surface Science **116**, 23-27 (1997).
3. "Proceedings of the 8th U.S. Compact Toroid Symposium and 9th U.S.-Japan Workshop on Compact Toroids," edited by A.W. DeSilva and G. C. Goldenbaum, College Park, MD, 4 June 1987.
4. M. Li and H. F. Dylla, *J. Vac. Sci. Technol.* A **13** (3), 571-575 (1995).

# SECTION 3

# BEAMS,
# STRONGLY-COUPLED PLASMAS,
# SPECIAL TOPICS

# Temperature Measurements of Laser-Cooled Ions in a Penning Trap

M. J. Jensen*, T. Hasegawa*† and J. J. Bollinger*

*National Institute of Standards and Technology, Boulder, CO 80305, USA
†Permanent address: Himeji Institute of Technology, Hyogo 678-1297, Japan

**Abstract.** Between $10^4$ and $10^6$ $^9$Be$^+$ ions are trapped in a Penning trap. The ions are laser-cooled to ~millikelvin temperatures, where they form ion crystals. This system is an example of a strongly coupled one-component plasma. By means of Doppler laser spectroscopy we have measured the temperature and heating rate of the plasma. Initially the heating rate is low, $60 \pm 40$ mK/s, but after about 100 ms the plasma heats up rapidly to a few kelvin. The onset of the rapid heating coincides with the solid-liquid phase transition.

## INTRODUCTION

Clouds of ions trapped in either Paul or Penning traps are routinely laser-cooled to temperatures sufficiently low for the cloud to enter the liquid or solid phases. These cold ion plasmas offer unique possibilities for studies of both atomic physics and strongly coupled plasma physics with related applications as diverse as cold antimatter, low-temperature chemistry, high-precision atomic and molecular spectroscopy, frequency standards, and quantum information processing [1]. Because the confinement in a Penning trap is provided by static electric and magnetic fields, the Penning trap is particularly well suited for trapping and cooling of large numbers of ions. We trap up to $\sim 10^6$ $^9$Be$^+$ ions in a Penning trap and laser-cool these ions to below 10 mK, where the ion cloud undergoes a phase transition to the solid phase and forms a crystal with an interparticle spacing of about 20 $\mu$m [2, 3].

A laser-cooled ion plasma in a Penning trap provides a clean, rigorous realization of a one-component plasma (OCP). An OCP is a system of a single species of classical point charges (ions) embedded in a uniform neutralizing background charge [4, 5, 6]. The thermodynamic state of an OCP is determined by the coupling parameter, $\Gamma$, which is defined as

$$\Gamma = \frac{1}{4\pi\varepsilon_0} \frac{q^2}{a_{WS} k_B T},$$ (1)

where $\varepsilon_0$ is the vacuum permittivity, $q$ is the ion charge, $k_B$ is Boltzmann's constant, $T$ is the temperature, and $a_{WS}$ is the Wigner-Seitz radius, given by the expression for the plasma density $n_0 = 3/(4\pi a_{WS}^3)$. $\Gamma$ is a dimensionless measure of the ratio of the potential energy between nearest-neighbor ions to the ion thermal energy. Strongly

CP692, *Non-Neutral Plasma Physics V*, edited by M. Schauer et al.
2003 American Institute of Physics 0-7354-0165-9/03/$20.00

coupled OCPs are believed to exist in dense astrophysical objects such as in the crust of a neutron star, where iron nuclei are embedded in a degenerate electron gas. They can also exist in less dense objects, however only if the temperature is correspondingly low. Cold trapped ions form a convenient low-density and low-temperature strongly coupled laboratory OCP. Indeed, it has been shown that ions trapped in a Penning trap represent a rigorous realization of an OCP where the uniform neutralizing background charge is provided by the trapping potential [6, 7]. OCPs have been subject to extensive theoretical and computational studies. For instance, the coupling parameter where the plasma undergoes the liquid-solid phase transition has been determined to high accuracy. In recent years, the calculations have converged to $\Gamma \simeq 172 - 174$ [5, 8]. There is no corresponding experimental determination of this value. In fact, the actual liquid-solid phase transition has never been directly observed in an OCP.

Due to the crossed electric and magnetic fields in a Penning trap, the trapped ion plasma will undergo a rotation about the magnetic field axis. In thermal equilibrium, this rotation is rigid [6]. At the low temperatures characterizing the present work, the plasma density $n_0$ is constant within the plasma boundary and can be expressed as $n_0 = 2\varepsilon_0 m \omega_{rot}(\Omega_c - \omega_{rot})/q^2$, where $m$ is the ion mass, $\omega_{rot}$ is the rotation frequency about the magnetic field axis, and $\Omega_c$ is the cyclotron frequency. Accordingly, the density is determined by the rotation frequency. At the plasma boundary, the density decreases from this constant value to zero over a distance comparable to the Debye length $\lambda_D = \sqrt{\varepsilon_0 k_B T / n_0 q^2}$ which for the present system is shorter than the interparticle spacing. Due to the quadratic trapping potential, the plasma is shaped as a spheroid, the aspect ratio of which is directly related to the rotation frequency $\omega_{rot}$. The stability region of the trap, in terms of rotation frequency, is $\omega_m < \omega_{rot} < \Omega_c - \omega_m$, where $\omega_m$ is the magnetron frequency.

Here, we present measurements of the temperature and heating rate of the trapped ions, i.e., the rate at which the temperature of the ions increases when no laser-cooling is applied. This work is motivated mainly by the prospects of creating many-particle entangled states. Entanglement can lead to improved spectroscopic precision. For an ensemble of non-entangled ions, the uncertainty on a given transition frequency scales as $1/\sqrt{N}$, where $N$ is the number of ions; whereas for entangled ions the scaling can be as strong as $1/N$ [9, 10]. Hence, by entangling the ions the spectroscopic precision can potentially be improved by orders of magnitude. Also, entanglement is one of the key components to quantum information processing. The creation of many-particle entangled states therefore has interesting implications for experiments on quantum simulation or even quantum computation. The main technical difference between the two is that quantum computation requires that the ions can be individually addressed with a laser beam.

A necessary condition in the proposed schemes for creating entangled states is that the ions are in the Lamb-Dicke limit while their internal and motional states are being manipulated to create the entanglement [11, 12, 13]. The Lamb-Dicke limit is the limit in which the amplitude of the ion motion in the propagation direction of the state-manipulating radiation is much less than $\lambda/2\pi$, where $\lambda$ is the radiation wavelength. This constraint is basically equivalent to imposing an upper limit to the ion temperature. In addition, it is not possible to carry out the desired state manipulations while directly laser-cooling the ions. Therefore, entanglement can be obtained only if the system is cold initially and has a heating rate low enough that the ions will remain in the Lamb-Dicke limit throughout the process. Here we do not consider the option of sympathetically cooling the ions [14].

The measurements of temperature and heating rate are not only necessary to investigate the prospects of creating entangled states, but also present an opportunity to study basic properties of OCPs, in particular the solid-liquid phase transition. It will be argued that the dominant cause of heating (due to external heat sources) is collisions with the residual gas. Since the kinetic energy of the room-temperature residual gas particles by far exceeds the kinetic energy of the trapped ions, the collision energy is the same in the solid and liquid phases. Furthermore, the collisions occur on a time scale much shorter than any time scales associated with the motional modes of the ion plasma. As a result, the energy input due to the collisions is independent of whether the plasma is in the solid or liquid phase. Hence, the energy input due to residual gas collisions is constant. Accordingly, a measurement of temperature as a function of time after turning off the laser-cooling is equivalent to measuring temperature as a function of internal energy. Because the latent heat must be supplied to melt the crystal, such a measurement may show direct evidence of the solid-liquid phase transition, in the form of a short interval where the temperature remains constant despite a continuous increase in internal energy.

## EXPERIMENT

A cylindrical Penning trap is used to confine between $10^4$ and $10^6$ $^9$Be$^+$ ions at densities on the order of $10^8$ cm$^{-3}$ [15, 16]. A schematic diagram of the setup is shown in Fig. 1. All data presented here were recorded at a pressure of $\sim 10^{-9}$ Pa. The magnetic field of 4.5 tesla is provided by a superconducting solenoid. The cyclotron frequency is $\Omega_c = 2\pi \times 7.6$ MHz, and with a typical trapping voltage $V_{trap} = 500$ V, the axial and magnetron frequencies are respectively $\omega_z = 2\pi \times 565$ kHz and $\omega_m = 2\pi \times 21$ kHz. Laser-cooling to $T < 5$ mK leads to a coupling parameter $\Gamma > 300$, and the plasma is therefore in the solid state. Cooling laser beams are sent through the center of the trap both parallel with and perpendicular to the magnetic field axis. These two cooling beams are derived from the same 313 nm beam which is produced by frequency-doubling a 626 nm beam from a dye laser. The parallel cooling beam has a wide waist of a size comparable to the diameter of the ion plasma, which is on the order of 1 mm. The waist of the perpendicular cooling beam is significantly narrower, and this beam, if offset from the center, can be used to apply a torque on the plasma and change its rotation frequency. All the diagnostics are based on fluorescence from the cooling laser beams. Real-space images of the plasma can be obtained both from the side, by use of a position-sensitive photomultiplier tube, and from the top, by use of a CCD camera equipped with an image intensifier. Phase-locked control of the rotation frequency is obtained by applying a rotating wall potential to a segmented ring electrode around the center of the trap [17]. Control of the rotation frequency implies control of the plasma density and aspect ratio.

Doppler laser-cooling was carried out on the $^2S_{1/2}$ $(m_I = +\frac{3}{2}, m_J = +\frac{1}{2}) \leftarrow {}^2P_{3/2}$ $(m_I = +\frac{3}{2}, m_J = +\frac{3}{2})$ transition (see Fig. 2). The parallel and perpendicular cooling beams

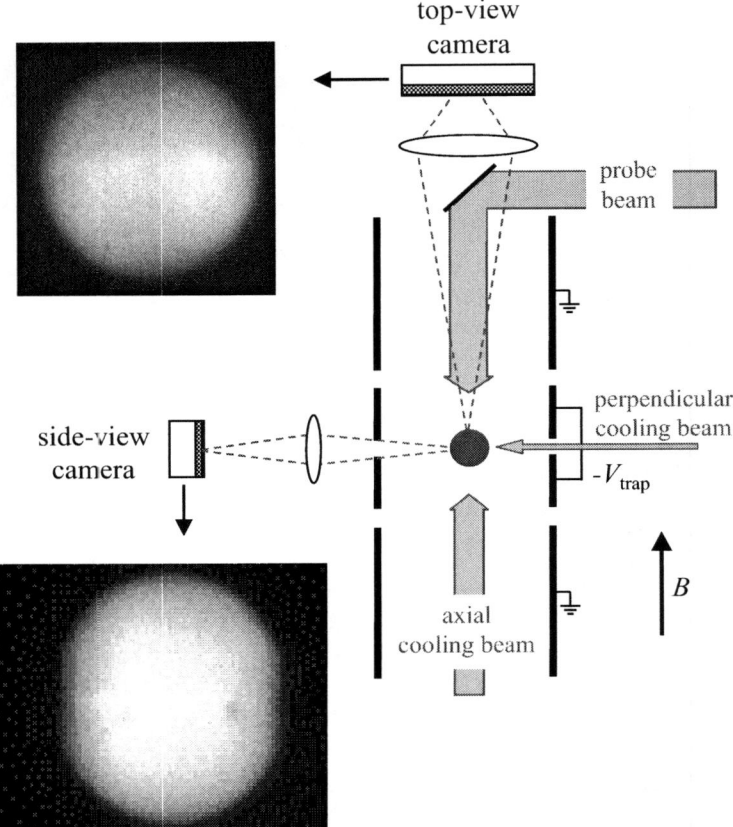

**FIGURE 1.** Schematic diagram of setup. Figure is not to scale. The trap diameter is 4 cm. The electrodes used to apply the rotating wall field are not shown. The vertical edges of the cloud visible in the side-view image are due to the presence of non-fluorescing impurity ions heavier than Be. The direction of the side-view light collection and the direction of the perpendicular cooling beam form a 60° angle in a plane perpendicular to the magnetic field axis.

cool the motion parallel and perpendicular, respectively, to the magnetic field axis. Due to the Coulomb interaction between the ions, there will be some coupling, and hence sympathetic cooling, between these two directions. The lowest achievable temperature using Doppler laser cooling, i.e., the Doppler cooling limit, is $T = \hbar\gamma/2k_B$, where $\gamma$ is the decay rate of the upper state. For the present transition $\gamma = 2\pi \times 18$ MHz, which leads to a Doppler cooling limit of 0.43 mK. This limit is reached when the laser is detuned below the resonance by the amount $\gamma/2$. The data presented here were recorded at a larger detuning, but the dependence of the cooling limit on the detuning is relatively weak at detunings larger than $\gamma/2$, and the actual cooling limit should still be below 1 mK.

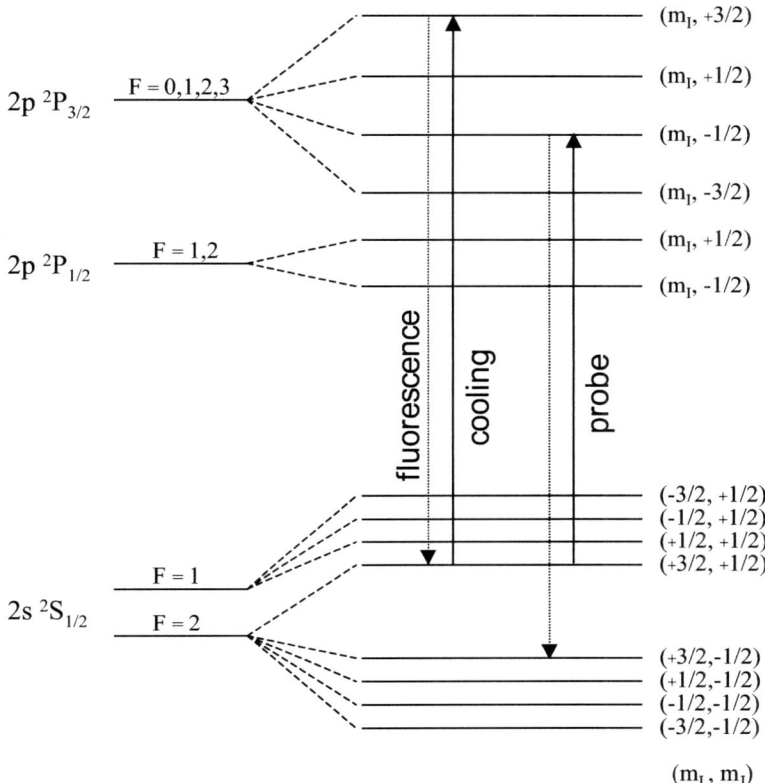

**FIGURE 2.** Energy-level diagram for $^9Be^+$ in high magnetic field (not to scale). The cooling and probe transitions are indicated. The wavelength of both laser beams is 313 nm. All diagnostics are based on the fluorescence from the cooling transition.

The temperature was probed by Doppler laser spectroscopy on the single-photon transition $^2S_{1/2}$ $(m_I = +\frac{3}{2}, m_J = +\frac{1}{2}) \leftarrow {}^2P_{3/2}$ $(m_I = +\frac{3}{2}, m_J = -\frac{1}{2})$ [18] (see Fig. 2). The probe beam was sent into the trap along the magnetic field axis, and consequently the measurement provides a determination of the axial temperature. Since the probe transition is 160 GHz below the cooling transition, a second frequency-doubled dye-laser system produces the probe beam. The upper state $^2P_{3/2}$ $(m_I = +\frac{3}{2}, m_J = -\frac{1}{2})$ decays with probability 1/3 to the $^2S_{1/2}$ $(m_I = +\frac{3}{2}, m_J = +\frac{1}{2})$ state and with probability 2/3 to the $^2S_{1/2}$ $(m_I = +\frac{3}{2}, m_J = -\frac{1}{2})$ state. Hence, the resonance can be observed as a decrease in fluorescence from the cooling laser due to a depopulation of the cooling laser cycle. The data-taking procedure is as follows. At $t = 0$, the cooling laser beams are turned off. After a desired amount of time, $t_{delay}$, the ions are exposed to a 10 ms probe laser pulse. Immediately thereafter, the cooling lasers beams are turned back on. The count rate in the side-view camera is measured before turning off ($y_{before}$) and after turning back on

($y_{\text{after}}$) the cooling beams, and the measured signal is defined as $y_{\text{after}}/y_{\text{before}}$. After such a measurement, the ions will be exposed to the cooling beams for about 25 s in order to allow for the system to fully repump into the original state. The cooling beams will slowly move the population trapped in the $^2S_{1/2}$ ($m_I = +\frac{3}{2}, m_J = -\frac{1}{2}$) state back to the cooling transition. This cycle is repeated for different probe frequencies until the line profile has been recorded. Mechanical shutters are used for turning on and off the laser beams.

The resonance is described by a Voigt profile

$$V(v) \propto \int_{-\infty}^{\infty} du \frac{e^{-u^2}}{\left[\frac{v-v_0}{\Delta v_D} - u\right]^2 + \frac{1}{4}\left[\frac{\gamma}{2\pi\Delta v_D}\right]^2}, \tag{2}$$

which is a convolution between a Lorentzian and a Gaussian. $v$ is the laser frequency, $v_0$ is the center frequency of the probe transition, and $\Delta v_D$ is the Gaussian width. The Lorentzian contribution, the magnitude of which is known from theory [19], is due to the natural linewidth of the transition, while the Gaussian contribution is due to Doppler broadening. A Voigt profile with a fixed Lorentzian width, $\Delta v_L = \gamma/2\pi = 18$ MHz, is fitted to a measured line profile and the Gaussian width $\Delta v_D$ is extracted from the fit. The temperature is found from the equation

$$\Delta v_D = \frac{v_0}{c}\sqrt{\frac{2k_BT}{m}}. \tag{3}$$

A correction is added to the fitting procedure to take into account saturation of the probe transition. The saturation is caused by the fact that the population in a given state is bound to be between 0 and 1. The depopulation rate of the monitored $^2S_{1/2}$ ($m_I = +\frac{3}{2}, m_J = +\frac{1}{2}$) state is proportional to the population in this state. Even at low probe laser power, we can approach the limit where almost all population is removed and the depopulation rate is significantly affected. This modifies the line profile. In the saturation correction, it is assumed that all ions are subject to the same degree of saturation. Because the waist of the probe beam is comparable to the size of the ion plasma, this assumption only holds true if the ions mix within the plasma boundary and thereby sample all possible probe laser powers. This is generally not the case, and certainly does not apply to the solid phase. However, a set of line profiles for very different probe laser powers were recorded and the Gaussian width extracted from the fits was independent of probe laser power. We therefore conclude that the error in analysis caused by the assumption of uniform saturation is negligible.

## RESULTS AND DISCUSSION

Fig. 3 shows two examples of measured line profiles and corresponding Voigt profile fits. The profile recorded at $t_{\text{delay}} = 0$ is, ignoring the 10 ms duration of the probe pulse, a measure of the temperature when laser-cooling the ions. From the fit we determined a temperature of $T = 1.6 \pm 0.6$ mK which is close to the Doppler-cooling limit. The

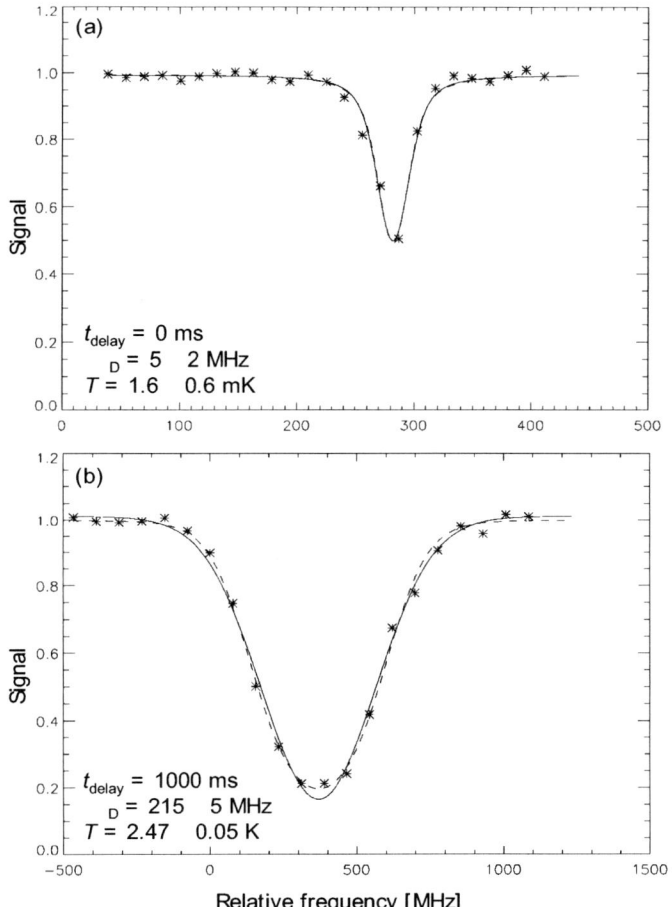

**FIGURE 3.** Line profiles recorded for a spherical plasma of 440,000 ions at (a) $t_{delay} = 0$ ms and (b) $t_{delay} = 1000$ ms. The dashed and solid lines are Voigt profile fits respectively with and without saturation correction. The Doppler widths and corresponding temperatures extracted from the fits with saturation correction are shown for comparison. For technical reasons, the frequency scales of figures (a) and (b) are shifted relative to each other. This shift is not due to an actual shift of the line center.

accuracy of this temperature measurement is limited by the fact that at the lowest temperatures the natural linewidth of the transition is significantly larger than Doppler width. At higher temperatures the Doppler width dominates (see Fig. 3(b)), resulting in much smaller relative uncertainties on the extracted temperatures.

Heating rate curves were obtained by measuring the temperature as a function of delay time $t_{delay}$ after shutting the cooling laser. Fig. 4(a) shows a heating-rate curve for a plasma of 440,000 ions. Initially, the heating rate is slow. A close-up on the first 80 ms shows that the short-time heating rate is $60 \pm 40$ mK/s, which is consistent with

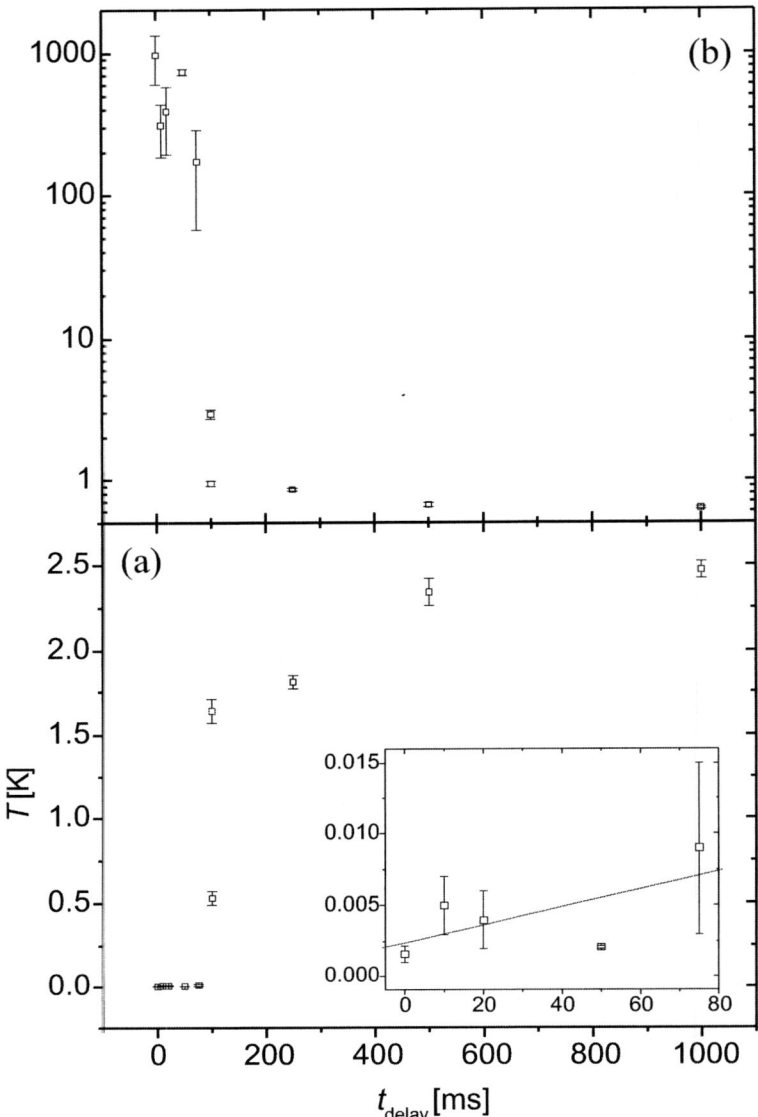

**FIGURE 4.** (a) Heating-rate curve (temperature as a function of $t_{delay}$) for a 440,000 ion cloud. A close-up of the short-time data is shown in the insert. A straight-line fit to the short-time data gives a short-time heating rate of $60 \pm 40$ mK/s. In the fit, we assumed equal weighting of all data points. (b) Same data set as in (a), but here plotted in terms of the coupling parameter, $\Gamma$, as a function of $t_{delay}$. According to theory, the solid-liquid phase transition occurs at $\Gamma \simeq 170$.

the expected heating rate due to collisions with the room-temperature residual gas. This heating rate is comparable to those observed in miniature radio-frequency traps typically used for pursuing ion-entanglement experiments, although here the source of heating is different [20]. Up to four ions have been entangled in a miniature radio-frequency trap. Here, we have the possibility of entangling many more ions. Preliminary estimates indicate that it may be possible to entangle as many as 1000 ions in the present Penning trap.

At $t_{delay} \simeq 100$ ms, the temperature increases to $T \sim 2$ K within about 100 ms. The onset of this rapid heating coincides with when the coupling parameter $\Gamma$ decreases to below 170 (see Fig. 4(b)), i.e., when the plasma reaches the solid-liquid phase transition. We therefore believe that the rapid heating is a signature of this transition. It is, however, not the signature expected from the latent heat. At present, we do not understand the rapid heating at the solid-liquid phase transition. We are investigating a number of ideas, with the goal of understanding the exact cause of this behavior.

# ACKNOWLEDGMENTS

This research was supported by the U.S. Office of Naval Research. We gratefully acknowledge Scott Robertson and Wayne M. Itano for useful comments on this manuscript.

# REFERENCES

1.  Anderegg, F., Schweikhard, L., and Driscoll, C. F., editors, *Non-Neutral Plasma Physics*, AIP Conference Proceedings 606, 2002.
2.  Itano, W. M., Bollinger, J. J., Tan, J. N., Jelenković, B., Huang, X.-P., and Wineland, D. J., *Science*, **279**, 686–689 (1998).
3.  Mitchell, T. B., Bollinger, J. J., Dubin, D. H. E., Huang, X.-P., Itano, W. M., and Baughman, R. H., *Science*, **282**, 1290 (1998).
4.  Ichimaru, S., *Rev. Mod. Phys.*, **54**, 1017 (1982).
5.  Ichimaru, S., Iyetomi, H., and Tanaka, S., *Phys. Rep.*, **149**, 91–205 (1987).
6.  Dubin, D. H. E., and O'Neil, T. M., *Rev. Mod. Phys.*, **71**, 87–172 (1999).
7.  Malmberg, J. H., and O'Neil, T. M., *Phys. Rev. Lett.*, **39**, 1333–1336 (1977).
8.  Dubin, D. H. E., *Phys. Plasmas*, **7**, 3895 (2000).
9.  Wineland, D. J., Bollinger, J. J., Itano, W. M., and Heinzen, D. J., *Phys. Rev. A*, **50**, 67–88 (1994).
10. Bollinger, J. J., Itano, W. M., Wineland, D. J., and Heinzen, D. J., *Phys. Rev. A*, **54**, R4649–R4652 (1996).
11. Cirac, J. I., and Zoller, P., *Phys. Rev. Lett.*, **74**, 4091–4094 (1995).
12. Sørensen, A., and Mølmer, K., *Phys. Rev. A*, **62**, 022311 (2000).
13. Leibfried, D., DeMarco, B., Meyer, V., Lucas, D., Barrett, M., Britton, J., Itano, W. M., Jelenković, B., Langer, C., Rosenband, T., and Wineland, D. J., *Nature*, **422**, 412–415 (2003).
14. Larson, D. J., Berquist, J. C., Bollinger, J. J., Itano, W. M., and Wineland, D. J., *Phys. Rev. Lett.*, **57**, 70–73 (1986).
15. Mitchell, T. B., Bollinger, J. J., Itano, W. M., and Dubin, D. H. E., *Phys. Rev. Lett.*, **87**, 183001 (2001).
16. Kriesel, J. M., Bollinger, J. J., Mitchell, T. B., King, L. B., and Dubin, D. H. E., *Phys. Rev. Lett.*, **88**, 125003 (2002).
17. Huang, X.-P., Bollinger, J. J., Mitchell, T. B., and Itano, W. M., *Phys. Rev. Lett.*, **80**, 73–76 (1998).

18. Brewer, L. R., Prestage, J. D., Bollinger, J. J., Itano, W. M., Larson, D. J., and Wineland, D. J., *Phys. Rev. A*, **38**, 859–873 (1988).
19. Yan, Z.-C., Tambaso, M., and Drake, G. W. F., *Phys. Rev. A*, **57**, 1652–1661 (1998).
20. Turchette, Q. A., Kielpinski, D., King, B. E., Leibfried, D., Meekhof, D. M., Myatt, C. J., Rowe, M. A., Sackett, C. A., Wood, C. S., Itano, W. M., Monroe, C., and Wineland, D. J., *Phys. Rev. A*, **61**, 063418 (2000).

# Simultaneous Trapping of Electrons and Anionic Clusters in a Penning Trap

L. Schweikhard, A. Herlert and G. Marx

*Institut für Physik, Ernst-Moritz-Arndt-Universität Greifswald, D-17487 Greifswald, Germany*

**Abstract.** The simultaneous storage of electrons and anionic clusters in a Penning trap influences the trapping conditions of the ions, e. g. it results in a shift of the motional frequencies. In addition, the overlap of the electron cloud with the stored anions leads to electron attachment to the already charged anionic clusters. The time dependence of the attachment process and the influence of the trapping potential are reviewed. These experimental results may be useful in the characterization of the stored electron ensemble.

## INTRODUCTION

Atomic and molecular clusters are particles consisting of a few up to a few hundred identical constituents, i. e. atoms or small molecules, respectively. They bridge the seemingly discrete research area of atomic and molecular physics/chemistry and the regime of condensed matter. Many gas-phase properties of clusters have been studied already and smooth transitions of these properties as well as discontinuities have been found. In particular the latter have attracted considerable interest. When the cluster size is increased by successively adding one atom after the other, magic numbers are found, i. e. cluster sizes with increased stability vs. dissociation, ionization energy, resistance against chemical reactions and the like [1].

There are two main causes for the appearance of magic numbers: (a) Geometrical compactness: When closed shells of atoms are formed, it is rather difficult to, e. g., break up the cluster and there are not many sites for further molecules to react at [2, 3]. (b) Electronic structure: In particular in the case of metal clusters, the atomic valence electrons are rather free to move in the cluster volume. The localized ionic cores of the atoms, as a whole, form a confining potential, in which the valence electrons find their states according to the Pauli principle. Similar to the case of filled shells of electrons in noble-gas atoms (and of nucleons in magic-number atomic nuclei) the clusters are particularly stable when electron shells are filled [4, 5]. The corresponding numbers differ from those of atoms since there is not just one center with positive charge (and they differ from the case of nuclei since there is no strong spin-orbit coupling).

The two causes for enhanced stability are different in nature and can be distinguished when the cluster properties are investigated not only as a function of cluster size but also as a function of charge state. An example is the investigation on small silver clusters [6, 7] where almost all increased abundances in mass spectra, i. e. enhanced stabilities, were found to be due to the electronic structure, i. e. the corresponding cluster size varied in agreement with the clusters' charge state. In contrast to the electronic structure,

CP692, *Non-Neutral Plasma Physics V*, edited by M. Schauer et al.
© 2003 American Institute of Physics 0-7354-0165-9/03/$20.00

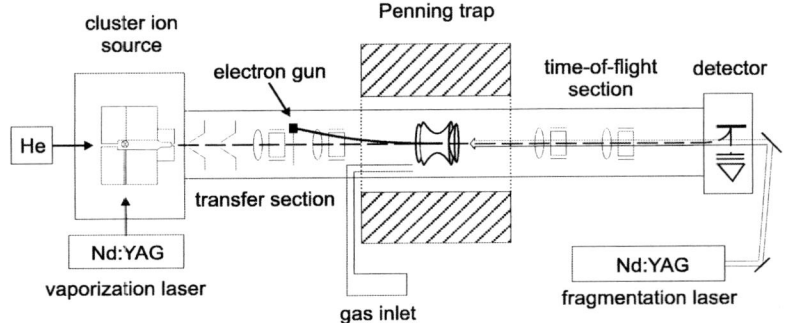

cluster ion
source

Penning trap

electron gun

time-of-flight    detector
section

He

transfer section

Nd:YAG

gas inlet

Nd:YAG

vaporization laser

fragmentation laser

**FIGURE 1.**  Overview of the experimental setup.

$Ag_{55}$ shows a strong signal independent of the charge state, i. e. its stability is due to its geometry. The number 55 is well known for the so-called Mackay icosahedra of noble gas clusters [2] (alternatively, it is also a magic number for fcc sphere packing). The silver cluster investigations were performed on cations, $Ag_n^{z+}$, $z = 1, 2, 3$, as most of the measurements on multiply charged atomic and molecular clusters.

As a matter of fact, until recently, no polyanionic metal clusters had been observed. However, in an effort to extend the cluster investigations, a method to increase the charge state of mono-anions stored in a Penning trap by attachment of further electrons has been invented [8, 9], were an electron ensemble is stored simultaneously with the clusters in a Penning trap. The ions are essentially "bathed in a sea of electrons". The yield of higher charge state ions is cluster size dependent [10] and after their formation, the polyanionic metal clusters can be further studied. The investigations include the intrinsic properties, i. e. stability and decay pathways after excitation [7, 11, 12], as well as the ions' motion, e. g. when they change their charge state [13]. These studies are in general performed after the electrons have been ejected, i. e. in a space charge free environment.

In this contribution, however, we focus on some aspects concerning the electron ensemble and its interaction with the clusters: (a) The ions can act as probes since the frequency of their cyclotron resonance is a function of the space charge. (b) Electron attachment is correlated to the kinetic energy of the electrons, which decreases as a function of time due to radiative cooling. A coupling of the different motional modes can be monitored by the study of the time dependence of the attachment process. (c) The electron attachment is also related to the depth of the trapping potential. First results show a distinct pattern in the relative abundance of produced dianions as a function of the trapping voltage. The observed dips are possibly an indication of the coupling of the electrons' and ions' magnetron and cyclotron motion in the trap.

## EXPERIMENTAL SETUP

The experimental setup has been described previously on several occasions [11, 12, 14, 15, 16, 17]. An overview is given in Fig. 1. In short, the cluster ions are produced in a laser-vaporization source and transferred by use of ion optical elements to a Penning trap with hyperbolically shaped electrodes. The ring electrode is segmented to allow the application of rf excitations. With a pulsed gas inlet system inert buffer gas (e. g. argon) may be introduced to the trap volume. In addition, an electron beam from an external electron gun can be guided through the trap. The ions (typically a few up to few tens) are detected and analyzed by single-ion counting time-of-flight mass spectrometry (TOF MS).

In case of the trapping of negatively charged particles, an additional electron bath may be created by use of argon gas that is pulsed into the trap volume and ionized with the electron gun ($E \approx 70\,\text{eV}$). The positively charged argon atoms leave the trap while the secondary electrons stay stored together with the previously captured anionic clusters. Further investigations of the clusters (not discussed in this contribution) includes, e. g., photoexcitation by use of a Nd:YAG laser [7, 12].

## RESULTS

### Ion-cyclotron-resonance frequency as a function of electronic space-charge density

The electron ensemble has not yet been further investigated, i. e. with respect to its spatial distribution, by any direct means like ejection onto a MCP, bolometric detection or the like. However, the ions can act as probes since the frequency of their cyclotron motion is a function of the space charge (of the electrons).

The clusters' (reduced) ion cyclotron frequency $v_+$ [18], is probed by dipolar radial excitation and by monitoring the number of cluster anions as a function of excitation frequency. In resonance, the anions are ejected from the trap volume which leads to a drop of the ion signal. In Fig. 2 a resonance curve for the dipolar excitation of $Au_{27}^{1-}$ is shown as an example (solid circles, $v_+ = 13367\,\text{Hz}$).

When a space charge is created, a shift $\Delta v_+$ of the ions' cyclotron frequency is observed (see Fig. 2). In addition, the resonance curves show an increased width. These preliminary results suggest that the ion cyclotron resonance frequency may be used to monitor the space charge density of the trapped electron ensemble.

The shifted frequency $\omega_{+,S} = 2\pi v_{+,S}$ of an ion with mass $M$ and charge $q$ that moves in a homogeneous and spherical charge distribution of density $n_S$ with charges $Q_S$ is given by [19]

$$\omega_{+,S} = \frac{\omega_c}{2} + \sqrt{\left(\frac{\omega_c}{2}\right)^2 - \frac{\omega_z^2}{2} - \omega_S^2} \tag{1}$$

with the cyclotron and the axial osciallation frequencies

$$\omega_c = \frac{qB}{M} \quad \text{and} \quad \omega_z = \sqrt{\frac{qU_0}{Md_0^2}}, \tag{2}$$

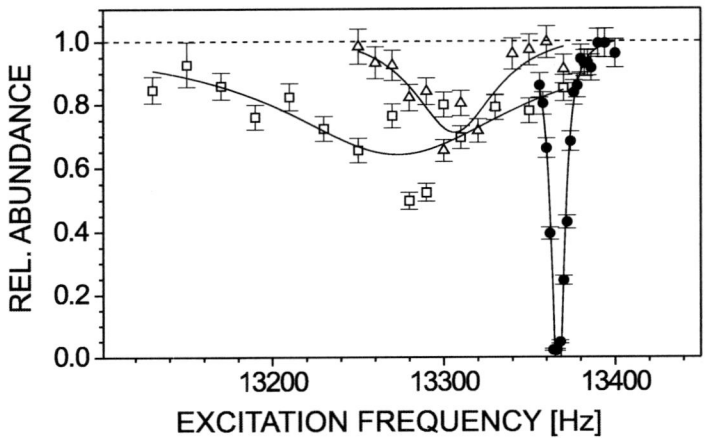

**FIGURE 2.** Relative abundance of gold clusters $Au_{27}^{1-}$ as a function of the dipole excitation frequency without (solid circles) and with electron production inside the Penning trap (open triangles and squares for 200-ms and 400-ms application of the electron beam, respectively).

respectively, and the additional term due to the space charge

$$\omega_S^2 = \frac{qQ_Sn_S}{3M\varepsilon_0}. \tag{3}$$

In the example of Fig. 2 the trap diameter is $2r_0 = 2\sqrt{2}d_0 = 40\,\text{mm}$, the ring potential $U_0 = 12.4\,\text{V}$ and the magnetic field $B = 5\,\text{T}$. From the frequency shifts of the center of the resonance curves (see Fig. 2) the charge density $n_S$ can be estimated to be about $n_S = 2.7 \times 10^5\,\text{cm}^{-3}$ (for $\Delta\nu_+ = -60\,\text{Hz}$) and $n_S = 4.2 \times 10^5\,\text{cm}^{-3}$ (for $\Delta\nu_+ = -95\,\text{Hz}$). The broadening of the resonance curves is probably due to the inhomogeneity of the space charge.

## Temporal behavior of electron attachment

Electron attachment occurs in the time range of seconds with an exponential behavior. In Fig. 3 the yield of dianionic clusters $Au_{27}^{2-}$ relative to all stored size-selected cluster anions ($Au_{27}^{1-}$ and $Au_{27}^{2-}$) is plotted as a function of the time of interaction with the electron ensemble. The open symbols denote measurements with about half the amount of gas during the application of the electron beam as in the case of the data symbolized with solid circles. The lines are exponential fits to the data, which yield constants of $\gamma = 0.50(5)\,\text{s}^{-1}$ (solid circles) and $\gamma = 0.83(11)\,\text{s}^{-1}$ (open circles).

The experimental results can be tentatively explained as follows: In order to attach further electrons to the clusters which are already negatively charged, the electrons need sufficient kinetic energy to overcome the repulsive Coulomb barrier. The initial radial (cyclotron) motion is cooled rapidly by radiation of the revolving electrons. For the

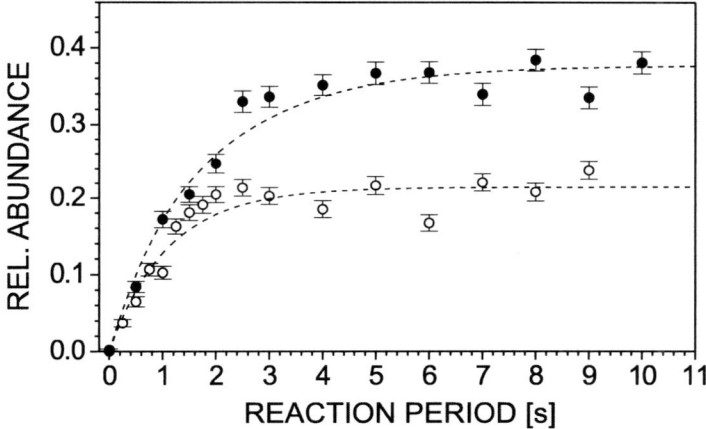

**FIGURE 3.** Relative abundance of dianions $Au_{27}^{2-}$ as a function of the reaction period in the electron bath. Solid and open symbols denote measurements for the application of 5 and 2 argon gas pulses, respectively, during the electron bombardment just before the reaction period.

present case of a 5-T magnetic field the corresponding radiative cooling rate is given by [18]:

$$\gamma_c = \frac{1}{4\pi\varepsilon_0}\frac{4e^2\omega_c^2}{3m_ec^3} \approx 10\,\mathrm{s}^{-1} \tag{4}$$

The axial motion is not cooled as fast by radiation ($\gamma_z \approx 7 \times 10^{-8}\,\mathrm{s}^{-1}$) but may be coupled to the radial motion. Thus, the experimental rates seem to be a measure of the coupling between the cyclotron and the axial modes.

## Electron attachment as a function of trapping potential

In addition to the temporal behavior a distinct pattern is found in the charge-state-conversion yield as a function of the axial trapping voltage. In Fig. 4 (a) the relative abundance of dianions $Au_{25}^{2-}$ is plotted as a function of the applied ring potential $U_0$. In a previous investigation [20] the rise of the relative yield at $U_0 \approx 5\,\mathrm{V}$ is found to be correlated to the Coulomb barrier height of the singly charged cluster. A more detailed study for values $U_0 > 12\,\mathrm{V}$ gave, at first glance, a seemingly irreproducible dependence of the relative yield of dianions from the potential well depth. A further systematic measurement reveals a characteristic structure as shown in Fig. 4 (b).

The analysis of the absolute number of anions shows that the observed dips are not the result of ion loss, e. g. of dianions. It seems, that under the given trapping conditions the electrons from the electron bath are less likely to attach to the singly charged cluster anions or that the electrons' motion is instable, i. e. they leave the trap before dianions are formed.

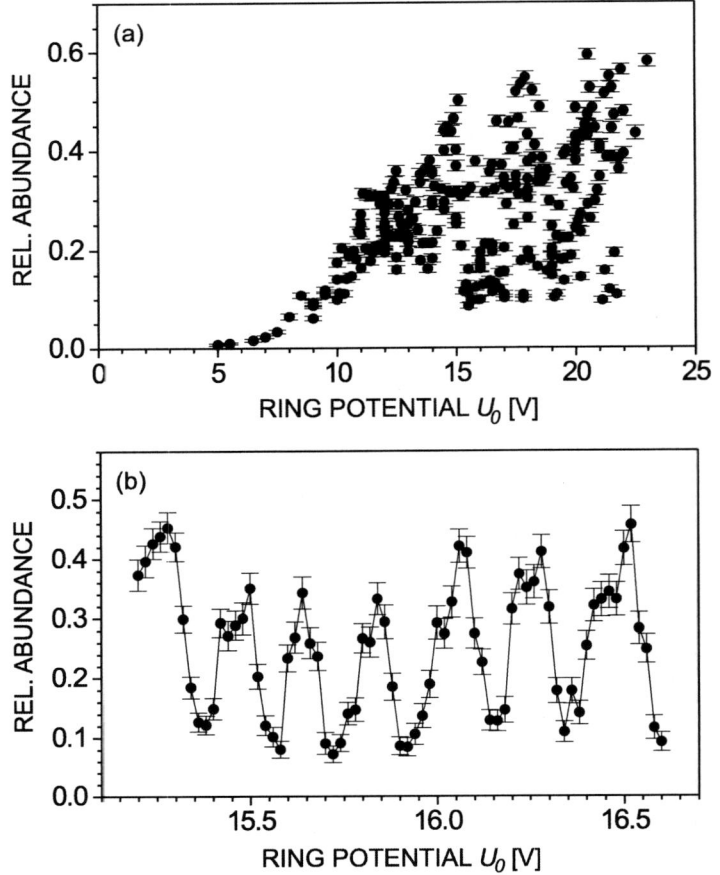

**FIGURE 4.** Relative abundance of dianions $Au_{25}^{2-}$ as a function of the applied ring potential $U_0$. (a) Full range from $U_0 = 0$ V to 25 V. (b) Detailed measurement for $U_0 = 15.1$ V to 16.7 V.

The stability of the trajectories of electrons in a Penning trap has been investigated recently by G. Werth and coworkers [21]. If the frequencies of the axial and/or the radial motional modes are related as

$$n_+ \omega_+ + n_- \omega_- + n_z \omega_z = 0 \qquad (5)$$

with $n_+, n_-$ and $n_z$ integral numbers, the electrons are lost after extended storage periods. In case of the observed dips for the electron attachment to stored clusters $Au_{25}^{1-}$ no agreement could be found. However, it seems that the attachment yield is related to the frequency ratio of the magnetron motion of electrons and the ions' cyclotron motion, possibly a coupling between the electrons' and ions' motional modes. Further investigations are in preparation, e. g. a direct detection of the electrons and monitoring their abundance as a function of storage time and trapping potential.

## CONCLUSION AND OUTLOOK

The preliminary results of this contribution have been obtained as a by-product of measurements intended to increase the yield of multiply charged metal clusters. They have not been extended as a subject of their own after this goal was achieved. Instead, we have concentrated on the intrinsic properties of the metal clusters [7, 10, 11, 12, 22]. However, it can be noted that a few ions stored simultaneously with an electron ensemble may be useful as a probe to investigate this ensemble: The interaction between ions and electrons leads to significant ion observables which allow conclusions on the electron ensemble.

The current investigations were performed on negative species. In analogy, cations may be used to investigate ensembles of stored positrons [23, 24, 25]. This should be possible for atomic ions, too, which can be converted to higher charge states by positron-electron annihilation, in contrast to the case of anions where only larger species can accommodate more than one surplus electron. Furthermore, it may be speculated as to whether ensembles of protons or other light ions, although several orders of magnitude heavier than electrons and positrons, can be probed by larger clusters, too.

## ACKNOWLEDGMENTS

This work was supported by the Deutsche Forschungsgemeinschaft and the EU network "CLUSTER COOLING".

## REFERENCES

1. de Heer, W. A., *Rev. Mod. Phys.* **65**, 611-676 (1993).
2. Echt, O., Sattler, K., and Recknagel, E., *Phys. Rev. Lett.* **47**, 1121-1124 (1981).
3. Martin, T. P., *Phys. Rep.* **273**, 199-241 (1996).
4. Knight, W. D., Clemenger, K., de Heer, W. A., Saunders, W. A., Chou, M. Y., and Cohen, M. L., *Phys. Rev. Lett.* **52**, 2141-2143 (1984).
5. Brack, M., *Rev. Mod. Phys.* **65**, 677-732 (1993).
6. Krückeberg, S., Dietrich, G., Lützenkirchen, K., Schweikhard, L., Walther, C., and Ziegler, J., *Eur. Phys. J. D* **9**, 169-172 (1999).
7. Schweikhard, L., Hansen, K., Herlert, A., Herráiz Lablanca, M. D., Marx, G., and Vogel, M., *Hyperfine Int.*, in press
8. Herlert, A., Krückeberg, S., Schweikhard, L., Vogel, M., and Walther, C., *Physica Scripta* **T80**, 200-202 (1999).
9. Herlert, A., and Schweikhard, L., *Int. J. Mass Spectrom.*, in press
10. Yannouleas, C., Landman, U., Herlert, A., and Schweikhard, L., *Phys. Rev. Lett.* **86**, 2996-2999 (2001).
11. Schweikhard, L., Hansen, K., Herlert, A., Herráiz Lablanca, M. D., Marx, G., and Vogel, M., *Int. J. Mass Spectrom.* **219**, 363-371 (2002).
12. Schweikhard, L., Hansen, K., Herlert, A., Marx, G., and Vogel, M., *Eur. Phys. J. D*, in press

13. Herlert, A., Schweikhard, L., and Vogel, M., *AIP Conf. Proc.* **606**, 652-657 (2002).
14. Schweikhard, L., Becker, St., Dasgupta, K., Dietrich, G., Kluge, H.-J., Kreisle, D., Krückeberg, S., Kuznetsov, S., Lindinger, M., Lützenkirchen, K., Obst, B., Walther, C., Weidele, H., and Ziegler, J., *Physica Scripta* **T59**, 236-243 (1995).
15. Becker, St., Dasgupta, K., Dietrich, G., Kluge, H.-J., Kuznetsov, S., Lindinger, M., Lützenkirchen, K., Schweikhard, L., and Ziegler, J., *Rev. Sci. Instrum.* **66**, 4902-4910 (1995).
16. Schweikhard, L., Krückeberg, S., Lützenkirchen, K., and Walther, C., *Eur. Phys. J. D* **9**, 15-20 (1999).
17. Schweikhard, L., Herlert, A., and Vogel, M., "Metal Clusters as Investigated in a Penning Trap," in *The Physics and Chemistry of Clusters (Proceedings of the Nobel Symposium 117)*, E. E. B. Campbell and M. Larsson (Eds.), World Scientific, Singapore, 2001, pp. 267-277.
18. Brown, L. S., and Gabrielse, G., *Rev. Mod. Phys.* **58**, 233-311 (1986).
19. Jeffries, J. B., Barlow, S. E., and Dunn, G. H., *Int. J. Mass Spectrom. Ion Processes* **54**, 169-187 (1983).
20. Herlert, A., Jertz, R., Alonso Otamendi, J., González Martínez, A. J., and Schweikhard, L., *Int. J. Mass Spectrom.* **218**, 217-225 (2002).
21. Paasche, P., Angelescu, C., Ananthamurthy, S., Biswas, D., Valenzuela, T., and ,Werth, G., *Eur. Phys. J. D* **22**, 183-188 (2003).
22. Schweikhard, L., Herlert, A., Krückeberg, S., Vogel, M., and Walther, C., *Philos. Mag. B* **79**, 1343-1352 (1999).
23. Greaves, R. G., and Surko, C. M., *Nucl. Instr. Meth. B* **192**, 90-96 (2002).
24. van der Werf, D. P., Jorgensen, L. V., Watson, T. L., Charlton, M., Collier, M. J. T., Doser, M., and Funakoshi, R., *Appl. Surf. Sci.* **194**, 312-316 (2002).
25. Gabrielse, G., Estrada, J., Tan, J. N., Yesley, P., Bowden, N. S., Oxley, P., Roach, T., Storry, C. H., Wessels, M., Tan, J., Grzonka, D., Oelert, W., Schepers, G., Sefzick, T., Breunlich, W. H., Cargnelli, M., Fuhrmann, H., King, R., Ursin, R., Zmeskal, J., Kalinowsky, H., Wesdorp, C., Walz, J., Eikema, K. S. E., and Hänsch, T. W., *Phys. Lett. B* **507**, 1-6 (2002).

# Modeling Intense Beam Propagation in the Paul Trap Simulator Experiment (PTSX)

Erik P. Gilson, Ronald C. Davidson, Philip C. Efthimion, Richard Majeski, and Edward A. Startsev

*Plasma Physics Laboratory, Princeton University, Princeton, New Jersey, 08543*

**Abstract.** The Paul Trap Simulator Experiment (PTSX) is a compact laboratory facility whose purpose is to simulate the nonlinear dynamics of intense charged particle beam propagation over large distances through an alternating-gradient magnetic transport system. The PTSX device is a 200 cm long, 20 cm diameter cylindrical Paul trap in which a 400 V, 100 kHz signal confines cesium ions to an rms radius of 1 cm. The one-component cesium plasmas can be confined for hundreds of milliseconds, which would correspond to an equivalent alternating-gradient transport system many kilometers long. The normalized intensity parameter $\hat{s} = \omega_p^2|_{r=0}/2\omega_q^2$, where $\omega_q$ is the average transverse focusing frequency, describes whether the plasma is emittance-dominated ($\hat{s} \ll 1$) or space-charge-dominated ($\hat{s} \to 1$). By increasing the amount of charge loaded into the trap, PTSX reaches values of $\hat{s} = 0.8$. Thus, the opportunity exists to study important physics topics such as: the conditions necessary for quiescent intense beam propagation over large distances, beam mismatch and envelope instabilities, collective mode excitations, the dynamics and production of halo particles, emittance growth, compression techniques, and the effects of the distribution function on stability properties. Results are presented that demonstrate the sustainment of the radial density profile over long times, and the ability of PTSX to reach large $\hat{s}$. The results of initial three-dimensional particle-in-cell simulations are also presented.

## INTRODUCTION

The Paul Trap Simulator Experiment (PTSX) is a cylindrical Paul trap whose purpose is to simulate the transverse dynamics of charged particle beams in alternating-gradient (AG) magnetic transport systems [1]. The PTSX device is able to simulate such charged particle beam transport systems because the transverse equations of motion including both the externally applied forces and the self-field forces are similar as long as the beam radius $r_b$ is small compared to the spatial periodicity of the AG system [1, 2]. This is because the magnetic forces that beam particles feel in the laboratory frame are Lorentz transformed into oscillatory electric forces in the beam frame. Because of the long confinement times of ions in PTSX relative to the oscillation frequency of the trap voltage, PTSX is capable of simulating the beam dynamics over equivalent propagation distances of many kilometers. Perhaps most importantly, the waveform of the trap voltage is controlled by an arbitrary function generator so that the waveform can be varied in order to study a wide variety of focusing field configurations.

The novel, essential feature of PTSX is its ability to study intense beams in which the space-charge is non-negligible compared to the confining force, and the normalized intensity parameter $\hat{s} = \omega_p^2|_{r=0}/2\omega_q^2$ [where $\omega_p(r)$ is the plasma frequency, and $\omega_q$ is the smooth focusing applied betatron frequency] approaches unity. Such intense beams

CP692, *Non-Neutral Plasma Physics V*, edited by M. Schauer et al.
© 2003 American Institute of Physics 0-7354-0165-9/03/$20.00

are of increasing interest due to their relevance to high energy and nuclear physics, heavy ion fusion, spallation neutron sources, tritium production, and nuclear waste transmutation [3, 4]. As new accelerator facilities use beams with increasing space-charge intensity, a basic understanding of the complex nonlinear interactions in the system becomes increasingly important. Our goal is to simulate the properties of high-intensity beams where the self-field effects can significantly alter beam equilibrium, stability and transport properties.

In this paper, a brief description of the machine is presented. A reanalysis of pre-viously presented data is shown to clarify the stability of the radial excursions of the particle trajectories. Data are presented that show the sustainment of the trapped plasma over long confinement times. Measurements demonstrating the ability of PTSX to reach intensities of $\hat{s} = 0.8$ are shown. The initial results of 3D particle-in-cell simulations that aid in the interpretation of the data and help guide experimental planning are also pre-sented. Finally, modifications that will improve the ion source performance and Faraday cup diagnostic are discussed.

## APPARATUS

References 5 and 6 contain detailed descriptions of the PTSX device, and so only the essential concepts are summarized here. The PTSX device consists of three cylindrical electrodes of radius $r_w = 10$ cm that are sliced into four 90° sectors as shown in Fig. 1. The central electrode has length $2L = 2$ m while the end electrodes are each 40 cm long. The trap confines charged particles radially by applying a periodic voltage $\pm V_0(t)$ to the four sectors, creating a ponderomotive force that points radially inwards. A dc voltage $+\hat{V}$ applied to the end electrodes confines the ions axially. The ratio $L/r_p$ is large in order to minimize finite-length effects that are outside of the analogy. When the plasma radius $r_p$ is small compared to $r_w$, then the fields from the walls are almost purely quadrupolar with corrections on the order of $(r_p/r_w)^4$. For PTSX, $V_{0\,\text{max}} \leq 400$ V, $\hat{V} = 150$ V, and the frequency $f$ of the oscillating voltage is less than 150 kHz. Presently, cesium atoms are used, although future experiments will use barium. With barium's atomic structure it will be straightforward to implement the laser-induced fluorescence (LIF) diagnostic planned for future experiments.

For the sinusoidal oscillations used to date, $\omega_q = 4eV_{0\,\text{max}}/\sqrt{2}m\pi^2 r_w^2 f$. Another im-portant system parameter is the vacuum phase advance $\sigma_v$ that describes what portion of a transverse oscillation a beam particle completes per focusing system period. Thus, the smooth-focusing vacuum phase advance is given by $\sigma_v^{sf} = \omega_q/f$. Values from 0° to beyond 90° are easily accessible by the proper choice of $V_{0\,\text{max}}$ and $f$. For the ex-periments and simulations presented here, $V_{0\,\text{max}} = 235$ V and $f = 75$ kHz, so that $\omega_q = 6.51 \times 10^4$ s$^{-1}$ and $\sigma_v^{sf} = 49.7°$.

The machine manipulates plasmas using a load—trap—dump cycle. During loading (dumping), the short electrodes on the source (diagnostic) end of the machine are made to oscillate with the same voltages $\pm V_0(t)$ as the long electrodes, which allows the ions to pass. When dumped, the ions strike a Faraday cup that has a collection aperture with a diameter, $2r_{ap} = 1$ cm. Either the charge or the average current can be measured with

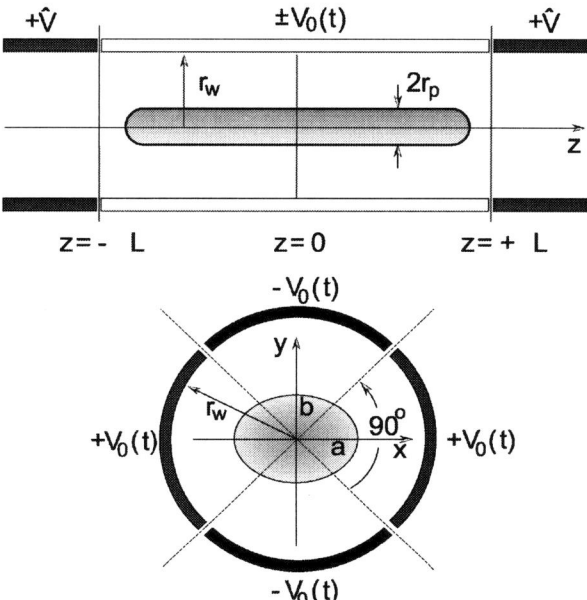

**FIGURE 1.** The PTSX device consists of three cylindrical electrodes with radius $r_w$, each sliced into four 90° sectors. An oscillating voltage $\pm V_0(t)$ confines the plasma radially to a radius $r_p$. Static voltages $+\hat{V}$ on end electrodes confine the ions axially.

an electrometer to determine the radial plasma density profile.

## EXPERIMENTAL RESULTS

Previously, it was reported that the single-particle orbits became unstable when the smooth-focusing vacuum phase advance $\sigma_v^{sf}$ exceeded 90° [7]. This claim was based on the mistaken assumption that the solutions to the Mathieu equation became unstable at a phase advance of $\sigma_v = 90°$ and that the deviation of $\sigma_v^{sf}$ from the exact vacuum phase advance $\sigma_v$ was small. However, the solutions to the Mathieu become unstable at $\sigma_v = 180°$. Further, at such large phase advances, the difference between $\sigma_v$ and $\sigma_v^{sf}$ becomes large. By extending the analysis of Davidson and Qin [3], it is found that as $\sigma_v$ approaches 180°, $\sigma_v^{sf}$ approaches 115.6° (see Fig. 2). Therefore, a more accurate description of the PTSX data in Ref. 7 is that the single-particle orbits become unstable when $\sigma_v = 180°$, or equivalently, when $\sigma_v^{sf} = 115.6°$. Since the difference between the curve for $\sigma_v^{sf} = 90°$ and $\sigma_v^{sf} = 115.6°$ is small, the error in the previously reported result is relatively small.

Radial density profiles are measured by dumping the plasma into a moveable Faraday

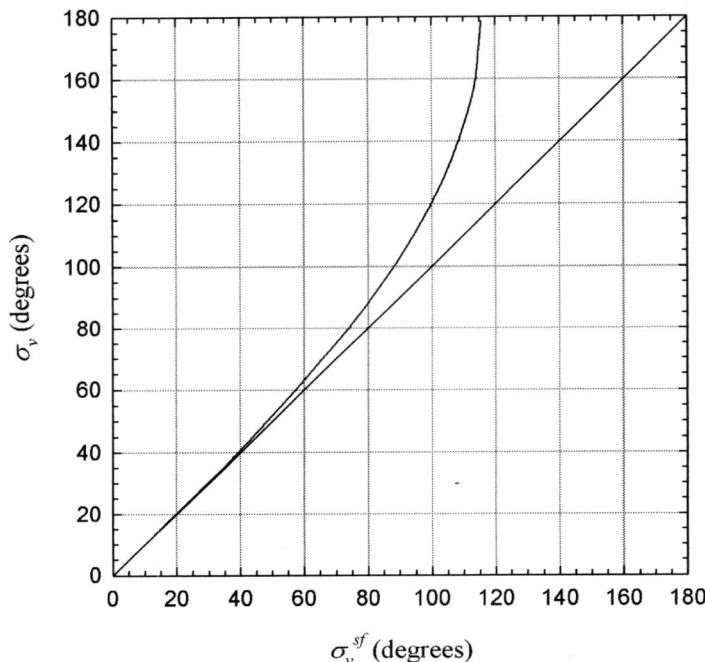

**FIGURE 2.** The exact vacuum phase advance $\sigma_v$ is obtained from numerically-determined matched-beam solutions and compared with the phase advance $\sigma_v^{sf}$ determined using $\omega_q$ as calculated in the smooth-focusing approximation [3].

cup. Then, as described in Ref. 6, $r_{ap}$ and an estimate of the length of the plasma column $L_p$ allow the measured charge profile $Q(r)$ to be recast as a density profile $n(r)$. The estimate of $L_p$ is taken from particle-in-cell simulations (see below) and the resulting uncertainty in $L_p$ contributes a systematic uncertainty to quantities such as $N$ and $\hat{s}$. Since the amount of charge contained in the plasma is small ($n \leq 10^6$ cm$^{-3}$), the charges from hundreds or even thousands of "shots" of the experiment are averaged. The averaging reduces the effect of electrometer noise and allows better charge resolution.

Measurements of the radial profile as a function of hold time were presented in Ref. 6 where it was shown that the line charge $N = \int_0^{r_w} n(r)2\pi r\,dr$ remains relatively unchanged for hundreds of milliseconds. Since the PTSX electrodes oscillate with frequencies in the tens of kilohertz range, it is seen that PTSX can simulate beam transport over equivalent distances of many kilometers. Several of the individual radial profiles used to show the constancy of $N$ are shown in Fig. 3, including profiles corresponding to hold times of 1 ms, 10 ms, and 100 ms. It is interesting to note that the profile apparently becomes somewhat taller and narrower over time, while $N$ (given by the area under radial density profile) remains relatively constant. The profiles are reasonably described by Gaussians as is expected for moderately warm plasmas in thermal equilibrium [3]. Note that an offset of 0.62 cm has been removed from the profiles in Fig. 3 in order to center the peaks about $r = 0$ cm. The origin of the offset is perhaps electrode asymmetries or distortion of the ion trajectories due to the Faraday cup.

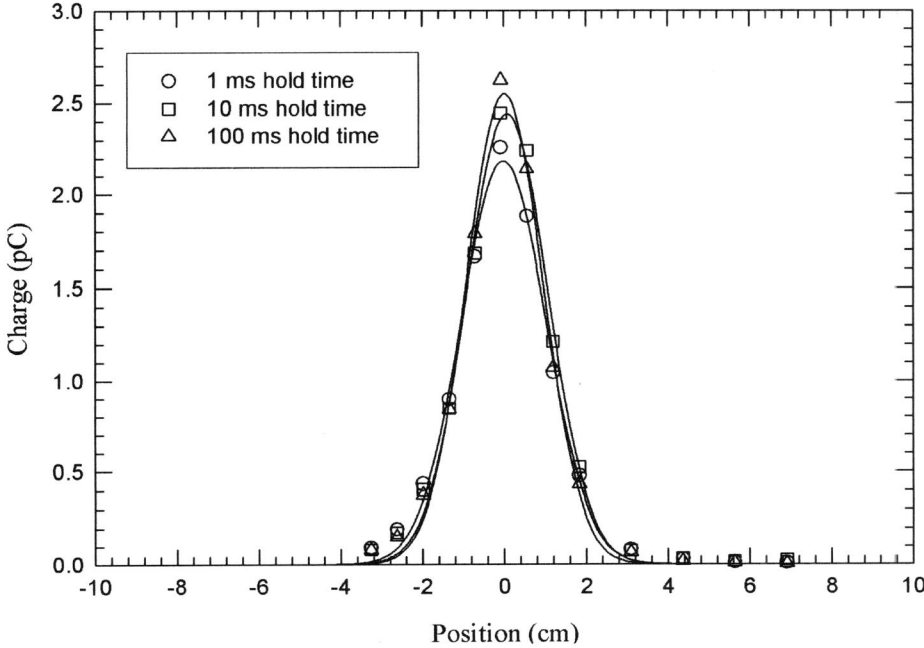

**FIGURE 3.** The radial profile of the plasma charge dumped into the Faraday cup is relatively unchanged after 100 ms. For $f = 75$ kHz and a spatial magnet period of 1 m, this corresponds to beam transport over an equivalent distance of 7.5 km.

An important feature of PTSX is the ability to perform experiments with plasmas for values of $\hat{s}$ approaching unity. By increasing the current injected into the trap, plasmas with $\hat{s} \leq 0.8$ have been observed. Figure 4 shows a radial profile for a plasma with $\hat{s} = 0.8$. Note that the apparent halo structure is likely due to measurement noise since repeated measurements do not reliably reproduce this feature. As with Fig. 3, the data in Fig. 4 have had an offset removed, 0.72 cm in this case.

## PARTICLE-IN-CELL SIMULATIONS

Three-dimensional particle-in-cell simulations employing the WARP-3D code [8] have been used to model the injection of plasmas into PTSX and investigate the properties of the trapped plasmas. The simulations include the detailed geometry of the ion source and the confinement electrodes. The code was run using a $48 \times 48 \times 512$ grid with $2.5 \times 10^5$ particles and a 0.66 $\mu$s timestep on 64 processors at NERSC. To compare a simulation to a set of experimental results, the simulation is started by adjusting the voltage on the acceleration grid to match the injection current in the simulation to that measured in the

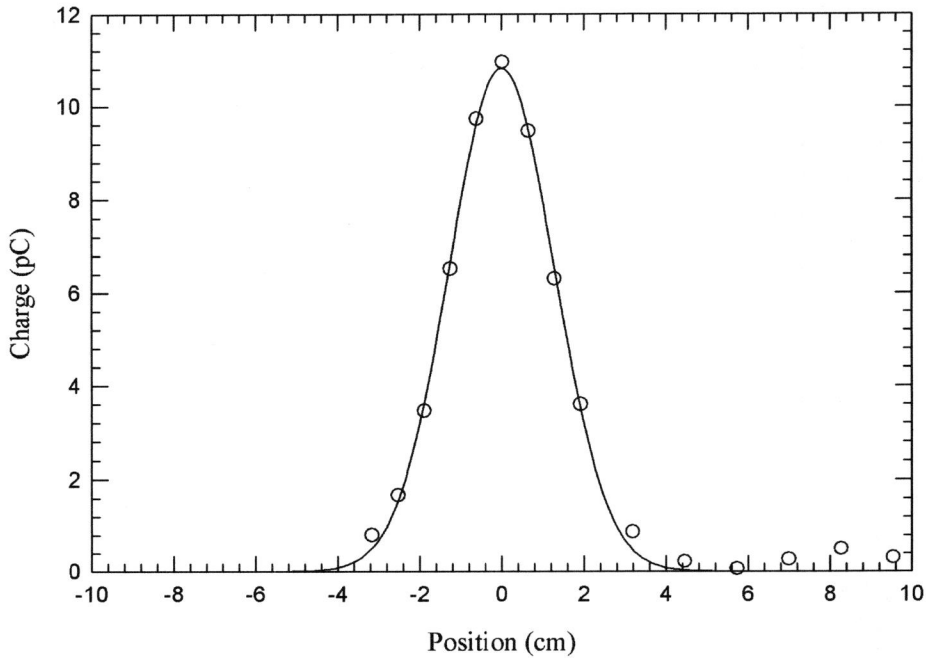

**FIGURE 4.**  By increasing the current injected into the trap, the charge collected in the Faraday cup is increased and $\hat{s} = 0.8$.

experiment. No other free parameters are necessary.

Simulations of the injection process are important because there is an inherent mismatch between the beam created at the ion source and the beam that we wish to simulate in PTSX. Matched beams have pulsating, elliptical cross sections that pulsate with the oscillation period of the transport system. The beam created at the ion source, however, has a circular cross section. Simulations confirm that this mismatch leads to some plasma heating and subsequent broadening of the plasma density profile.

Figure 5 shows the simulation particles and the equipotential surfaces in $(x, z)$ space immediately after the closure of the injection electrodes. An important result is an improved estimate of the plasma length $L_p$. To lowest order, one may assume that $L_p$ is the same as the length of the confinement electrodes, which is 200 cm. However, the axial-trapping voltages $\hat{V}$ "leak" into the central trapping region as can be seen in Fig. 5. For a variety of parameters, it is found that $L_p$ is approximately 170 cm. As evident from Fig. 5, the sudden change in the end-electrode potential in only one timestep has launched a density perturbation that propagates axially, back and forth. The Faraday cup cannot detect the presence or absence of this disturbance as it only measures $z$-averaged quantities. A simulation in which the trapping voltage is applied more gradually shows that the plasma is heated in the transverse direction rather than in the axial direction.

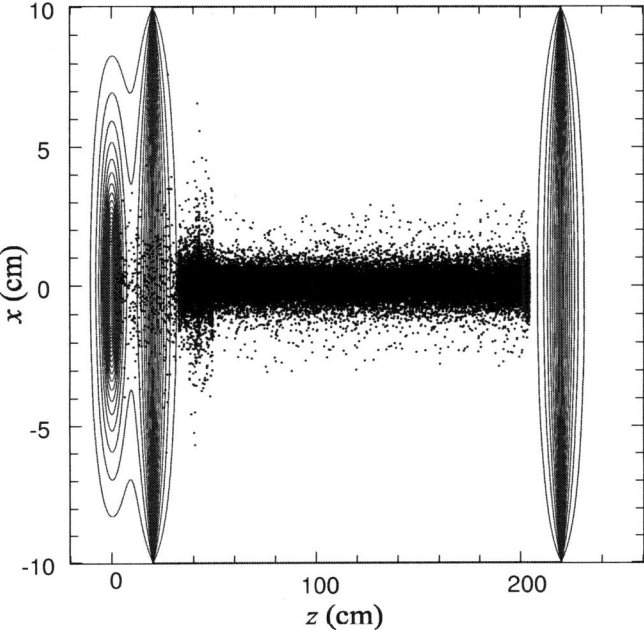

**FIGURE 5.** WARP 3D simulations showing the $(x,z)$ phase-space projection at $t = 5.033$ ms, right after the closure of the injection electrodes at $t = 5$ ms. Note that phase-space projections are generated every 50 timesteps. The equipotential contours exclude the transverse quadrupolar component and include only the self-field potential created by the ions and the applied end-electrode potentials.

## MACHINE MODIFICATIONS

The two systems that are being improved in PTSX are the ion source system and the Faraday cup system. After adding a Pierce electrode to the original ion source design to inject smaller diameter plasmas, the ion source has been redesigned to ensure a geometry that is more nearly planar and to ensure azimuthal symmetry. Deviations from azimuthal symmetry lead to electric field perturbations that increase the transverse temperature of the injected plasma. In addition to the shielding added to the original Faraday cup system, a faceplate that spans the entire PTSX cross section now provides a conductive boundary that remains fixed even as the Faraday cup moves radially.

Engineering constraints required that the overall diameter of the previous ion source installation be 1.25 inches whereas the emitting surface itself is a 0.6 inch diameter disk. This allowed a Pierce electrode and grids where the difference between the outer radius and inner radius was only 0.325 inches. Moreover, electrical leads and grounded support structures passed through the Pierce electrode, degrading the azimuthal symmetry. By redesigning the ion source to be mounted directly on the vacuum-side of a large flange, transverse size constraints are removed and all source structures now extend to a diameter of four inches. The Pierce electrode and grids now open to a diameter of 2.75 inches

(Fig. 6). The increased clearances allow for easier assembly and maintenance and for the addition of a type K thermocouple to the rear of the ion source. The thermocouple will allow the temperature and thus emission of the source to be better controlled.

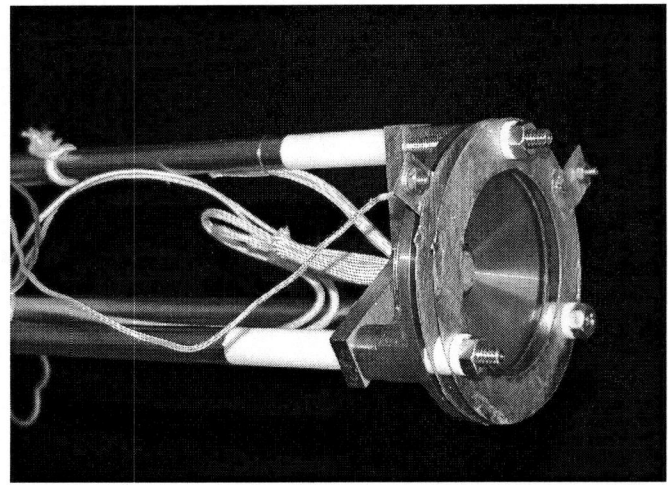

**FIGURE 6.** The ion source has an overall diameter of four inches while the grids and the conical surface of the Pierce electrode are 2.75 inches in diameter. The acceleration grid is placed 1 cm from the emitting surface and the spacing between the acceleration grid and deceleration grids is 0.5 cm.

The previous Faraday cup design resulted in a variable boundary condition that changed as the Faraday cup was moved to different radial locations. The Faraday cup enclosure was a rectangular copper box that measured 1.25 inches by four inches. Thus, measurements of the radial density profile at the origin and at large radii occurred under different experimental conditions and may not be directly comparable. To ensure that the boundary condition instead remains fixed, the end of the dump electrodes has been fitted with a pair of copper disks (Fig. 7). The upstream disk has a slot (not visible in Fig. 7) that runs radially over three-quarters of the disk's diameter. The Faraday cup and its rectangular enclosure sit behind this disk so that the dumped plasma sees an equipotential boundary regardless of the position of the Faraday cup. Because of the finite size of the vacuum chamber and the scope of the Faraday cup's linear-motion feedthrough, when the Faraday cup is at large radius, ions passing through the slot near the axis can escape. Therefore, a second disk sits behind the Faraday cup to collect stray ions. The aperture in front of the Faraday cup has been reduced in size to a diameter of 0.5 cm to improve spatial resolution. The copper plates will also provide a direct measurement of the total charge of the plasma.

## CONCLUSIONS

By demonstrating that PTSX can create and trap plasmas that are well-behaved for hundreds of milliseconds, the opportunity exists to perform experiments in which the

**FIGURE 7.** A pair of copper disks sandwiches the Faraday cup. A slot (not visible here) in the forward disk allows ions to enter the Faraday cup as the cup moves radially. The second disk acts as a backstop to catch ions that may pass through the slot but miss the Faraday cup.

voltage waveform applied to the confinement electrodes is varied to simulate a wide range of beam transport problems. Equally as important as trapping plasmas for long equivalent distances is the creation of plasmas with intensity parameter $\hat{s} = 0.8$. This allows PTSX to explore nonlinear beam phenomena in regimes where intense space-charge effects cannot be neglected. Computer simulations help to validate the PTSX results and also help to guide the design of future modifications. Improvements to the ion source and to the Faraday cup diagnostic will give finer control of the injected plasma and more precise measurements of the trapped plasma.

## ACKNOWLEDGMENTS

This work was supported by the United States Department of Energy.

## REFERENCES

1. Davidson, R. C., Qin, H., and Shvets, G., *Phys. Plasmas*, **7**, 1020 (2000).
2. Okamoto, H., and Tanaka, H., *Nucl. Instrum. and Methods A*, **437**, 178 (1999).
3. Davidson, R. C., and Qin, H., *Physics of Intense Charged Particle Beams in High Intensity Accelerators*, World Scientific, Singapore, 2001.
4. Reiser, M., *Theory and Design of Charged Particle Beams*, Wiley, New York, 1994.
5. Gilson, E. P., Davidson, R. C., Efthimion, P. C., Majeski, R., and Qin, H., *Laser and Particle Beams*, in press (2003).
6. Gilson, E. P., Davidson, R. C., Efthimion, P. C., Majeski, R., and Qin, H., *Proceedings of the 2003 Particle Accelerator Conference*, in press (2003).

7. Celata, C. M., Bieniosek, F. M., Henestroza, E., Kwan, J. W., Lee, E. P., Logan, G., Prost, L., Seidl, P. A., Vay, J.-L., Waldron, W. L., Yu, S. S., Barnard, J. J., Callahan, D. A., Cohen, R. H., Friedman, A., Grote, D. P., Lund, S. M., Molvik, A., Sharp, W. M., Westenskow, G., Davidson, R. C., Efthimion, P., Gilson, E. P., Grisham, L. R., Majeski, R., Qin, H., Startsev, E. A., Bernal, S., Cui, Y., Feldman, D., Godlove, T. F., Haber, I., Harris, J., Kishek, R. A., Li, H., O'Shea, P. G., Quinn, B., Reiser, M., Valfells, A., Walter, M., Zou, Y., Rose, D. V., and Welch, D. R., *Phys. Plasmas*, **10**, 2064 (2003).
8. Friedman, A., Grote, D. P., and Haber, I., *Phys. Fluids B* (1992).

# Coherent Structures in low Energy Electron Beams in ELTRAP

G.Bettega*, F. Cavaliere*, M.Cavenago†, F. De Luca*, A. Illiberi*, I. Kotelnikov**, C. Maroli*, V. Petrillo*, R. Pozzoli*, M. Romé*, L. Serafini* and Yu. Tsidulko**

*I.N.F.M., I.N.F.N., Dipartimento di Fisica, Università degli Studi di Milano, Milano, Italy
†I.N.F.N Laboratori Nazionali di Legnaro, Legnaro, Italy
**Budker Institute of Nuclear Physics, Novosibirsk, Russian Federation

**Abstract.** The experimental investigation of the formation of coherent structures in an electron beam of very low energy produced in a Malmberg-Penning trap is presented. A rapid development of the structures is observed when, by increasing the emission current of the cathode, a sharp transition to a space charge dominated regime occurs, where a region unaccessible to beam electrons is found around the axis, and is characterized by a sharp density gradient at the edge. The transport of the structures along the beam is studied by varying the strength of the axial magnetic field. A 3D PIC simulation code is used to interpret the results. A modification of the configuration is under way, in which a pulsed electron source is used, and plasma effects in the beam transport, of relevance for X-ray SASE-FELs, can be investigated.

## STRUCTURES IN A BEAM

We have investigated the development of coherent structures in a beam produced in a Malmberg-Penning trap, where the cylindrical electrodes are grounded and a low energy electron plasma continuously flows from the emitting thermionic cathode to a charge collector (phosphor screen), held at a fixed potential (few kV). In this case, the life time of single electrons in the trap is very short with respect to the time scale of drift motion. We have found a direct evidence of the transition to a space charge dominated regime with a rapid development of coherent structures. The experimental investigation has been performed by means of a CCD camera diagnostic, which allows to extract the density distribution of the electrons reaching the phosphor screen. Each image corresponds to stationary conditions of the system. Depending on the external parameters, a simple translation of the source image, or a fast evolution of coherent structures can be observed on the screen. Observations of structures in beams using screens for different physical conditions date back to many years [1, 2, 3]. Interesting characteristics of configurations where the electron plasma continuously flows from the source and is reflected by a potential barrier have been recently investigated [4].

The schematic of the experimental apparatus, the ELTRAP Malmberg-Penning trap [5, 6, 7], is shown in Fig. 1. The length of the beam is about $1.2m$, the uniform magnetic field $B$ can be varied in the interval $0.002 - 0.2T$. The electron source is a spiral cathode whose characteristics have been already investigated in the literature [8].

CP692, *Non-Neutral Plasma Physics V*, edited by M. Schauer et al.
© 2003 American Institute of Physics 0-7354-0165-9/03/$20.00

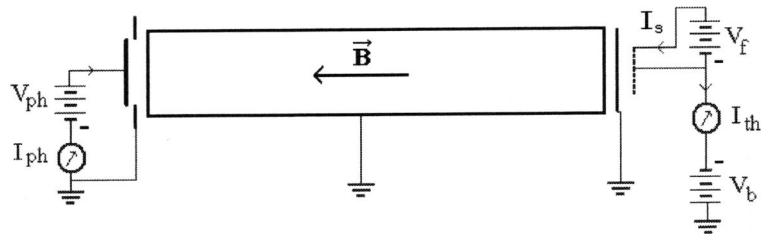

**FIGURE 1.** Schematic of the device for beam investigation

The acceleration grid is grounded; the spiral center is held at the bias potential $V_b$, the potential at the outer end of the spiral is $V_f$; the current flowing in the cathode, towards the spiral center is $I_s$ and the emission current $I_{th}$. By adjusting $V_b$ and $I_s$ different spatial distributions of the electrons reaching the grid can be obtained. Fig. 2 shows a CCD image of the phosphor screen obtained with a source emission current of $0.05mA$. When the emission current is increased up to about $1mA$ similar images are obtained, showing increasing brightness and deformation due to azimuthal rotation (Fig. 3). If the emission current is further varied of less than one percent a sharp change of the picture is observed (Fig. 4) where only two rings of the spiral are apparent, and are characterized by the presence of well developed coherent structures. The transition is characterized by a sharp drop of the electron flux reaching the phosphor screen, as shown in Fig. 5 , points $A$ and $B$. If the emission current is further increased starting from the transition and is then decreased only the low current regime is found. The beam remains in this regime also when the emission current is decreased below the transition value.

A cold fluid, drift-Poisson description of the plasma seems appropriate to the performed observations. In this model the fluid velocity is written as $\mathbf{v} = \mathbf{v}_\perp + w\mathbf{e}_z$, where $\mathbf{v}_\perp = -(c/B)\nabla_\perp\phi \times \mathbf{e}_z$, being $\mathbf{e}_z$ the unit vector parallel to $\mathbf{B}$ (along the axis of the device). For a stationary beam, the following system is obtained:

$$\mathbf{v}_\perp \cdot \nabla_\perp n + \frac{\partial}{\partial z}(nw) = 0, \tag{1}$$

$$m\mathbf{v}_\perp \cdot \nabla_\perp w + \frac{\partial}{\partial z}(mw^2/2 - e\phi) = 0, \tag{2}$$

$$\nabla^2\phi = 4\pi en. \tag{3}$$

A solution with cylindrical symmetry, i.e. with $\phi(r,z)$ independent of the polar angle $\theta$, and characterized by a purely azimuthal $\mathbf{E} \times \mathbf{B}$ drift, may exist for this system, and obeys the following equation:

222

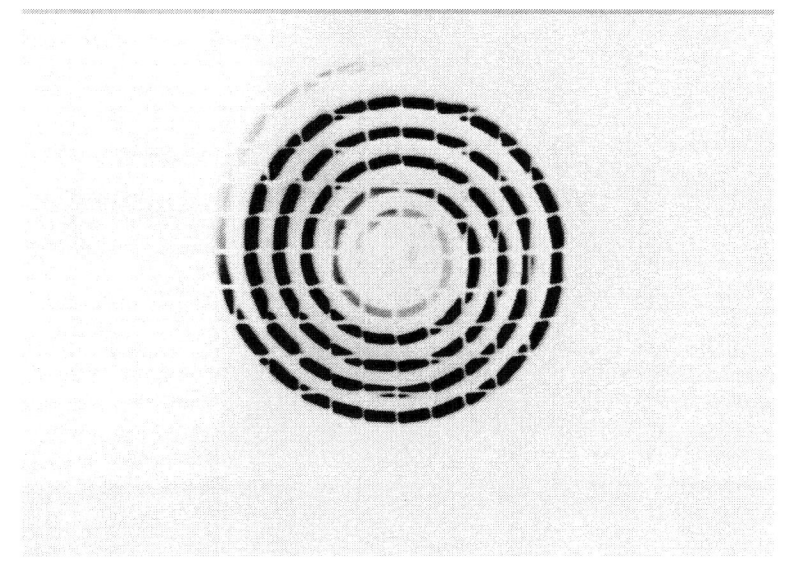

**FIGURE 2.** CCD image for a cathode emission current $I_{th} = 0.05mA$; bias potential $V_b = -10V$; $I_s < 0$; magnetic field $B = 0.066T$

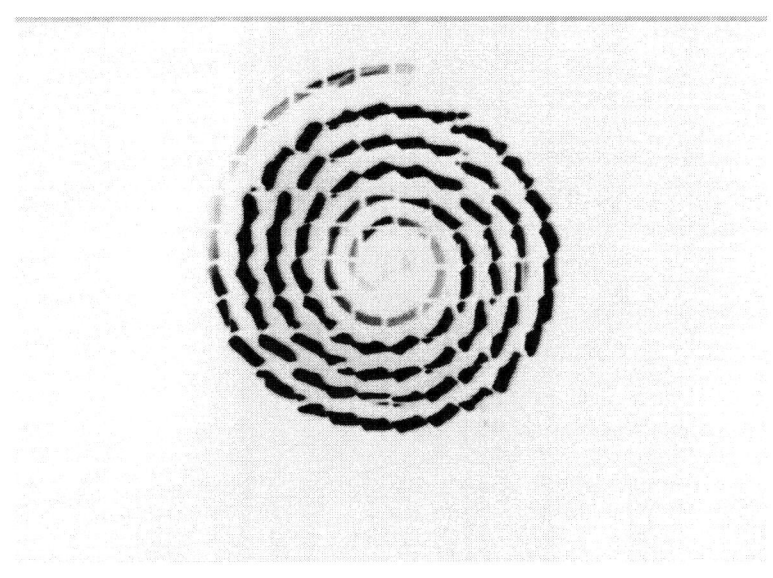

**FIGURE 3.** Same as in Fig. 2 but with $I_{th} = 1.02mA$, corresponding to point $A$ of Fig. 5

**FIGURE 4.** Same as in Fig. 2 but with $I_{th} = 1.01mA$, corresponding to point $B$ of Fig. 5

$$\nabla^2\phi(r,z) = 4\pi e\frac{n_0(r)w_0(r)}{\sqrt{(2/m)(e\psi_0(r) + e\phi(r,z))}},\tag{4}$$

where $e\psi_0 = mw_0^2/2 - e\phi_0$, and the quantities with index 0 represent the values at $z = 0$ (at the grid, with $\phi_0 = 0$), or at $z = L$ (at the charge collector, with $\phi_0 = V_{ph}$).

As unit of energy we introduce $e\bar{\psi} = m\bar{w}_0^2/2$, a value of the energy of electrons reaching the grid; as unit length $\bar{x} = \sqrt{\bar{\psi}/(4\pi e\bar{n}_0)}$ with $\bar{n}_0$ a value of density at the grid. In dimensionless form Eq. 4 reads:

$$\tilde{\nabla}^2\tilde{\phi}(\tilde{r},\tilde{z}) = \frac{\alpha(\tilde{r})}{\sqrt{\beta(\tilde{r}) + \tilde{\phi}(\tilde{r},\tilde{z})}},\tag{5}$$

where $\tilde{x} = x/\bar{x}$, $\tilde{\phi} = \phi/\bar{\psi}$, $\alpha = (n_0 w_0)/(\bar{n}_0\bar{w}_0)$, $\beta = \psi_0/\bar{\psi}$. Note that $\bar{\psi}, \bar{n}_0$, and the functions $\alpha$ and $\beta$ can be estimated from the settings of source parameters. The denominator in the fraction at the right hand side of Eq. 5 is the local axial beam velocity $\tilde{w} = w/\bar{w}_0$; the whole right hand side is the local plasma density $\tilde{n} = n/\bar{n}_0$, and, depending on the external parameters, may exhibit sharp radial and longitudinal variations. This behavior of the solution is related to the problem of space-charge-limited emission, which has been recently rivisited in the literature [9, 10, 11]. In our case the region dominated by space charge phenomena extends along the electron beam for a considerable length (up to about thirty times the source diameter) and allows the development of the vortices which characterize our observations. A typical behavior of the solution of Eq. 5 in a space charge dominated case is characterized by the presence of the surface $\tilde{\phi} + \beta = 0$

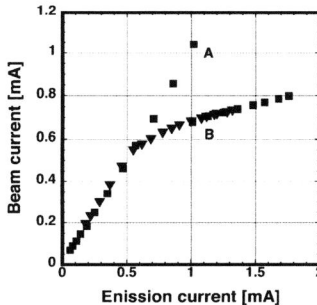

**FIGURE 5.** Electron current through the phosphor screen versus the source emission current. Squares indicate increasing emission current; triangles indicate decreasing emission current.

which encloses a spatial region not accessible by the electrons emitted from the source. Close to this surface the plasma density exhibits sharp variations.

To account for the well developed coherent structures (characterized by a vorticity $\zeta$ larger than the inverse electron transit time: $\zeta > w/L$) observed at the transition we note that, along the beam, close to the edge of the shadow region, shortly after the "nose", the equilibrium beam is nearly parallel to the axis, and the radial part of the Laplacian becomes dominant with respect to longitudinal: then, the plasma vorticity becomes proportional to plasma density, and the plasma dynamics is close to 2D Eulerian, described by the system:

$$w\frac{\partial}{\partial z}n+\mathbf{v}_\perp\cdot\nabla_\perp n=0, \qquad (6)$$

$$\nabla_\perp^2\phi = 4\pi en. \qquad (7)$$

The analysis presented above refers to $\theta$ independent stationary conditions, and justifies the conditions where transition occurs. To investigate the time evolution of the system where the electron flow, including its spatial and velocity distribution, is given at the grid, a recently developed 3D PIC code[12], which solves the cold drift-Poisson system (i.e. Eqs. (1, 2, 3) with the time derivatives included) has been used. The obtained results confirm the overall picture given above. In a space charge dominated regime a part of the plasma near the axis is reflected back to the grid while an annulus of high density is formed, and gives rise to the development of vortices. Note that the higher

**FIGURE 6.**   Same as in Fig.4 but with $B = 0.040T$

density region is inside the beam, while close to the phosphor, held at high potential, the electron density decreases.

## BEAM TRANSPORT

We have considered the transport of the observed structures along the beam, i.e. when the length of the central part of the beam (not close to the ends) is varied. Note from Eqs. (6,7) that a change of the magnetic field from $B_0$ to $B_1$ corresponds to a modification of the beam length of the ratio $B_0/B_1$, when the other parameters are kept fixed. This is true as far as the axial part of the Laplacian can be neglected with respect to the transverse and the variation of the emission flow due to change of the magnetic field is compensated.

Thus, in the limit of validity of the above system, the transport along the plasma column can be studied using a fixed phosphor screen and varying the confinement field $B$. A real time "temporal" evolution of the distribution of transverse plasma density following the beam can obtained by slowly decreasing (with respect to the rotation time scale) the magnetic field. Typical processes of 2D dynamics as merger and filamentation can be seen to occur in the system. Fig. 7 shows the image of the screen obtained starting from the condition of Fig. 4 and decreasing the magnetic field by a factor 2.44.

**FIGURE 7.** Same as in Fig.4 but with $B = 0.027T$

## SPACE CHARGE EFFECTS IN ELECTRON BEAMS

As a development of the research described above, a different ELTRAP configuration is under way (ELETRASP experiment, I.N.F.N.), where a pulsed (4 ns, 5 keV, 0.1 nC) electron source is used [13], and the CCD diagnostic allows the investigation of plasma effects (e.g., wave-breaking) in the beam. These phenomena, in relativistic laminar beams produced by photo-injectors at higher energy (150 MeV) severely limit the brightness achievable, which is of great relevance in many applications, like X-ray SASE-FELs [14]. Their experimental investigation is very difficult owing to the small width of the beam. We aim to measure such effects in ELTRAP, on beams with quite lower energy (few keV), but with the same properties (as far as the transverse dynamic is concerned) of the relativistic beams: this is possible if a scale parameter, defined below, is kept constant while the current is properly scaled with the beam energy and spot size, so that the beam envelope follows the same behavior with a larger spot size.

For a circular beam the envelope perturbations $\delta r$ around a matched beam are described by the equation

$$\partial^2(\delta r)/\partial s^2 + 4\kappa(1 - \omega_p^2/\gamma\omega_c^2)\delta r = 0, \tag{8}$$

where $\kappa = (\omega_c/2\gamma\beta c)^2$ and $s$ is the axial coordinate. The same behavior is obtained if $\omega_p^2/\gamma\omega_c^2 \propto I/(\beta\gamma B^2 r^2)$ is kept constant and the length is scaled by factor $\kappa^{-1/2}$. With reference to the injector of the SPARC project, for the development of a high brightness electron beam source to drive a SASE-FEL experiment [15], the following parameters are considered $\gamma\beta = 10; I = 100A; r \simeq 1mm; B \simeq 0.3T$. For the new ELTRAP configuration $\gamma\beta \simeq 0.2$; $I = 20mA$. Using spot sizes with radii of few $mm$ and magnetic fields of about $100gauss$ good experimental conditions can be found.

## ACKNOWLEDGMENTS

The authors are grateful to G. Malvezzi for his invaluable help in the construction of various mechanical parts of the device. This work was partially supported by the Italian Ministry of Education and Scientific Research.

## REFERENCES

1. Cutler, C., *J. Appl. Phys.*, **27**, 1028 (1956).
2. Kyhl, R., and Webster, H., *IRE Trans. Electron Devices*, **3**, 172 (1956).
3. Kapetanakos, C., Hammer, D., Striffler, C., and Davidson, R., *Phys. Rev. Lett.*, **30**, 1303 (1973).
4. Pasquini, T., and Fajans, J. (2002), F. Anderegg, L. Schweikhard and C. F. Driscoll, editors, *Non-Neutral Plasma Physics IV*, AIP 606, page 453, New York, 2002, American Institute of Physics.
5. Pozzoli, R., and Ryutov, D., *Electromagnetic Waves and Electronic Systems*, **3**, 12 (1998).
6. Amoretti, M., Bettega, G., Cavaliere, F., Cavenago, M., De Luca, F., Pozzoli, R., and Romé, M. (2002), F. Anderegg, L. Schweikhard and C. F. Driscoll, editors, *Non-Neutral Plasma Physics IV*, AIP 606, page 603, New York, 2002, American Institute of Physics.
7. Amoretti, M., Bettega, G., Cavaliere, F., Cavenago, M., De Luca, F., Pozzoli, R., and Romé, M., *Rev. Scient. Instr.* (2003).
8. Kriesel, J. M., and Driscoll, C. F., *Phys. Plasmas*, **5**, 1265 (1998).
9. Lau, Y., *Phys. Rev. Lett.*, **87**, 278301 (2001).
10. Akimov, P., Shamel, H., Kolinsky, H., Ender, A., and Kuznetsov, V., *Phys. Plasmas*, **8**, 3788 (2001).
11. Luginsland, J., Lau, Y., Umstattd, R., and Watrous, J., *Phys. Plasmas*, **9**, 2371 (2002).
12. Tsidulko, Y., Pozzoli, R., and Romé, M. (2003), *"A new 3D PIC code for the simulation of the dynamics of a non-neutral plasma"*, in: *Non-Neutral Plasma Physics V*, 2003, American Institute of Physics, to be published.
13. Feldman, D., Valfells, A., Neumann, J., Harris, J., Beaudoin, B., O'Shea, P., and Virgo, M. (2001), Proceedings of the 2001 Particle Accelerator Conference, PACS 2001, vol. 3, p. 2132, 2001, IEEE.
14. Serafini, L., and Rosenzweig, J. B., *Phys. Rev. E*, **55**, 7565 (1997).
15. Alesini, D., and et al. (2003), PAC 2003, 11-17 May 2003, Portland, Oregon.

# Numerical Non-neutral Plasmas

Ross L. Spencer, Grant W. Mason, and S. Neil Rasband

*Department of Physics and Astronomy, Brigham Young University, Provo, Utah 84602*

**Abstract.** For the past 15 years, or so, a set of computational tools for studying non-neutral plasmas has been developed at Brigham Young University. These codes, which include equilibrium codes (for both static plasmas and solitons), radial eigenvalue codes, 2-d eigenvalue codes, and both 2-d and 3-d particle-in-cell simulation codes, will be discussed, along with their applications to problems of interest to the non-neutral plasma physics community.

## INTRODUCTION

For many years we have been developing at Brigham Young University a set of computational tools for studying non-neutral plasmas, tools which are now being used by several different groups around the world. This paper is a brief review of these tools, the computational methods they use, and the situations in which they might appropriately be used. Since we usually work with students, we have tried to keep the methods that we use as simple as possible. In practice this means that we mostly use finite-difference methods on uniform grids, although we have done some work with non-uniform grids using finite-element methods. The codes are all written in FORTRAN, either using embedded graphics via the PLPLOT package from the magnetic fusion group at the University of Texas, Austin, or using MATLAB to produce graphics after the codes have produced data files. The all-important electrostatic field solve is handled in three different ways, depending on the number of computational dimensions. For one-dimensional problems we simply use a standard tri-diagonal Gauss elimination algorithm. For two-dimensional problems we use a banded matrix solver, similar to the tridiagonal Gauss elimination algorithm, but with a much larger bandwidth. We number the points on the grid by moving along the smallest dimension first (the $r$ direction in $(r, z)$ problems in the standard axially elongated non-neutral plasma geometry) so that the bandwidth of the sparse matrix that results from finite-differencing the Laplacian is $2N + 1$, where $N$ is the number of grid points in the shortest direction. On modern computers this produces a manageable matrix problem. In three-dimensions we use a multi-grid algorithm.

In the following sections each of the codes in our suite is described.

## DRIFTK

This code[1] is a 1-dimensional radial eigenvalue code based on the drift-kinetic equation, which means that the dynamics perpendicular to the magnetic field is governed

CP692, *Non-Neutral Plasma Physics V*, edited by M. Schauer et al.
© 2003 American Institute of Physics 0-7354-0165-9/03/$20.00

by the $\mathbf{E} \times \mathbf{B}$ drift, while parallel to the field we use kinetic theory, as embodied in the plasma dispersion function. The plasma equilibrium is assumed to be infinitely long and axisymmetric so that we may write

$$\phi_1 = \phi_1(r) \exp(ikz + im\theta - i\omega t) , \tag{1}$$

and the mode equation based on this model is

$$\frac{1}{r}\frac{d}{dr}\left(r\frac{d\phi_1}{dr}\right) - \frac{m^2}{r^2}\phi_1 - k^2\phi_1 - \frac{\omega_p^2(r)}{v_{th}^2}W\left(\frac{\omega - m\omega_0(r) - kv_b(r)}{kv_{th}}\right)\phi_1$$

$$+ \frac{mq}{\varepsilon_0 Br(\omega - m\omega_0(r) - kv_b(r))}\frac{d}{dr}(n_0(r))\left[W\left(\frac{\omega - m\omega_0(r) - kv_b}{kv_{th}}\right) - 1\right]\phi_1 = 0 , \tag{2}$$

where $k$ is the axial wave number, $m$ is the azimuthal wave number, $\omega_p^2(r)$ is the plasma frequency, which varies with radius because the density $n_0(r)$ varies with radius, $v_{th} = \sqrt{k_B T/m}$ is the thermal velocity, $q$ is the particle charge, $B$ is the magnetic field strength, $\varepsilon_0$ is the permittivity of free space, $\omega_0(r)$ is the equilibrium drift $\mathbf{E} \times \mathbf{B}$ rotation frequency, and where $v_b$ is the net velocity of the plasma in the $z$-direction. The function $W(z)$ is defined by

$$W(z) = \frac{1}{\sqrt{2\pi}} \int_{-\infty}^{\infty} \frac{xe^{-x^2/2}}{x - z}dx , \tag{3}$$

analytically continued from the upper-half $z$ plane, and the boundary conditions are that $\phi_1$ vanish at the conducting wall and that $\phi_1 = 0$ for $m \geq 1$ or that $d\phi_1/dr = 0$ for $m = 0$. It should also be mentioned that the code also allows for multiple species.

The algorithm to solve this equation starts by finite-differencing the differential equation on a radial grid. Since the resulting system of linear equations is homogeneous a straightforward solve will just give $\phi_1 = 0$ for any choice of $\omega$. To solve this problem we remove the equation corresponding to some radius $r_0$ from the system and replace it with the equation

$$\phi_1(r_0) = 1 , \tag{4}$$

which now gives a system for which there is a solution for any value of $\omega$. There is, however, a kink in the solution at $r = r_0$ unless $\omega$ is one of the mode frequencies.

The algorithm now varies the mode frequency $\omega$ until the removed equation is satisfied and the kind is removed (this technique is called matrix shooting.) The user chooses values for $k$ and $m$, then supplies the code with an initial guess for $\omega$. The algorithm depends sensitively on this initial guess and often blows up when the guess is inappropriately chosen, so the code also allows the user to scan $\omega$ either along the real axis or across a region of the complex $\omega$ plane to find approximately where the mode frequency lies. When an approximate value found by scanning is used as an initial guess, the code usually converges.

This code has been used to study damped quasi-modes (spatial Landau damping) and is routinely used by the experimenters at Brigham Young University and other places to identify the modes observed in experiments.

# DRIFT2D

We (S. Neil Rasband) have also developed a 2-dimensional eigenvalue code[2] for use with the $(r,z)$ cold-fluid $v_b = 0$ version of Eq. (2). In this case the perturbed potential is taken to be of form

$$\phi_1 = \phi_1(r,z)\exp(im\theta - i\omega t) \tag{5}$$

and the mode equation involves both $r$ and $z$ derivatives. This code uses a finite-element algorithm on a non-uniform rectangular mesh in the $r,z$ plane. After performing the many moment integrals of the finite element method a large linear algebra problem of the form

$$L(\omega)_{ij}\phi_j = 0 \tag{6}$$

is obtained. To avoid the trivial solution $\phi = 0$ we replace the zero on the right-hand side by some suitably chosen vector $r_i$ to obtain

$$L(\omega)_{ij}\phi_j = r_i \,, \tag{7}$$

which is directly solved by LU decomposition using a banded matrix solver. The unknown mode frequency $\omega$ is then varied until the solution vector $\phi_j$ becomes numerically infinite, indicating that a mode has been found. (The mode frequencies are the values of $\omega$ that make the operator $L$ singular.)

This code is used to find frequency shifts and eigenfunction changes due to finite length, equilibrium shape, and induced charges on nearby walls in non-neutral plasma modes.

# EQUILSOR

This code[3] is our axisymmetric equilibrium code which finds the equilibrium potential $\phi(r,z)$ and density $n(r,z)$ given electrode shapes and applied voltages. The code can model both rings at the outer edge of the cylindrical computing volume, or electrode segments specified as line segments in the $(r,z)$ plane within the cylindrical computing volume. By choosing many such segments any axisymmetric electrode arrangement may be modeled. When the electrode line segments cross the grid lines we use a short-legged version of the finite-difference approximation to the Laplacian to minimize the relatively large errors produced by a "stair-step" approximation.

The code computes three types of equilibria, corresponding to the following three forms of the non-linear equilibrium equation:

**Mid-plane density profile:**

$$\nabla^2\phi = -\frac{q}{\varepsilon_0}n_{mid}(r)\exp\left(-q(\phi(r,z) - \phi(r,0))/kT\right) . \tag{8}$$

This form allows the user to specify the radial density profile $n_{mid}(r)$ at the plasma midplane $(z = 0)$ then allow the plasma to adjust its density along the magnetic field lines until equilibrium is reached.

**∫ndz profile:**

$$\nabla^2\phi = -\frac{q}{\varepsilon_0}n_{mid}(r)\exp\left(-q(\phi(r,z)-\phi(r,0))/kT\right). \tag{9}$$

This form allows the user to specify a radial profile of the line-integrated density $\int ndz$, and then the algorithm, as part of the equilibrium solve, adjusts the mid-plane density profile $n_{mid}(r)$ until the computed profile of $\int ndz$ matches the profile supplied by the user. So this form is just a slight modification of the mid-plane profile form.

**Global thermal equilibrium:**

$$\nabla^2\phi = -\frac{q}{\varepsilon_0}n_0\exp\left(-q(\phi(r,z)-\phi(0,0))/kT + Cr^2\right). \tag{10}$$

This version adjusts both the radial and axial profiles using the above form of the density obtained from the condition for global thermal equilibrium[4]. The constant $C$ is adjusted as the algorithm proceeds to achieve the mid-plane radius specified by the user.

The convergence rate of the code is determined by the ratio $\lambda_D/L_p$, where $\lambda_D$ is the Debye length and where $L_p$ is a measure of the size of the plasma. The code converges very quickly when this ratio is of order 1, but slows down as this ratio becomes small. In addition to the convergence problem, small values of this ratio also cause the exponential factor on the right-hand side of the equilibrium equation to blow up. This can be controlled by starting the iteration sequence at a larger value of the plasma temperature than desired, then slowly decreasing it toward the target value. This allows the equilibrium to slowly adjust to a decreasing Debye length, controlling the potentially large value of the exponential function when the temperature is small.

This code provides the input for DRIFT2D, RATTLE, and INFERNO (discussed below) and is routinely used by experimenters to turn profiles of $\int ndz$ into pictures of what the plasma equilibrium looks like.

# RATTLE

This code[5, 6] is an axisymmetric particle-in-cell simulation in $(r,z)$ geometry. Note that axisymmetry means that each "particle" is actually a ring of charge. The Larmor radius is assumed to be infinitesimally small, so that the radial position of each particle remains unchanged. For simplicity, particles are constrained to lie on the $z = $ const grid lines of the simulation, so that the interpolations that produce the density $n(r,z)$ and the axial electric field $E_z(r,z)$ only take place in $z$. This code reads an equilibrium file produced by EQUILSOR and uses the density $n(r,z)$ from this file to load the particles onto the simulation grid. RATTLE has the same varied electrode shape capability as EQUILSOR. The code has a time step constraint given by the usual electrostatic Courant condition:

$$\omega_p\tau < 1. \tag{11}$$

For most plasmas this constraint means that the code is useful for modeling plasma loading, plasma dumping, and Gould-Trivelpiece waves, e.g., processes that occur on a timescale of $1/\omega_p$ or $L/v_{th}$, where $L$ is the system size and $v_{th}$ is the thermal velocity. The code is not very useful for studying transport processes because the collisions are constrained to axial exchanges of energy along the field lines (which means that the axial distribution function $f(v_z)$ can't evolve due to collisions) and because transport processes are slow.

The basic cycle that occurs during each time step is:

(1) The particles are moved using the standard leapfrog algorithm in which velocities and positions are known at different times:

$$v_{n+1/2} = v_{n-1/2} + \frac{q}{m}E_z(\mathbf{r}_n)\tau \quad ; \quad z_{n+1} = z_n + v_{n+1/2}\tau , \tag{12}$$

with $E_z(\mathbf{r})$ found by linear or quadratic interpolation on the grid.

(2) The particle positions are then used to create density at each grid point, using a matching form of interpolation to the one used in step (1). If they don't match, a single particle will exert a force on itself.

(3) Using the density from step (2) the potential $\phi(r,z)$ is found by LU-decomposition and back-substitution on the large banded matrix that represents the Laplacian operator. This direct solve is quite manageable on modern PCs. For instance, a $100 \times 500$ grid makes a banded matrix of size $[100 \times 500, 201]$ which requires only 80 Megabytes of memory.

This code is used to interpret the meaning of observed mode frequencies in experiments, to simulate plasma loading and dumping, to study nonlinear waves (including solitons), etc..

# INFERNO

This code[7] (Grant Mason) is a 3-dimensional particle-in-cell simulation in which the motion in $z$ is governed by Newton's second law, but the motion in the $x,y$ plane is governed by the $\mathbf{E} \times \mathbf{B}$ drift. We compute in a cylindrical geometry, but use a Cartesian grid in $(x,y)$. The reason for this somewhat awkward choice is that the pain of finite differencing the Laplacian near a circular boundary that crosses Cartesian grid lines is easier to bear than that associated with the $r = 0$ singularity in a cylindrical coordinate system. In addition, the small errors in $\phi$ caused by the mismatched boundary are of very short wavelength, and hence decay exponentially from the wall into the plasma. This code is suitable for simulating non-axisymmetric Gould-Trivelpiece modes, diocotron modes, loading and dump effects, etc.. Due to the same limitations on the modelling of particle collisions as in RATTLE, and due to the increased run time required with a 3-dimensional grid, it is not very useful for studying transport problem.

INFERNO has the same basic 3-step cycle as RATTLE, except that particles must be moved in the $(x,y)$ plane as well as in $z$. To follow the drift motion in the $(x,y)$ plane we use a predictor-corrector algorithm discussed by Tajima[8].

The field solve must also be handled differently because in 3-dimensions the banded matrix form of the Laplacian is too large for LU decomposition, so we use a multi-grid algorithm for the field solve.

This code has been used to study the instability of the $m = 1$ diocotron mode for hollow density profiles, and has been used more recently to simulate the trapped-particle asymmetry modes discovered by Kabantsev, *et al*[9].

## · CONCLUSION

These codes occupy a middle ground between experiment and theory, and are therefore useful for connecting the two. They can be used to interpret experimental signals, to test theoretical ideas, and can be usefully thought of as highly diagnosable simplified experiments in which the laws of physics can be adjusted to see the relative importance of various effects. They are available upon request for use by anyone who is interested.

## REFERENCES

1. Grant W. Mason and Ross L. Spencer, Phys. Plasmas **9**, 3217 (2002).
2. S. Neil Rasband and Ross L. Spencer Phys. Plasmas **10**, 948 (2003).
3. R. L. Spencer, S. N. Rasband, and Richard R. Vanfleet, Phys. Fluids B, **5**, 4267, (1993).
4. S. A. Prasad and T. M. O'Neil, Phys. Fluids **22**, 278 (1979).
5. M. D. Tinkle, R. G. Greaves, C. M. Surko, R. L. Spencer, and G. W. Mason, Phys. Rev. Lett. **72**, 352 (1994).
6. G. W. Mason, R. L. Spencer, and J. A. Bennett, Phys. Plasmas **3**, 1502 (1996).
7. Grant W. Mason and Ross L. Spencer Phys. Plasmas **9**, 3217 (2002).
8. *Computational Plasma Physics, with Applications to Fusion and Astrophysics*, Toshiki Tajima, (Addison-Wesley, New York, New York, 1989), Chapter 7.
9. A. A. Kabantsev, C. F. Driscoll, T. J. Hilsabeck, T. M. O'Neil, and J. H. Hu, Phys. Rev. Lett. **87**, 225002 (2001).

# Collisional Cooling of Pure Electron Plasmas Using $CO_2$

W. Bertsche* and J. Fajans*

*Physics Department, U. C. Berkeley

**Abstract.**
Inelastic collisions with $CO_2$ buffer gas cool a pure electron gas in a Penning-Malmberg trap at low magnetic fields. 0.6 eV electrons are cooled by down to 30% of their original temperature.

## INTRODUCTION

Pure electron plasmas can be held indefinitely by applying a rotating electric potential. Such "rotating walls" work only if there is some mechanism that cools the plasma to counteract the heating from the rotating wall. Experiments using this technique [1] have relied on electron cyclotron cooling in large B-fields to counter this heating. However, at substantially lower fields, cyclotron cooling is not fast enough to balance the heat-input from the wall drive, therefore requiring another cooling method to achieve confinement.

Inelastic electron collisions with buffer gas can transfer energy from electrons to gas molecules through excitation of molecular vibrational modes, acting to cool electrons, while elastic collisions tend to generate outward radial transport [2]. Good candidate buffer gases should have an inelastic electron scattering cross-section larger than its elastic cross-section for the energy range of interest, in addition to having vibrational modes with energies similar to the electron thermal energy. Motivated by an examination of spectroscopic data, calculations of electron-$CO_2$ scattering cross-sections [3], work done with gas-cooling in positron experiments [4] and ease of use, $CO_2$ seemed to be a good candidate for an initial investigation.

## EXPERIMENT

Cooling and preliminary compression measurements were carried out in a Penning-Malmberg trap. Typical plasmas used in the experiment measured approximately 30 cm. in length with a radius of 1 cm, and an approximate number density of $10^7$ cm$^{-3}$. The longitudinal B-field was 1500 gauss.

Temperature measurements were performed by gradually lowering the end confinement potential while measuring charge leaving the trap and inferring a $T_{\parallel}$ from a reconstruction of the tail of the parallel velocity distribution function[5]. Electrons were assumed to be in approximate thermal equilibrium giving $T_{\parallel} \approx T_{\perp}$. The plasma temperature with no buffer gas (base pressure $5 \cdot 10^{-10}$ torr $to$ $8 \cdot 10^{-10}$ torr) was approximately

CP692, *Non-Neutral Plasma Physics V*, edited by M. Schauer et al.
© 2003 American Institute of Physics 0-7354-0165-9/03/$20.00

0.6 eV for these experiments.

The buffer gas was 99.9% dry $CO_2$. Gas was introduced into the vacuum chamber via a leak valve and monitored with an ion gauge and residual gas analyzer. For measurements discussed in this paper, the $CO_2$ was contaminated with water from the gas transport line to a level between 2.5% and 5% partial pressure. Later experimental enhancements eliminated this contamination.

Measurements were made using the standard inject-hold-dump sequence. Hold times were varied and temperature measurements made at the dump cycle. Figure 1 shows cooling curves at several $CO_2$ pressures.

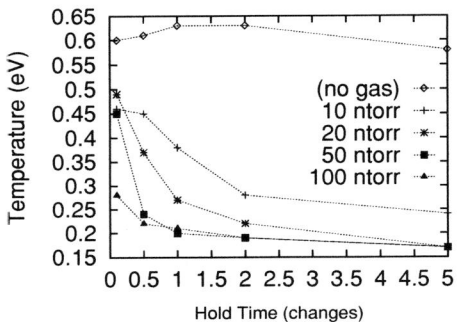

**FIGURE 1.** Plasma temperature as a function of hold time. Pressures listed are pressure of $CO_2$ plus base pressure (.8 ntorr).

Preliminary work involving a rotating wall drive was carried out using an annular gate in the center of the plasma column divided into four 90° sectors. A sinusoidal potential was applied to each gate with successive 90° phases with the same rotation sense as E$x$B rotation. The drive frequency was swept linearly for a period of time (chirp time) and held at the end frequency for the remainder of the hold cycle. Measurements of the RMS plasma radius and total number of electrons were made during the dump cycle by dumping the electrons on a phosphor screen and analyzing the image. Figures 2 and 3 show results for one set of parameters.

# DISCUSSION

As shown in Fig. 1, the $CO_2$ buffer gas successfully cools the electron gas. In general, higher $CO_2$ pressure increased cooling rates, however cooling saturated at temperatures between 0.2 and 0.3 eV. It was uncertain if this effect is due to saturation of the cooling process or is a limitation in the temperature diagnostic.

Higher partial pressures of $CO_2$ resulted in substantial radial transport and some $CO_2$ ionization – two effects opposing the goal of infinite plasma confinement. Coupling this system with a rotating wall potential produced some radial compression to counteract this elastic collisional transport.

Figure 2 demonstrates moderate compression using a chirped, rotating wall potential. Maximal compression was achieved with a frequency near the first Trivelpiece-Gould

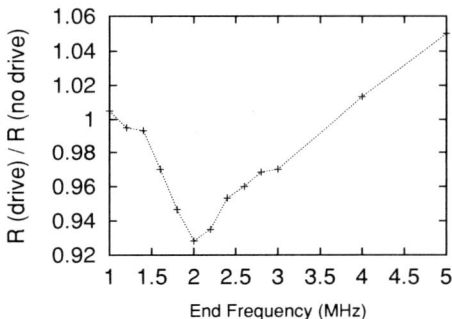

**FIGURE 2.** Normalized RMS Plasma radius as a function of wall drive end frequency. Wall Drive Voltage: 0.1 $V_{pp}$ Wall Start Frequency: 660 kHz Chirp Time: 20 sec. Total Hold Time: 30 sec. $CO_2$ Pressure:6 ntorr.

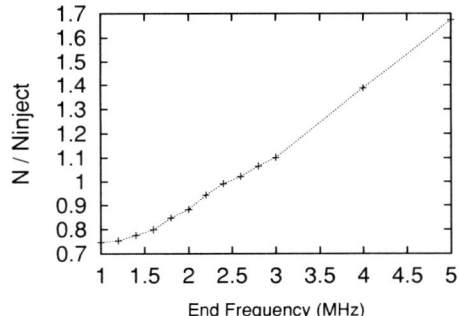

**FIGURE 3.** Total trapped charge as a function of wall drive and end frequency. Normalized to the amount of charge measured in an inject with no hold time and no buffer gas. System parameters the same as in figure 2.

mode of the system (1.9 MHz), as expected[4]. Unfortunately, this compression was met with much gas ionization, as evident by an increasing amount of total charge in the trap with increasing end frequency (figure 3).

In conclusion, this experiment shows that $CO_2$ buffer gas cools pure electron plasmas. Thus far, only modest plasma compression was achieved using a rotating wall potential, accompanied with ionization. Whether this is a parameter regime of $CO_2$ pressure and drive potential in which plasma columns are indefinitely confine remains an open question.

# REFERENCES

1. Huang, X. P., Anderegg, F., Hollman, E. M., Driscoll, C. F., and O'Neil, T. M., *Phys. Rev. Lett.*, **78**, 875 (1997).
2. Douglas, M. H., and O'Neil, T. M., *Phys. Fluids*, **21**, 920 (1978).
3. Morrison, M. A., and Greene, A. E., *J. Geophys. Rsrch.*, **83**, 1172 (1978).
4. Greaves, R. G., and Surko, C. M., *Phys. Rev. Lett.*, **85**, 1883 (2000).
5. Eggleston, D. L., Driscoll, C. F., Beck, B. R., Hyatt, A. W., and Malmberg, J. H., *Phys. Fluids B*, **4**, 3432 (1992).

# High Voltage Amplifier Designs for Penning and Radio-Frequency Traps

B. T. Chang and T. B. Mitchell

*Department of Physics and Astronomy, University of Delaware, Newark, Delaware, 19716*

**Abstract.** Two economical designs for high voltage amplifiers are described, and the performance of constructed units is characterized. The first amplifier is based on a design originating from the non-neutral plasma group at UC San Diego, and is useful for driving capacitive loads such as the containment rings of Penning traps. The second is based on a design published by the Jones group of the University of Utah Chemistry Department, and is designed to power radio-frequency (rf) guides and traps. Complete design specifications including schematics, PC board files and circuit simulation inputs are available for those interested in building either amplifier [1].

Charged-particle traps require confinement fields in order to provide potential wells. In this contribution we describe two economical solutions for providing electric fields for use with Penning and radio-frequency (rf) traps.

With Penning traps, radial confinement is provided by an axial magnetic field while axial confinement is provided by electric fields. In a typical electron plasma experiment, large voltage amplitudes of ~100 V applied to trap electrodes are needed to provide confinement of electrons with densities of ~$10^7 \text{cm}^{-3}$ and length ~20 cm [2]. The voltage waveforms should be readily adjustable to permit fine control of the injection process [3] and later shaping of the electron distribution [4, 5]. The bandwidth of the amplifier should be as wide as possible to permit fast density measurements [6]. Total harmonic distortion should be minimized to allow the amplifier to be used to apply large perturbations such as can be needed for rotating wall frequency stabilization [7]. These requirements can be met with a DC-coupled high voltage amplifier with a gain of 15 (to allow control with computer-controlled DAQ cards with ± 10V outputs).

With rf traps, an oscillator modulates the potential applied to a trap electrode(s) at typical frequencies of 1 MHz and greater, and amplitudes of 20 $V_{p-p}$ to 900 $V_{p-p}$. The time-dependent electric field can result in a pondermotive force which, depending on amplitude and frequency, will trap particles of particular charge to mass ($q/m$) ratios while ejecting others [8]. The ideal rf oscillator can be switched off rapidly (to facilitate ion ejection), is stable and tunable in both frequency and amplitude, and is well coupled to the trap.

## High Voltage DC-Coupled Amplifier

Our high voltage amplifier design utilizes the Apex PA84A high voltage power operational amplifier, and is based on a circuit designed by Bob Bongard of UCSD. The

CP692, *Non-Neutral Plasma Physics V*, edited by M. Schauer et al.
© 2003 American Institute of Physics 0-7354-0165-9/03/$20.00

amplifier accepts up to $\pm 9.5$ V input, has a voltage gain of -15, can output $\sim\pm 25$ mA, and has a 2000$\Omega$ input impedance. It is designed to drive a capacitive load $C_L$ on the order of $\sim$300 pF. The difficult aspect of the circuit design is accomplishing this while avoiding oscillations. The method used is to avoid poles in the characteristic polynomial of the closed loop transfer function, and to introduce zeros to cancel unavoidable ones.

The current limits of the op-amp limit the maximum slew rate possible. A 1 k$\Omega$ output isolation resistor $R_{iso}$ creates a zero to eliminate a pole caused by the internal PA84A output resistance and the capacitive load. This $R_{iso}$–$C_L$ network establishes the slew rate and the bandwidth (as measured by the frequency at which the output is 3 dB down). With no load and maximum output amplitude, the amplifier slew rate is 245 $V/\mu s$ and its bandwidth is 320 kHz. These figures reduce linearly with load capacitance, and are 135 $V/\mu s$ and 185 kHz with a load of 500 pF.

The amplifier's total harmonic distortion (THD) is roughly independent of the load capacitance, but increases with frequency. We characterized the performance by driving the amplifier with a sinusoidal input with a THD of 0.35%, and measuring the THD of the output with an amplitude near the maximum value achievable of 270 $V_{p-p}$. At 20 Hz, the output was similar to the input with a THD of $\sim$0.30%, but at 50 kHz we measured THD $\sim$0.70%.

In the full version of the amplifier, analog switches, potentiometers and op amps are used in the front end along with panel mounted controls to provide a great deal of flexibility for pulse generation. TTL voltage levels are used to select one of two inputs for the amplifier, either of which can be supplied by either external signals or user-controlled DC levels. These advanced features can be bypassed by a jumper to provide a simplified amplifier with a single input.

Complete design specifications are available at [1]. This website includes Gerber 274X images and Excellon drill files which can be used to fabricate a two-layer fiberglass printed circuit board. It also includes CAD drawings for the amplifier housing, electronic schematic diagram files, PC board design files and some SPICE models. (The parts of the circuit for which SPICE models were available were simulated in Electronics Workbench's Multisim 2001 to verify that the circuit would work.) An accompanying parts lists indicates suggested vendors and costs of components. One full amplifier costs about $600 to build. Simplified amplifiers use the same board but require fewer components and cost about $450.

Although the amplifier circuit is intended to benefit from the application of negative feedback, every attempt has been made to eliminate sources of positive feedback, which can result in uncontrollable oscillations. Potential sources of positive feedback include modulation on DC supply lines that is coupled back into the amplifier, non-zero ground voltages reaching the non-inverting amplifier input which is also grounded, near 360° phase shift at unity loop gain, parasitic capacitance on the op amp input which must be compensated, and improper placement of feedback components and paths near high amplitude signals and ground returns.

# Radio Frequency High Voltage Amplifier

In this section we describe the design of a high voltage 2.50 MHz frequency rf oscillator capable of sinusoidal output with linear amplitude controllable from 100 $V_{pp}$ to 900 $V_{pp}$. It can be switched off from a 6146 vacuum tube anode voltage of 900 $V_{pp}$ to 0 $V_{pp}$ in 30 nanoseconds. It is intended for use with an ion trap for trapping and time-of-flight experiments. Because the ion trap serves as a capacitive load that participates in an inductor-capacitor tank of the rf oscillator, no impedance matching transformer is needed.

The basic oscillator design is motivated by the work of the Jones group of the University of Utah Chemistry department [9, 10, 11]. Two 6146 vacuum tubes are connected in an astable multivibrator configuration, with the output anode of each tube driving the grid input of the other tube. With a parallel LC tank circuit connected between the two anodes, the circuit never comes to rest in a stable configuration, and instead freely oscillates. The frequency of oscillation is fixed, but can be adjusted from a few hundreds kHz to about 25 MHz by changing the inductance of the tank circuit.

Based upon recommended gains, current and voltage levels in data sheets for 6146 tubes used as rf oscillators, and upon Jones' circuit designs [11], we constructed a prototype oscillator but encountered parasitic oscillations when we sought to improve the amplitude and frequency stability. To investigate these, we used Electronic Workbench's Multisim 2001 circuit modelling software package. This software accurately predicted that the resonant frequency of the rf oscillator would be the measured value of $\omega_0 = 1.57 \cdot 10^7$ rad/sec, and be sensitive to mutual reflected impedance between the halves of the tank inductor, with the center-tap capacitively bypassed to ground. (Although seemingly each tube drives a parallel LC circuit of a $15\mu$H inductor half and a 131pF 4.1 ft. cable with grounded braided shield from its anode and ground, the reflected mutual impedance of M = 11.55 $\mu$H affects the current flow in the other half of the inductor, as well.)

We then wrote a second software simulation to gain insight into the feedback nature of the circuit, after the working design had been attained. Wolfram Research Inc.'s Control Systems Professional and Mathematica 5.0 software packages were used to translate a set of Kirchoff current and voltage law equations, applying the non-conventional two-loop positive feedback in astable multivibrators to them, to derive state space representations of the circuit. The software translated this into a frequency response which shed light on how the circuit equalizes its effort to minimize energy dissipated in the 6146 plate resistance, while maximizing the current flowing into the tank capacitance, as it ought to at resonance.

Bode plots from the second simulation giving magnitude (Fig. 1) and phase (Fig. 2) of the rf oscillator of the anode 1 voltage, divided by the component of anode 1 current furnished by the DC power supply, show that the transfer function has high gain and, counterintuitively, that the supply does not deliver current to the tube zero degrees out of phase to the voltage, but rather that the voltage lags the supply current by 1.6 degrees at resonance.

At the resonant frequency $\omega_0$, we found that the anode current and the voltage were out of phase due to a lag effect created by a 313pF capacitance and the 20K$\Omega$ plate resistance. A mathematical model found a current flowing **out** of the plate resistor equal

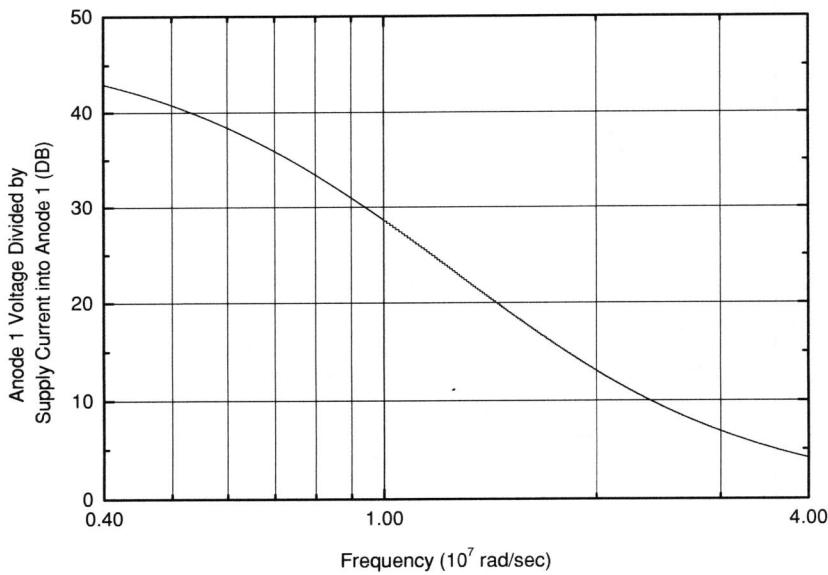

**FIGURE 1.** Bode plot of the magnitude of the transfer function relating rf oscillator Anode-1 voltage divided by the component of anode 1 current furnished by the DC power supply.

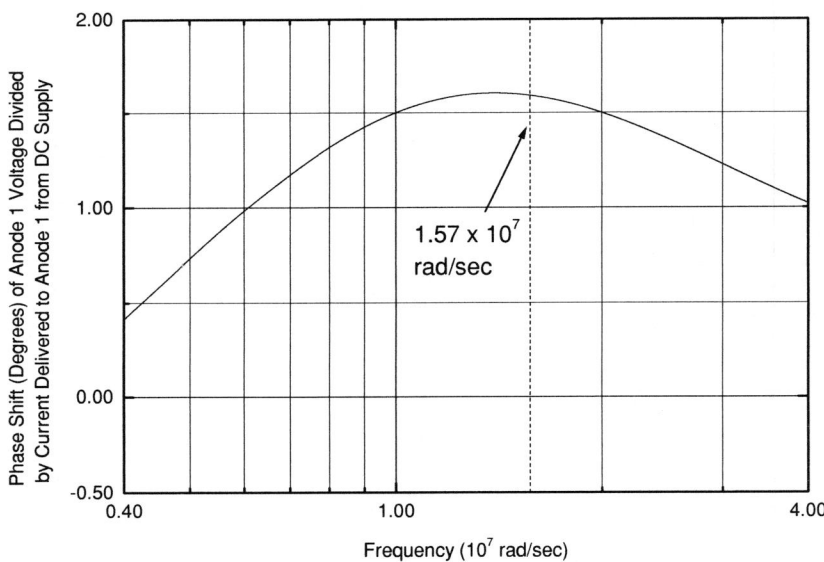

**FIGURE 2.** Bode plot of the number of degrees the $V_{A1}$ leads the component of anode 1 current furnished by the DC power supply. Positive angles imply $V_{A1}$ leads the current.

to $V_{A1}(2.749 \ 10^{-3}\cos(\omega_0 \ t - 0.02834))$, where $V_{A1}$ is the anode voltage amplitude. A current $V_{A1}(27.7 \ 10^{-6}\cos(\omega_0 \ t + 0.0279))$ flows from the DC supply **into** the resistor. The phases and directions are such that the DC supply furnishes a current whose real part minimizes that of the current in the resistor due to the tube, minimizing the energy lost in it. The imaginary part of current flowing out of the resistor is maximized. It flows into the capacitor, the correct destination for resonant current.

For harmonics other than the $\omega_0$, the tank becomes heavily inductive, forcing the the current from the power supply to lead the anode voltage. This reduces the phase angle, as seen in Fig. 2.

In the remainder of this contribution, we discuss some changes we have made to the Jones oscillator design.

**Noise Reduction** The harmonic-content power dissipated in the non-ideal resistance of the tank inductor can be reduced by mixing in a sinusoidal signal having one pure frequency. Phase-different voltages will cancel, and the spectral purity of the output will improve. The mixing was accomplished by adding a **B** field by driving a current into two loops of AWG 22 wire wrapped around one half of the center-tapped tank inductor. A relative clean function-generator-derived 2.5MHz oscillating current of $100 \ mA_{p-p}$ was produced using an emitter-follower high current transistor. The improvement from this mixing can be seen in Fig. 3, where the top shows an FFT of the frequency stabilized oscillator's fundamental and first harmonic, and the bottom shows an FFT of a freely running oscillator.

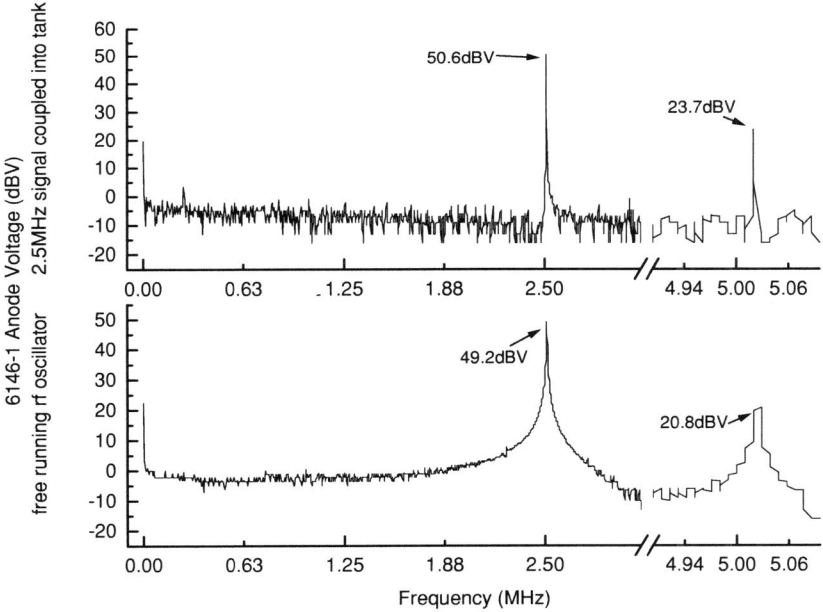

**FIGURE 3.** Fourier spectrum of the rf oscillator output; freely running (bottom) and stabilized by a 2.50 MHz source (top).

**Amplitude control** Incorporating a design from Ref. [12], we have added an automatic gain control circuit which provide a linear control of a single anode's voltage in the 100 $V_{pp}$ to 900 $V_{pp}$ range. Amplitudes from 22 $V_{pp}$ to 100 $V_{pp}$ range are also attainable, but are not proportional to the control voltage.

We control anode amplitudes by varying the 6146 gains; namely, their grid-cathode voltage dependent transconductance. This is done by varying the cathode current, which then controls the grid-cathode DC bias and also the excursion range of plate-cathode voltages.

To regulate current, a power transistor is connected between the cathode and ground of each tube. Their resistances are variable, and increase to limit current flow. When the the transistors are nearly turned off, corresponding to a small anode voltage output range, the cathodes of the tubes are pulled high. Effectively, the voltage divider pulls the cathode high with a low plate-cathode resistance, while the cathode-ground resistance due to the transistor is very high. This also decreases grid-cathode bias, reducing the gain of the tubes. When the transistors allow strong current flow, their resistance is low, and the cathode-plate potential has a wide range. Grid-cathode voltage is made more positive by a less positive grid. This increases the tube amplification.

**Fast turn-off** The rf oscillator can be switched on and off rapidly. When shut off, it ramps to zero volts in 30 nanoseconds. We use International Rectifier IR2213 combination high side and low side driver integrated circuit, and IRFPG50 high voltage, high current MOSFET transistors. We use four of the MOSFETs as on-off switches of very large voltages, controlled by small levels.

The high side driver is needed to apply a large voltage to the gates of a MOSFET, supplying or denying high 475$V_{DC}$ voltages to our load, the DC power supply line of the rf oscillator. The low side driver grounds this line rapidly when needed.

We use a third and fourth IRFPG50, each connected to a 6146 tube, with their drain connected to the anode, and their source to the ground. When the TTL control is low this second pair conducts, discharging the tubes and the tank circuit and shutting down the oscillations. When the TTL control goes high, this second pair opens.

A very fast high voltage diode, reverse-connected across the DC power supply line is essential to clip turn-off noise transients to levels which would not trigger the oscillator spuriously. These transients are evident in Fig. 4 which plots the control and output voltages during the time turn-off occurs. The figure indicates the sudden turn-off achievable as well as the relative cleanliness of the output.

# ACKNOWLEDGMENTS

This research was supported by the National Science Foundation. We wish to thank T. Pham for software assistance, G. Robinson and L. Shulman for analog design tips, J. Poirier for technical assistance, C. F. Driscoll for sharing his group's amplifier design, and D. Schaeffer for constructing amplifiers and measuring their performance.

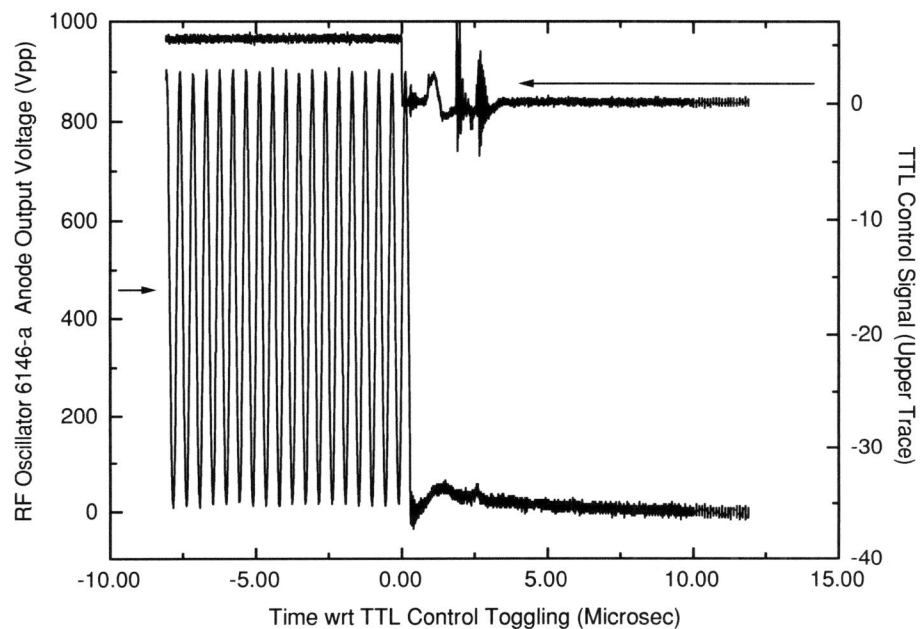

**FIGURE 4.** Input TTL control signal and oscillator output at turn-off. Note the noise due to inductive kickback and the sudden 900 V transition in 30 nanosec.

# REFERENCES

1.   www.physics.udel.edu/~mitchell/projects.
2.   Prasad, S. A., and O'Neil, T. M., *Phys. Fluids*, **22**, 278 (1978).
3.   Kriesel, J. M., and Driscoll, C. F., *Phys. Plasmas*, **5**, 1265–1272 (1998).
4.   Driscoll, C. F., *Phys. Rev. Lett.*, **64**, 1528–1543 (1990).
5.   Fine, K. S., Driscoll, C. F., Malmberg, J. H., and Mitchell, T. B., *Phys. Rev. Lett.*, **67**, 588 (1991).
6.   Malmberg, J. H., and Driscoll, C. F., *Phys. Rev. Lett.*, **44**, 654–657 (1980).
7.   Huang, X.-P., Bollinger, J. J., Mitchell, T. B., and Itano, W. M., *Phys. Rev. Lett.*, **80**, 73–76 (1998).
8.   Ghosh, P. K., *Ion Traps*, Clarendon, Oxford, 1995.
9.   Jones, R. M., Gerlich, D., and Anderson, S. L., *Rev. Sci. Instrum.*, **68**, 3357 (1997).
10.  Jones, R. M., and Anderson, S. L., *Rev. Sci. Instrum.*, **71**, 4335 (2000).
11.  www.chem.utah.edu/chemistry/staff/jones/rfgen.
12.  O'Connor, P. B., Costello, C. E., and Earle, W. E., *J. Am. Soc. Mass Spectrom.*, **13**, 1370 (2002).

# Electrostatic Confinement of a Reflecting Ion Beam

J. R. Correa and C. A. Ordonez

*Department of Physics, University of North Texas, Denton, Texas 76203*

**Abstract.** The confinement of ions in a purely electrostatic, three-electrode configuration is predicted. The electric potential within the trap is calculated by numerically solving Laplace's equation using a sequential over-relaxation method. The classical trajectory of a proton is computed by numerically solving the equations of motion. Parameters that result in particle confinement are provided.

## INTRODUCTION

Charged particles that experience a periodically changing electric potential can be considered to be acted on by an effective force, which can confine the particles. The electric potential can change either periodically in time, as in a Paul trap [1], or periodically in the spatial region through which the charged particles travel, as in an ion-trap resonator [2]-[5]. An ion-trap resonator employs a static electric field and no magnetic field to confine charged particles within a long, cylindrical, axially symmetric confinement region. An electric potential well provides axial particle confinement. Radial particle confinement occurs because the particles travel axially through a spatially periodic electric potential that provides an effective force directed radially inward. Two modes of operations have been reported for the ion-trap resonator. For each, an ion bunch is initially caught in the trap having an axial length smaller than the axial length of the confinement region. In one mode of operation, referred to as synchronization, the length of the ion bunch remains smaller than the axial length of the confinement region [5]. In the other, the ion bunch quickly spreads out axially during a time period that is short compared to the particle confinement time [4], and the particle velocity distribution is that associated with a reflecting ion beam. Even with a non-Maxwellian velocity distribution, long particle confinement times, in excess of two seconds, have been reported. A confinement analysis presented in Ref. [4] is consistent with the following explanation for the long confinement times: The effect of periodic particle collisions among themselves, which tends to cause the velocity distribution of the trapped particles to become Maxwellian, is counteracted by the effect of periodic particle encounters with focusing-field regions, which tends to cause the velocity distribution of the trapped particles to remain non-Maxwellian. In the work presented here, a classical trajectory simulation is used to explore another possibility for electrostatic confinement of a reflecting ion beam. Preliminary results suggest that a reflecting ion beam can be confined within a simple configuration consisting of three cylindrical electrodes that are aligned end to end.

CP692, *Non-Neutral Plasma Physics V*, edited by M. Schauer et al.

# TRAP CONFIGURATION

The configuration considered is shown schematically in Fig. 1. A Cartesian coordinate system is defined such that the coordinate origin is located at one corner of the trap. A rectangular slab geometry is used to model three electrodes, which are infinite in extent in the z direction. For each electrode, $x_{max} = 10$ mm. The two end electrodes, held at $V_1 = 1$ V potential, are 10 mm long. One is located at $0 < y \leq y_1$ and the other at $y_2 < y \leq y_{max}$. The center electrode is $y_2 - y_1 = 20$ mm long and is grounded, $V_0 = 0$. The separation between the electrodes is assumed to be very small compared to the trap length. In the computation, the end electrodes are considered as being capped, which serves to simulate the effect of very long end electrodes without having to extend the computation region to an equally large length.

# COMPUTATION OF ELECTROSTATIC POTENTIAL WITHIN THE TRAP

The electric potential $\phi$ in the absence of charge obeys Laplace's equation

$$\nabla^2 \phi = 0. \tag{1}$$

The solution is uniquely defined if the boundary conditions are met [6]. For the configuration considered, the boundary conditions are as follow:

$$\phi = 0 \text{ for } x = 0 \text{ or } x = x_{max}, \text{ and } y_1 < y \leq y_2 \tag{2}$$

$$\phi = 1 \text{ for } x = 0 \text{ or } x = x_{max}, \text{ and } 0 < y \leq y_1 \text{ or } y_2 < y \leq y_{max} \tag{3}$$

$$\phi = 1 \text{ for } y = 0 \text{ or } y = y_{max} \tag{4}$$

The electric potential within the configuration was found using a sequential over-relaxation finite-differences method.

# PARTICLE TRAJECTORY COMPUTATION

For a two-dimensional analysis, the particle's motion can be described in terms of the $x$ and $y$ coordinates and the corresponding velocity components $(v_x, v_y)$. These phase space coordinates are equivalently expressed in terms of $(x, y, E, \alpha)$, where $E$ is the kinetic energy of the particle and $\alpha$ is an angle defined by $\alpha = arctan(v_x/v_y)$. Suppose that at an initial instant in time, a particle is located at phase space coordinates $(x, y, E, \alpha) = (x_0, y_0, E_0, \alpha_0)$ with $y_0 = y_{max}/2$. Note that the particle starts at the configuration midplane (at $y_0 = y_{max}/2$). It is helpful to define two types of resulting trajectories. The first will be referred to as a "confinement" trajectory. Here, the particle travels a path length that is much larger than the axial length of the trap. The particle experiences a large number of axial reflections, and it appears the particle trajectory will continue indefinitely without intersecting the electrode wall. In contrast, the second kind of trajectory, referred to as a "loss" trajectory, is one in which the particle encounters an electrode boundary.

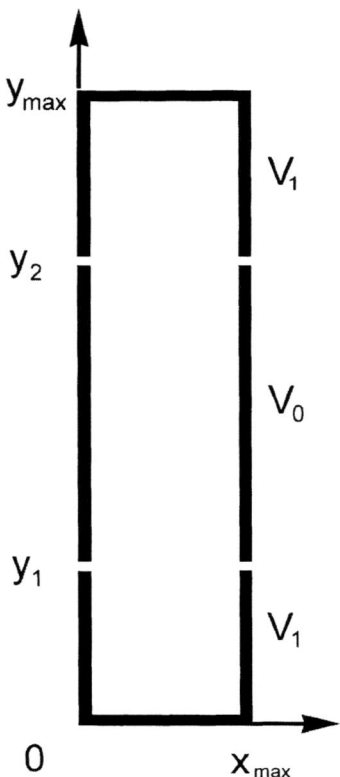

**FIGURE 1.** Electrode configuration. Three electrodes are aligned end-to-end. The center electrode is held at $V_0 = 0$ V, while the two end electrodes are held at $V_1 = 1$ V. The dimensions chosen are $x_{max} = 10$ mm, $y_1 = 10$ mm, $y_2 = 30$ mm, and $y_{max} = 40$ mm.

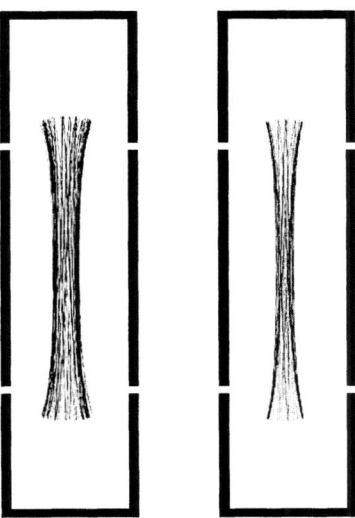

**FIGURE 2.** Confinement trajectories. For the figure on the left, $x_0 = 6$ mm, $y_0 = y_{max}/2 = 20$ mm, $E_0 = 0.63$ eV, and $\alpha_0 = 0$. For the figure on the right, $x_0 = 5$ mm, $y_0 = y_{max}/2 = 20$ mm, $E_0 = 0.64$ eV, and $\alpha_0 = 3$ degrees.

## RESULTS

Simulations indicate that single-particle confinement in the configuration presented in Fig. 1 is possible. A large number of confinement trajectories were found, where the particle remained trapped for times larger than the time it takes to bounce axially about 100 times (examples are shown in Fig. 2). A parameter study was conducted. The

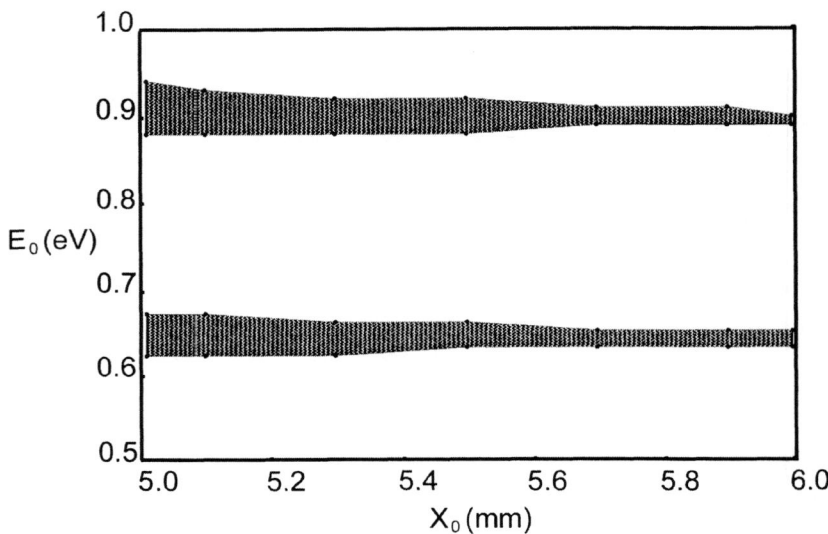

**FIGURE 3.** Results of parameter study with $y_0 = y_{max}/2 = 20$ mm and $\alpha_0 = 0$. The gray area indicates the range of values for $x_0$ and $E_0$ that result in confinement trajectories.

particle's initial position $x_0$ and energy $E_0$ were the variable parameters, while the other two parameters were not changed, $y_0 = y_{max}/2$ and $\alpha_0 = 0$. The results indicate that particle confinement occurs for a range of initial conditions. The results of the parameter study are presented in Fig. 3, where the shaded regions indicate the initial values that result in particle confinement. For the configuration considered, the results indicate that confinement is possible for two separate energy regions. It should be noted that the proton energy, the configuration dimensions, and the applied voltage were chosen rather arbitrarily. It is speculated that three-electrode confinement configurations can be designed for confining low energy (e.g., meV) ions and high energy (e.g., keV) ions.

## DISCUSSION

The results presented here may be considered to apply to a cylindrical system consisting of coaxial electrodes that are used to confine an annular ion beam in the limit that the beam thickness is much smaller than the beam radius. In future work, we plan to carry out a study that applies to axially symmetric ion beams that are not hollow. The future study will be similar (but more detailed) than the present study. The present study may be considered to provide preliminary results that suggest a reflecting ion beam can be confined within a simple axially symmetric configuration consisting of three cylindrical electrodes that are aligned end to end.

It may be possible to trap two oppositely signed particle species, with one species being a reflecting ion beam. One possibility is to form a nested well configuration similar to one of those described in Ref. [7]. Another possibility is to combine traps such that two or more reflecting beams intersect each other. A virtual anode (cathode) may be expected to form where two or more positive (negative) beams intersect, and a three-dimensional well may form within which electrons (positrons) can be confined.

The possibility exists of using the configuration for atomic and molecular physics studies of trapped ions in the absence of a magnetic field. For such studies, the ions located near the geometric center of the trap would be in a region having a small or negligible externally produced electric field. However, the ions do experience an electric field near their turning points.

Other possible applications of electrostatically trapped, reflecting ion beams include antihydrogen and fusion energy production. For antihydrogen production, the detrimental effect that a magnetic field can have on the three-body recombination rate [8] can be made negligible. This could be important for research associated with experiments at CERN, which includes studying antihydrogen that is formed by achieving simultaneous confinement of positrons and antiprotons such that recombination takes place [9, 10]. For fusion energy production, a large number of reflecting beams could be arranged to intersect a central location, and the inertial electrostatic confinement concept could be implemented in the "star" mode of operation [11], with the possibility of enhanced ion confinement.

## ACKNOWLEDGMENTS

This material is based upon work supported by the National Science Foundation under Grant No. PHY-0099617 and the Texas Advanced Research Program under Grant No. 3594-0003-2001.

## REFERENCES

1. W. Paul, Rev. Modern Phys. **62**, 531 (1990).
2. D. Zajfman et al., Phys. Rev. A **55**, R1577 (1997).
3. H. B. Pedersen et al., Phys. Rev. Lett. **87**, 055001 (2001).
4. H. B. Pedersen et al., Phys. Rev. A **65**, 042703 (2002).
5. H. B. Pedersen et al., Phys. Rev. A **65**, 042704 (2002).
6. J. D. Jackson, *Classical Electrodynamics*, 3rd Ed., New York: John Wiley & Sons, Inc., 1999, ch. 1, pp. 34 and 37-38.
7. K. F. Stephens II et al., in *Non-Neutral Plasma Physics III*, edited by J. J. Bollinger, R. L. Spencer and R. C. Davidson, AIP Conf. Proc. **498**, 451 (1999).
8. M. E. Glinsky and T. M. O'Neil, Phys. Fluids B **3**, 1279 (1991).
9. M. Amoretti et al., Nature **419**, 456 (2002).
10. G. Gabrielse et al., Phys. Rev. Lett. **89**, 213401 (2002).
11. Y. Gu and G. H. Miley, IEEE Trans. Plasma Sci. **28**, 331 (2000).

# Penning-Malmberg and Minimum-B Trap Compatibility: the Advantages of Higher-Order Multipole Traps

J. Fajans, A. Schmidt

*Dept. of Physics, U.C. Berkeley, Berkeley CA 94720*

**Abstract.** The ATHENA and ATRAP collaborations have recently created large numbers of un-trapped anti-hydrogen atoms. The most commonly suggested scheme for trapping the anti-hydrogen is to use a Minimum-B trap. Unfortunately, the Minimum-B fields are very likely to destroy the confinement of the anti-hydrogen constituents; the positrons and anti-protons, which are themselves held in double-well Penning-Malmberg traps. The reasons for the loss of confinement, and modifications to the Minimum-B trap that may alleviate this problem, are discussed in this paper.

## INTRODUCTION

Both ATHENA and ATRAP have had remarkable success generating anti-hydrogen ($\bar{\text{H}}$) [1, 2, 3]. The two experiments differ in details, but both experiments employ double-well, Penning-Malmberg traps to hold and cool the anti-hydrogen constituents: positrons ($e^+$), and anti-protons ($\bar{\text{p}}$). In the experiments considered here, temperature effects or "sloshing" are used to make the anti-hydrogen constituents overlap, and, occasionally, form anti-hydrogen through radiative or three-body recombination. Being neutral, the anti-hydrogen is not confined in the Penning-Malmberg trap. Consequently, the anti-hydrogen lasts only until it is carried into the trap wall by its initial momentum.

Because the anti-hydrogen is generated inside the Penning-Malmberg trap, the Minimum-B fields must be superimposed on the Penning-Malmberg fields. Unfortunately, simple Minimum-B fields (see Fig. 1) are likely to destroy the confinement of the anti-hydrogen constituents, $e^+$ and $\bar{\text{p}}$, in the Penning-Malmberg traps; anti-hydrogen may not have enough time to be formed before the positrons and anti-protons are lost. However, using high-order multipole fields may minimize the deleterious effects of the Minimum-B fields on the constituent confinement.

## ATHENA AND ATRAP PARAMETERS

For both ATHENA and ATRAP, the positron Debye length is much less than the positron column radius or length; thus the positrons are in the collective, or plasma regime. The anti-proton Debye lengths are only somewhat smaller (factors of two to seven) than the anti-proton column lengths and radii. Thus, the anti-protons are also in the plasma regime, but less robustly than the positrons. Note that both the positrons and the anti-

CP692, *Non-Neutral Plasma Physics V*, edited by M. Schauer et al.
© 2003 American Institute of Physics 0-7354-0165-9/03/$20.00

protons are far from the Brillouin limit, and diamagnetic effects are unimportant.

## EFFECT OF THE MIRROR FIELD

Axial $\bar{H}$ confinement in Minimum-B traps is provided by a "mirror" field that peaks the field at the trap ends. Such fields can be generated by two end coils, as shown in Fig. 1. Both the $e^+$ and $\bar{p}$ equilibria and transport will be affected by the mirror fields. Equilibria in mirror fields have been studied experimentally [4] and theoretically [5]. The equilibria are substantially more complicated than those in a simple solenoidal field. They may not be as advantageous. For example, the constituent densities will be highest at the ends of the trap, underneath the mirror coils, not in the center, where the $\bar{H}$ generation takes place.

A more serious consequence of the mirror field is increased transport. Kabantsev and Driscoll [6] have shown that transport increases dramatically with axial magnetic field variations. The transport is thought to be due to "Trapped Particle Scattering:" [7] scattering across trapped/untrapped particle separatrices. Kabantsev and Driscoll measured a five-fold increase in the transport for a 0.1% field variation. Given that a satisfactory Minimum-B trap must have at least a 100% variation in the field, this increase could be devastating; a linear extrapolation predicts that the transport would increase by a factor of 5000. It is unlikely that such high transport could be controlled by the rotating wall technique [8]. However, a 100% field variation is beyond the linear regime, and the nonlinear transport scaling is currently unknown.

Moving the mirror field coils out beyond the extent of the Penning-Malmberg trap would prevent this transport. This solution is not ideal as the volume into which the $\bar{H}$ is trapped would increase.

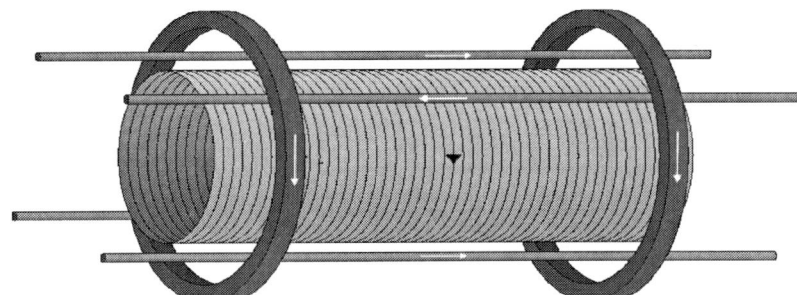

**FIGURE 1.** A quadrupole Minimum-B/Malmberg-Penning trap. The solenoid produces the axial field for the Penning-Malmberg trap, the two end coils generate the mirror field that confines $\bar{H}$ axially, and the four quadrupole wires generate the field that confines $\bar{H}$ radially. This configuration is similar to a Joffe configuration. To trap $\bar{H}$, the mirror coils must be positioned so that the minimum in the field extends over the overlap region.

# EFFECT OF THE MULTIPOLE FIELD

Radial neutral-particle confinement requires a magnetic field that increases with radius. Such fields can be generated by quadrupole or higher-order magnetic multipoles. The equilibria and transport in multipole fields are quite complicated. The magnetic field lines associated with the quadrupole superimposed on the Penning-Malmberg trap are shown in Fig. 2. The shape of the magnetic field will create elliptically shaped plasma, which may be unstable.

As with the mirror field, the most significant effect of the multipole field is likely to be increased transport. Experiments have demonstrated sharply increased transport with very small quadrupole fields; quadrupolar fields at the plasma edge 4000 times lower than solenoidal field have been shown to double the outward diffusion [9]. Preliminary data indicates that the diffusion increases linearly with quadrupole field strength when the quadrupole is strong [10]. To form a significant Minimum-B well, the quadrupole field would have to be comparable to the solenoidal field. Thus, the quadrupole might enhance diffusion by a factor of 4000.

The quadrupolar induced transport is strongest in the neighborhood of a resonance. Experimental data [9, 11] suggests that the resonance occurs when the ratio $nL/Bv_T$ remains constant, where $n$ is the plasma density, $L$ is the plasma length, $B$ is the solenoidal field strength, and $v_T = \sqrt{kT/m}$ is the thermal velocity. This scaling is consistent with an orbital resonance; if a particle rotates by 90 degrees during the time it takes to make one trap transit, then it can be on a trajectory that goes ever outwards or inwards [9]. While such resonant particles would be lost, there are too few precisely-resonant particles to induce significant transport. However, there is a broad class of particles that are near resonance, which undergo large radial excursions. If these particles collide, their excursions will cause them to diffuse. The detailed theory of this Resonant Particle Transport is difficult to construct, nonetheless, a back-of-the-envelope theory

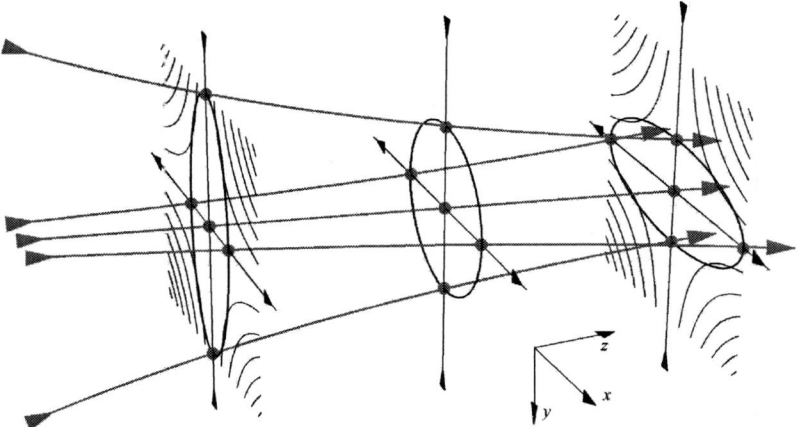

**FIGURE 2.**    Net field lines in the presence of a quadrupole and a solenoid.

predicts transport of the right order of magnitude [11]. However, the observed position of the resonance is off by a factor of two from the expected position. This discrepancy has lead Kabantsev and Driscoll to suggest [6] that the transport is more properly described by their Trapped Particle Scattering mechanism.

In the slowly rotating regime relevant here, the observed diffusion is inversely proportional to $f_R^2$. If the ATHENA and ATRAP plasmas were off resonance by a factor of more than 50, the diffusion would be down to tolerable levels. Unfortunately, they are much closer than this to resonance, and the diffusion is likely to be very large.

## COLLISIONS

Both Trapped Particle Scattering Transport and Resonant Particle Transport occur only when the plasma particle orbits are disturbed by collisions. Unfortunately, the transport theories are not sufficiently well developed to predict the scaling with collision frequency. Moreover the collision frequencies are complicated to predict because of O'Neil's collisional adiabatic invariant [12, 13]. Nonetheless, the collision frequencies appear to be easily great enough to induce transport [14].

A recent paper [15] contends that, because particles in Minimum-B fields orbit on unique, stable trajectories, transport will be insignificant. But the collisions described here disturb the orbits; proof of the stability of individual particle orbits is largely irrelevant to the issues of transport and loss.

## HIGH-ORDER MULTIPOLE ADVANTAGES.

As shown above, quadrupole fields pose profound problems for $e^+$ and $\bar{p}$ confinement. These problems may be alleviated by using higher-order multipole fields. The field from an infinitely-long multipole of order $s$ scales with radius $r$ as

$$|B| = B_{max} \left( \frac{r}{R_w} \right)^{s-1} , \tag{1}$$

where $B_{max}$ is the field at the wire radius $R_w$. For large order multipoles, the field is very small at the center. If a trap can be configured so that the $e^+$ and $\bar{p}$ column radii are small compared to the wall radius, the constituents would be largely unaffected by the weak multipole field at the center. The resulting $\bar{H}$ would still be trapped by the strong multipole field at the wall. For example, if the constituent radii were one third of the wall radius, the multipole field magnitude were equal to the solenoidal field magnitude at the wall, and the multipole order was ten, the maximum field experienced by the constituents would be twenty thousand times less than the solenoidal field. Such multipole fields would be on the same order as the fields used in the quadrupole experiments [9], a level that is probably tolerable. Finite length effects and the addition of the mirror field complicate this picture somewhat, and the fields are best calculated numerically. The fields from a typical trial configuration are shown in Fig. 3.

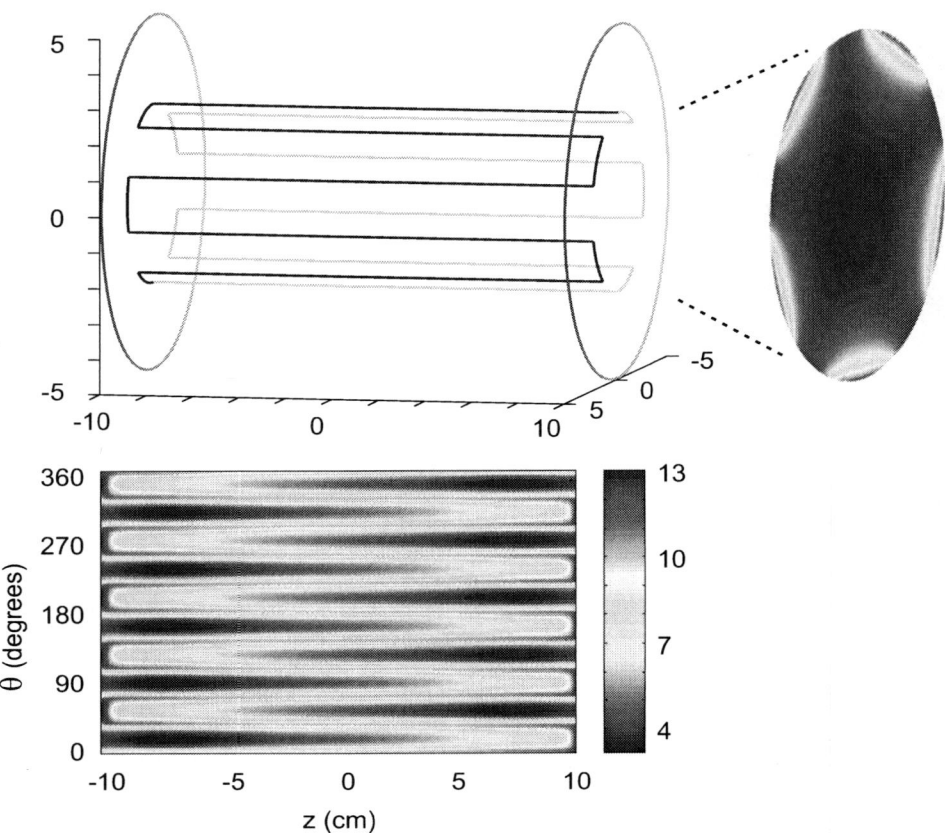

**FIGURE 3.** Boundary magnetic field values for the configuration shown at top left. All magnetic field values are normalized to the field at the center. The multipole is fifth order with wires at radius 2.5 cm. The mirror coils are of radius 5 cm. The ratio of current in the multipole to current in the mirror coils is 1.45. In addition, there is an infinite solenoid, oriented along the configuration axis, which contributes 80% of the field in the center of the trap. The plot at the left shows the field at the end, and the bottom plot shows the "unwrapped" field along on the sides at 90% of the radius of the multipole wires, i.e. a cylinder of radius 2.25 cm. The minimum field occurs about 14% of the way in from the end on the side, and is of magnitude 3.16.

One might worry that the greater degree of cancellation found in higher order multipoles would require larger currents for the same field increase. An elementary calculation shows that this is not true. Assuming a sinusoidal current distribution, the required current density $J_\lambda$, in $(A/m)$, is

$$J_\lambda = \frac{2B_{\text{max}}}{\mu_0} \sin s\theta, \qquad (2)$$

where $B_{\text{max}}$ is in Tesla. On gathering the current into $2s$ wires, the current in each wire drops proportional to $1/s$, a favorable outcome. As an added benefit, the magnetic forces on each wire also decrease proportionally. Nonetheless, the currents, and hence the forces, are quite large; a tenth order multipole that produces $1\,T$ at wires arrayed at $R_w = 1\,cm$ requires currents on the order of $4000\,A$. It may be easier to create the required fields with permanent magnets; fortuitously, high-order multipole fields are easier to create with permanent magnets than are quadrupole fields.

Increasing the order of the multipole moves the multipole resonance to proportionally higher velocities. This can be favorable or unfavorable depending on the plasma parameters; it is likely to be favorable for ATHENA $e^+$, but unfavorable for ATRAP $e^+$.

A disadvantage of higher order multipoles is that the $\bar{H}$ trapping volume is relatively large; the multipole fields are not significant until the $\bar{H}$ move far off the axis. This problem could be solved by turning on an auxiliary quadrupole to more tightly confine the $\bar{H}$ once a sufficient number have been created and trapped in the high order multipole.

## CAVEATS

The loss of confinement predicted here is based on significant extrapolations from the current experimental data. The transport could saturate. The experiments and theory to date have studied mirror and quadrupole configurations separately. It is conceivable, but unlikely, that each will ameliorate the effects of the other. The model for the multipole transport is based on orbital resonances. The effects of the magnetron fields are difficult to calculate. Further, particularly for the $\bar{p}$, the rotation frequency is difficult to determine in the $\bar{p}$-$e^+$ overlap region. There may be very odd effects there. Finally, Kabantsev and Driscoll suggest that the proper mechanism for quadrupole induced transport is their Trapped Particle Scattering, not the Resonant Particle Diffusion mechanism presented here. However, the extrapolations are based on observational scalings, not theoretical predictions.

## CONCLUSIONS

The tentative plans to trap $\bar{H}$ in ATHENA and ATRAP need to be refined. The proposed mirror and quadrupole field are very likely to destroy the confinement of the $\bar{H}$ constituents, $e^+$ and $\bar{p}$. Pushing the mirror field out beyond the constituent confinement region, and using a high order multipole, may alleviate the $\bar{H}$ constituent confinement problems. More exotic schemes are also available [16].

## ACKNOWLEDGMENTS

We thank Prof. Dima Budker, Prof. Gerry Gabrielse, Dr. Erik Gilson, Dr. Rolf Landau and Prof. Tom O'Neil. This work was supported by the Office of Naval Research and by the National Science Foundation.

# REFERENCES

1.  Amoretti, M., Amsler, C., Bonomi, G., Bouchta, A., Bowe, P., Carraro, C., Cesar, C. L., Charlton, M., Collier, M. J. T., Doser, M., Filippini, V., Fine, K. S., Fontana, A., Fujiwara, M. C., Funakoshi, R., Genova, P., Hangst, J. S., Hayano, R. S., Holzscheiter, M. H., Jørgensen, L. V., Lagomarsino, V., Landua, R., Lindelöf, D., Rizzini, E. L., Macri, M., Madsen, N., Manuzio, G., Marchesotti, M., Montagna, P., Regenfus, H. P. V. C., Riedler, P., Rochet, J., Rotondi, A., Rouleau, G., Testera, G., Variola, A., Watson, T. L., and der Werf, D. P. V., *Nature*, **419**, 456 (2002).

2.  Gabrielse, G., Bowden, N., Oxley, P., Speck, A., Storry, C., Tan, J., Wessels, M., Grzonka, D., Oelert, W., Schepers, G., Sefzick, T., Walz, J., Pittner, H., Haensch, T., and Hessels, E., *Phys. Rev. Lett.*, **89**, 213401 (2002).

3.  More precisely, ATHENA and ATRAP are actually creating highly excited $\vec{E} \times \vec{B}$ guiding center atoms. See C. F. Driscoll, comment submitted to Phys. Rev. Lett., and M. E. Glinsky and T. M. O'Neil, Guiding center atoms: Three-body recombination in a strongly magnetized plasma, 3 (1991) 1279.

4.  Gopalan, R., *Studies of Cryogenic Electron Plasmas in Magnetic Mirror Fields*, Ph.D. thesis, University of California, Berkeley (1998).

5.  Fajans, J., *Phys. Plasmas*, **10**, 1209 (2003).

6.  Kabantsev, A., and Driscoll, C., *Phys. Rev. Lett.*, **89**, 245001 (2002).

7.  Kabantsev, A., Driscoll, C., Hilsabeck, T., O'Neil, T., and Yu, J., *Phys. Rev. Lett.*, **87**, 225002 (2001).

8.  Huang, X. P., Fine, K. S., and Driscoll, C. F., *Phys. Rev. Lett.*, **78**, 875 (1997).

9.  Gilson, E., and Fajans, J., *Phys. Rev. Lett.*, **90**, 01501 (2003).

10. E. Gilson, unpublished.

11. Gilson, E., *Quadrupole Induced Resonant Particle Transport in a Pure Electron Plasma*, Ph.D. thesis, U.C. Berkeley (2001).

12. O'Neil, T. M., *Phys. Fluids*, **26**, 2128 (1983).

13. Beck, B. R., Fajans, J., and Malmberg, J. H., *Phys. Rev. Lett.*, **68**, 317 (1992).

14. J. Fajans and A. Schmidt, *Malmberg-Penning and Minimum-B trap compatibility: the advantages of higher-order multipole traps*. Submitted for publication to Nuc. Inst Meth. A.

15. Squires, T., Yesley, P., and Gabrielse, G., *Phys. Rev. Lett.*, **86**, 5266 (2001).

16. Dubin, D. H. E., *Phys. Plasmas*, **8**, 4331 (2001).

# A Magnetic Nozzle and Diverter Electrode to Improve Penning Fusion Efficiency

Carl C. Dietrich and Raymond J. Sedwick

*MIT Space Systems Laboratory, 77 Massachusetts Ave., Cambridge, MA 02139*

**Abstract.** Inertial Electrostatic Confinement of fusion ions in a modified Penning trap is further modified to include a mechanism for low-power recirculation of electrons via a diverter electrode placed in the low-field, cusp region of the trap. The locally divergent magnetic field lines act as a magnetic nozzle to extract energy bound up in the angular momentum of the larmor gyrations of the electrons, enabling the diverter potential to be close to that of the emitter while still collecting scattered electrons. In theory, the diverter recirculates a large fraction of the scattered core electron population back to the emitter through a much smaller potential difference, thereby reducing the overall power consumption for a given collisional diffusivity in the core region and improving overall system efficiency. Modeling of the general system is presented which suggests a power density that is too low to be practical for power generation unless ion density enhancement via a POPS type mechanism is realized. However, this technology is suggested as a potential candidate for an experimental plasma target, neutron source.

## INTRODUCTION

The Penning Fusion Reactor (PFR) is a type of Inertial Electrostatic Confinement (IEC) fusion device in which high energy ions are confined electrostatically by a virtual cathode produced by the space charge of confined electrons in a modified Penning Trap. This concept was first developed by Barnes et. al. at Los Alamos National Laboratory (LANL) [1,3,4,6,7,8]. In order to produce the potential well for ion confinement, a typical cylindrically symmetric Penning Trap is modified by the inclusion of an internal cavity in the anode structure to allow the space charge of the confined, non-neutral plasma to depress the potential inside the anode cavity. The elimination of the cathode wire grid in this type of IEC device may improve ion lifetime over more conventional, gridded, IEC devices thereby allowing greater potential for high-Q operation [2]. Confinement of ions in this type of virtual cathode has been recently demonstrated at LANL [3].

The PFR has a favorable scaling [4]. This scaling shows that the unit anode "cell" size for this type of reactor will typically be measured on the order of a few millimeters to a few tens of millimeters (not including power supplies, magnets, vacuum equipment etc). The small size allows for low-cost experimentation similar to other forms of IEC fusion. The combination of small unit size, low-cost experimentation, and the potential for high efficiency operation prompted interest in examining some of the practical issues associated with the construction of a PFR experiment. This work examines some of the issues that would be involved in the design and construction of a Penning Fusion Reactor experiment.

Figure (1) shows a finite element model of the potential structure with a line charge simulating a constant density reflected electron beam. The formation of an electrostatic potential well is clearly visible from the space charge of the beam. Of note is the small scale of the device and the high electrostatic potential differences that are necessary to confine thermonuclear fusion ions.

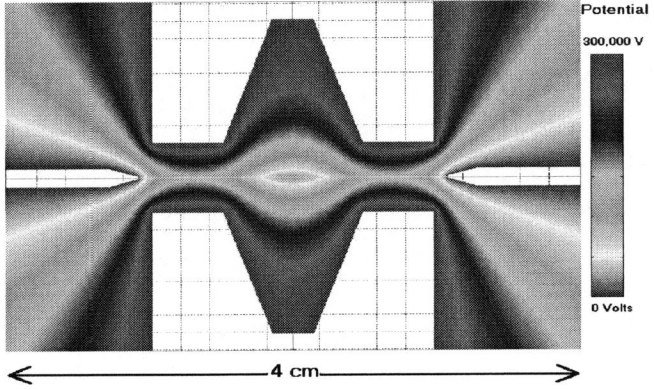

**Figure 1.** Constant potential contours of a cross-section of a single cell Penning Fusion Reactor with cold cathode emitters

## MODELING OF A PENNING FUSION REACTOR

In order to model this type of PFR, a typical single-cell geometry is presented for clarity, but it is assumed that a multi-cell, "string of pearls" configuration as described by Barnes [4] will likely be preferable in an operational reactor.

The maximum density that can be achieved in a magnetically confined, pure electron, non-neutral plasma column of uniform density is the Brillouin limit [5]

$$n_B = \frac{\varepsilon_o B^2}{2m_e} \tag{1}$$

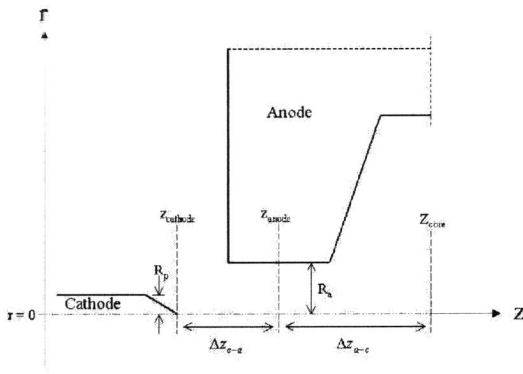

**Figure 2.** Geometry near cathode with labels used for modeling

Existing superconducting solenoid technology can achieve field strengths of up to 20 Tesla, which yields an upper density limit of nearly $2 \times 10^{21} \text{m}^{-3}$. Recent work suggests

260

that the optimum fractional neutralization, $f$, of the core is ~0.2 [2]. Ion number densities could exceed those of typical Tokamak devices (~$10^{20}$ m$^{-3}$) yielding favorable power densities. In addition, excitement of a spherical harmonic resonance in the potential structure via POPS (Pulsed Oscillating Plasma Sphere) operation [6] could allow core densities to exceed the Brillouin limit in a type of pulsed operating mode. There is evidence that these limits can be exceeded with a non-thermal electron distribution [7,8], but the analysis presented here will assume that the Brillouin limit will apply in general.

There is also a limiting radius of a constant density non-neutral plasma column at the Brillouin limit. This radius is defined by the location at which ExB drift would equal the speed of light. If the radial electric field exceeds $cB$, the electrons will not follow a confined trajectory. The location of this radius can be obtained by a simple substitution of the Brillouin density into Poisson's equation. The result is given here

$$R_{p_{max}} = \frac{4m_e c}{eB} \tag{2}$$

If we make the assumption that the core size is $O(R_{pmax})$, the core volume will scale like $B^{-3}$. This relationship would appear to detract from the concept, however the ion density will scale like $B^2$ (if we assume a constant fraction of the Brillouin limit). In addition, if we also impose a constant potential well depth for all scales, the power density will go like $B^4$ resulting in a net power output that is proportional to B. If we make the logical assumption that the size of the device will be related to the maximum confined plasma radius, the unit cell size will also scale like $B^{-1}$. This results in the interesting characteristic that as the magnetic field increases the power output per cell goes up but the size goes down. The ultimate limit will be determined by the magnetic field that can be achieved (ignoring the potential for POPS enhancement).

## 1D Modeling of Beam and Virtual Cathode

We start the analysis with the 1D assumptions that the beam density is constant in r and the variation in potential across the beam is negligible when compared to the axial variation in potential so that the potential can be approximated as a pure function of z. We will also continue to assume a cold electron population ($T_e$~$T_{emitter}$) and $\phi_{cathode}$=0. Now it is possible to estimate the maximum allowable reflected current density of the beam based on assuming the Brillouin limit density at the cathode (when the velocity is minimum, $O(v_{thermal})$ in the z-direction).

$$j_{max} = en_B v_{z_{min}} = \frac{e\varepsilon_o B^2}{2m_e} \sqrt{\frac{8kT_e}{\pi m_e}} \tag{3}$$

The electron reflexed current density, j, is constant at all z in this steady state model. Using the Child-Langmuir relationship for space charge limited currents (equation (7)) and the fluid continuity equation, it is possible to estimate the core well potential by imposing this constant current density constraint at all z along the reflexed beam.

$$\phi_{core} = C(z = z_{core}) = C(z = z_{anode})\left(1 - \sqrt{1-f}\frac{\Delta z_{a-c}}{\Delta z_{c-a}}\right)^{4/3} \tag{4}$$

where

$$\Delta z_{a-c} = \left| z_{anode} - z_{core} \right| \tag{5}$$

$$\Delta z_{c-a} = \left| z_{cathode} - z_{anode} \right| \tag{6}$$

Although a 1D analysis might seem simplistic, it appears to be reasonably accurate when compared to the well depths calculated by Barnes [3]. Equation (4) has only real solutions when the core-to-anode axial distance is less than or equal to the anode-to-cathode axial distance for zero ion fraction in the core. The limit where the distances are equal would correspond to the maximum achievable potential well depth, $\phi_{core}=\phi_{cathode}=0$. In that "equal distance" design scenario, if the ion fraction was near the optimal 0.2 identified by Chacon [2] instead of zero, the core potential would be approximately 5% of the centerline potential at the anode. It thereby follows that if a zero potential core is desired for an ion fraction of 0.2, the core-to-anode distance should be at least 1.12 times the anode-to-cathode distance.

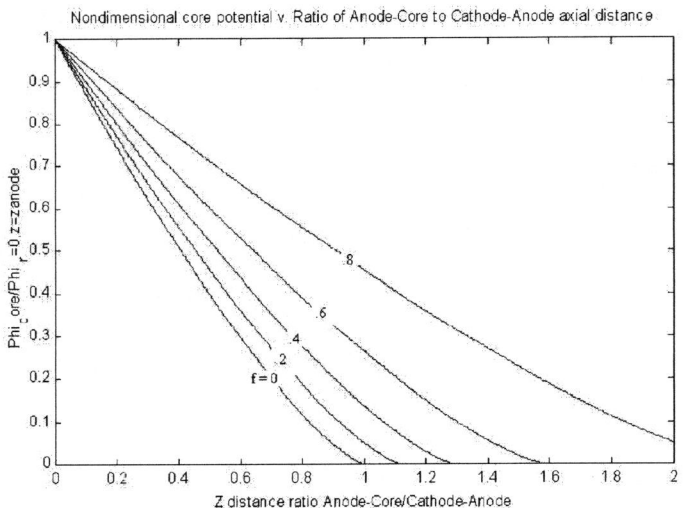

**Figure 3.** Nondimensional Core Potential v. Axial Distance Ratio

Generally, for large $\Delta z_{c-a}$, the magnitude of the space-charge-limited current density is uniquely determined by the 1D Child-Langmuir law

$$j = \frac{4\sqrt{2}}{9}\varepsilon_o \sqrt{\frac{e}{m_e}} \frac{C^{3/2}(z = z_{anode})}{\Delta z_{c-a}^2} \tag{7}$$

as long as this calculated current density is less than the maximum allowed by the radial, Brillouin limit in equation (3). It is important to note that Jory and Trivelpiece present the fully relativistic Child-Langmuir relationship [10], but it does not deviate substantially from the non-relativistic approximation for most anode potentials of interest, and the elliptic integrals involved in the relativistic calculation require numerical solution, so the non-relativistic approximation is used in this analysis.

When

$$\Delta z_{c-a} \leq \frac{2}{3} \frac{\sqrt{m_e}}{B} \left( \frac{\pi}{ekT_e} \right)^{\!\!1/4} C^{3/4} (z = z_{anode}) \qquad (8)$$

the current density will be limited by the Brillouin limit near the cathode. In that situation $j$ would be approximated by (3). Equations (8) and (4) can be used as a design guidelines, but it is important to note that the imposition of a constant ion fraction along the beam is inherently flawed because the ion density profile along the z-axis is independent of this simple, 1D electron based model. The actual variation of ion density is beyond the scope of this paper, but the purpose of the above analysis is to show that any $f>0$ will require a larger anode void radius than the cathode-anode gap in order to maximize the potential well depth.

Since high core densities are desirable, it is likely that the potential well will be quite deep so the axial electron velocity in the core is low. As can be seen by examination of (4), when fractional core neutralization is taken into account, the actual ratio $\Delta z_{a-c}/\Delta z_{c-a}$ will equal or even exceed unity while still maintaining a core potential somewhat above that of the cathode.

In order to estimate the electron density in the column at the anode for the calculation of $C(z=z_{anode})$, it is necessary to approximate $v_z(z=z_{anode})$. For the purposes of this approximate (not fully self-consistent) analysis, we will assume

$$v_z(z = z_{anode}) = c \sqrt{1 - \frac{1}{\gamma_{max}^2}} \qquad (9)$$

Now if $\Delta z_{c-a}$ satisfies the relationship in (8), j will be given by (3), and the density of the electron column at the anode can be approximated by

$$n_e(z = z_{anode}) = \frac{\varepsilon_o B^2}{2m_e c} \sqrt{\frac{8kT_e}{\pi m_e \left(1 - \frac{1}{\gamma_{max}^2}\right)}} \qquad (10)$$

When substituted into a cross-sectional expression for the potential that assumes constant beam density (ref. [13]), the centerline potential at the anode is given by

$$C(z = z_{anode}) = \phi_{anode} - \frac{eR_p^2 B^2}{m_e^{3/2} c} \sqrt{\frac{kT_e}{8\pi \left(1 - \frac{1}{\gamma_{max}^2}\right)}} \left(1 + 2\ln\!\left(\frac{R_a}{R_p}\right)\right) \qquad (11)$$

If the design calls for $R_p = R_{pmax}$, the dependence on the magnetic field drops out and the centerline potential at the anode is only a function of the anode potential, the radius ratio, and the thermal velocity of the electrons, $v_{the}$.

$$C(z = z_{anode}, R_p = R_{pmax}) = \phi_{anode} - \frac{4m_e c v_{the}}{e\sqrt{\pi \left(1 - \frac{1}{\gamma_{max}^2}\right)}} \left(1 + 2\ln\!\left(\frac{R_a}{R_p}\right)\right) \qquad (12)$$

For typical possible operating parameters with a cold cathode DD or DT system ($T\sim1000K$, $\phi_{anode}=300kV$, $R_a/R_p=2$), equation (12) gives about a 0.7% ($\sim2kV$) potential difference between the anode and the centerline at the anode. Clearly, in this situation the axial velocity estimation in (9) is good and the simple model is close to the actual, self-consistent solution. At high electron temperatures and non-

263

relativistic (low) anode potentials, the assumptions break down.

We now conclude the description of this electron beam model with an approximate expression for the core density again invoking the maximum current density from equation (3). Similar to (10) we get

$$n_e(z = z_{core}) = \begin{cases} \dfrac{\varepsilon_o B^2}{2m_e c} \sqrt{\dfrac{8kT_e}{\pi m_e \left(1 - \frac{1}{\gamma_{core}^2}\right)}}, \gamma_{core} > 1 \\ \\ \dfrac{\varepsilon_o B^2}{2m_e}, \gamma_{core} = 1 \end{cases}$$

(13)

where

$$\gamma_{core} = 1 + \frac{e\phi_{core}}{m_e c^2}$$

(14)

In the candidate 300kV system with an "equal distance" design and a neutralization fraction of 0.2, the core potential is ~15kV yielding a core electron density of $n_e = 5.4 \times 10^{18} m^{-3}$ and a core ion density of $n_i = 1.1 \times 10^{18} m^{-3}$. These densities are very sensitive to the dimensions of the gaps in the device. For example, to achieve the Brillouin limit density in the core (zero core potential) the ratio of the distances is changed from unity to 1.12 at f=0.2. This 12% dimension change results in a core electron density change by a factor of nearly 400, from $5 \times 10^{18} m^{-3}$ to $2 \times 10^{21} m^{-3}$. Careful design of the system to maintain a desired ion fraction will be important as will the dimensional stability and design tolerances of an operating PFR.

## System Power Balance

It is assumed that the ions trapped in the potential well equilibriate to a nearly maxwellian distribution with a characteristic temperature $T_i >> T_e$. This thermal distribution is likely to evolve because the average ion-ion collision time is significantly shorter than either the fusion time or the average ion confinement time and heated electrons are immediately removed from the system because of the high electron bounce frequency.

It is interesting to note that a PFR system with a truly monoenergetic electron distribution could be designed in which the rate of net energy transfer from ions to electrons is zero in the high-density core. This occurs when the thermal velocity of the ions is close to the velocity of the electrons. Specifically, for deuterium and a monoenergetic electron population in the core, a core potential of 52 Volts would result in a quasi-equilibrium with 20keV deuterons [11]. In practice, this quasi-equilibrium would likely be very hard to achieve because of the effects of electron thermalization on the timescale of the electron-electron collision frequency [12].

The reacting volume can be reasonably approximated as a sphere of radius $R_p$. The total power output of the cell is given by

$$P_{fus} = f^2 n_e^2 (z = z_{core}) \langle \sigma v \rangle_{fus} \frac{4}{3} \pi R_p^3 E_{fus}$$

(15)

Based on this model and a 20 Tesla maximum field design with a 20% ion fraction, a single cell could produce up to $10^8$ D-D neutrons/second or up to $10^{10}$ D-T neutrons/second. The D-T cycle would correspond to fusion output of at most 27mW per cell (without POPS enhancement). This power density may be too low to be competitive with alternative power generation technologies, but the neutron generation rates are comparable with commercially available sources.

Mitigation of the electron power loss can be accomplished by the use of a shaped magnetic field in combination with a low potential diverter. This field is arranged so that magnetic surfaces located between $R_p$ and $R_a$ intersect the diverter electrode in a region where the field magnitude is significantly lower than the field strength near the plasma column. The following figure illustrates a candidate geometry for this design concept.

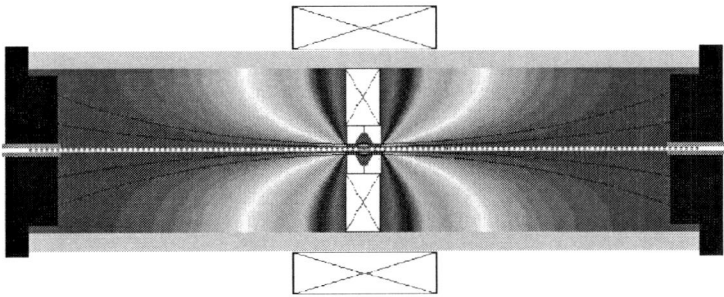

**Figure 4.** Single cell PFR with magnetic nozzle and diverter electrode. The finite element model of the electrostatic potential structure is shown and the "edge" magnetic surfaces of the magnetic nozzle are superimposed in black. The diverter electrodes are the black endcaps of the system.

It is hypothesized that if the gap between the anode and the core electron column (Ra-Rp) is larger than the maximum cyclotron radius (the classical diffusion step size), the high axial bounce frequency ($\sim10^{10}$ Hz) relative to the electron-electron collision frequency ($\sim10^8$ Hz) should allow a large fraction of the electron current to be recirculated through the low voltage diverter circuit instead of through the anode power supply.

The total power input to a Penning Fusion Reactor (discounting magnet, ion injection and vacuum systems) can be expressed by

$$P_{in} = I_{tot}(df V_{Diverter} + (1-df)V_{anode})  \qquad (16)$$

where $I_{tot}$ is the total current scattered out of the electron column by collisions and $df$ is the fraction of current that goes to the diverter. It is clear that the total input power can be minimized by diverting a large fraction of the total current to the diverter. Estimating the diverter fraction is beyond the scope of this paper.

Rider has shown that operation of any device, which relies on the maintenance of a trapped particle population with an energy distribution far from Maxwellian, requires a high recirculated-power fraction [12]. This effect of electron thermalization will fundamentally limit both the Q and the achievable power density of this type of reactor system. In addition, the magnetic nozzle and diverter configuration appears to require a "single chain" design as opposed to Barnes' proposed "multi-chain, stacked fuel rod" design. So while the nozzle-diverter would likely improve the efficiency of the system, it would further reduce the system power density because of the large area ratio required in the magnetic nozzle. This power density is further reduced because each chain would require its own high-field magnet.

## CONCLUSION

Although the system efficiency could be substantially improved by the implementation of such a magnetic nozzle and diverter, initial calculations suggest that the power density would likely be substantially lower than typical chemical battery power densities. This low power density makes such a system unlikely to be practical for power production with current magnet technology unless POPS type mechanisms can be implemented to further enhance the core density. The presented "non-POPS" design may be useful, however, to improve the efficiency and lifetime of low-yield, plasma-target, neutron sources ($10^6$-$10^{10}$ neutrons/second).

An experimental program to evaluate the potential for building an operational PFR with improved efficiency and lifetime may be justifiable given the heightened demand for neutron sources for bomb detection and medical isotope generation. In addition, work on POPS mechanisms for enhancing core ion densities should be pursued if there is interest in the development of PFR systems for power production.

## ACKNOWLEDGMENTS

This work is not presently funded. The authors wish to recognize helpful conversations with D. Barnes, M. Schauer, T. Rider, T. McGuire, and G. Carosi.

## REFERENCES

1. D. C. Barnes, L. Turner, R. A. Nebel, and T. N. Tiouririne, *Plasma Phys. Controlled Fusion,* **35**, 929 (1993).
2. L. Chacon, "Fokker-Planck Modeling of Spherical Inertial Electrostatic Virtual Cathode Fusion Systems," Nuclear Engineering Ph.D. Thesis, UIUC, (2000).
3. M. M. Schauer, D. C. Barnes, K. R. Umstadter, "Ion Trapping in the Virual Cathode of the Penning Fusion eXperiment-Ions," submitted May 6, 2002.
4. PFX-I web site: http://ext.lanl.gov/projects/pfxi/
5. L. Brillouin, *Phys. Rev.* **67**, 260 (1945).
6. R. A. Nebel, D. C. Barnes, *Fusion Tech.* **34**, 28 (1998).
7. D. C. Barnes, T. B. Mitchell, M. M. Schauer, *Phys. Plasmas*, **4**, 1745 (1997).
8. T. B. Mitchell, M. M. Schauer, D. C. Barnes, *Phys. Rev. Let.* **78**, 58 (1997).
9. R. C. Davidson, *Physics of Non-neutral Plasmas,* Imperial College Press (2001).
10. H. R. Jory, A. W. Trivelpiece, *J. Appl. Phys.* **40**, 3924, (1969).
11. J. D. Huba, *NRL Plasma Formulary.* NRL/PU/6790-98-358, Washington DC: Naval Research Laboratory (1998).
12. T. H. Rider, "Fundamental Limitations on Plasma Fusion Systems not in Thermodynamic Equilibrium," Electrical Engineering Ph.D. Thesis, MIT (1995).
13. C. C. Dietrich, R. J. Sedwick, *39ᵗʰ JPC Proceedings,* AIAA Paper 2003-4827, (2003).

# Interdisciplinary Issues Associated With Generating a Fully Non-Neutral Fusion Plasma

C. A. Ordonez

*Department of Physics, University of North Texas, Denton, Texas 76203*

**Abstract.** Various interdisciplinary issues are identified that are associated with the prospect of producing an electron-free fusion plasma.

A fully non-neutral plasma is considered here for application as a fusion plasma. A fully non-neutral plasma is defined at present as a plasma consisting of charged particles having the same sign of charge, while a fusion plasma is defined as a plasma within which nuclear fusion reactions are taking place at a useful rate. An advantage of using a fully non-neutral plasma as a fusion plasma is that plasma energy loss mechanisms would not include processes dependent on the presence of electrons, such as electron thermal conductivity and electron cyclotron and bremsstrahlung radiation. A disadvantage with using a fully non-neutral plasma as a fusion plasma is that the average fusion reactivity can be considerably limited. The average fusion reactivity is defined as the number of fusion reactions that occur per unit time per unit volume of the plasma confinement apparatus. In the work presented here, interdisciplinary issues that can lead to a limited average fusion reactivity are identified. Interdisciplinary issues are defined as involving disciplines outside of plasma physics. Disciplinary issues are not considered here. An example of a disciplinary issue is whether the Brillouin density limit can be exceeded for a magnetically confined fully non-neutral fusion plasma [1, 2]. It should be noted that this paper is intended to be somewhat pedagogical in simplicity [3], and only issues specific to fully non-neutral plasmas are identified. The issues identified should be considered for initiating any new study on the prospect of producing an electron-free fusion plasma. Although many of the issues identified here are likely known to researchers who have already undertaken such studies, the present paper is intended to provide a systematic compilation of possible interdisciplinary issues.

A general analysis is done by considering forces that might be combined to obtain a plasma confinement equilibrium. The forces considered are electric, magnetic, magnetic gradient, periodic potential, and centrifugal. The electric and magnetic fields that would be used to produce the first four of these are assumed to be time independent for the analysis. The forces are now defined for clarity. The electric and magnetic forces have the usual definitions. A plasma particle of charge $q$ within an electric field $E$ experiences an electric force $F = qE$. A plasma particle moving through a magnetic field $B$ with velocity $V$ experiences a magnetic force $F = qV \times B$. The magnetic gradient force is defined to act upon a plasma particle when the plasma particle is within a magnetic field having a magnitude gradient. In the guiding center approximation, the plasma particle

CP692, *Non-Neutral Plasma Physics V*, edited by M. Schauer et al.
© 2003 American Institute of Physics 0-7354-0165-9/03/$20.00

experiences the force, $F = -\mu\nabla B$, where $\mu$ is the magnetic moment associated with its cyclotron motion [4]. The magnetic gradient force arises from the magnetic force. However, for convenience, the magnetic gradient force is considered at present as a separate force. Charged particles that experience an electric potential that periodically changes can be considered to be acted on by an effective force, which can confine the particles [5]. The effective force, which is referred to here as the periodic potential force, arises from the electric force. However, for convenience, the periodic potential force is considered at present as a separate force.

Various model geometries are considered, although specific ways in which more than one force would be combined to achieve a force balance for each are not considered. Five geometries are illustrated in Fig. 1, of which the first three are specifically considered. The five model geometries are referred to as the spherical, cylindrical, planar, toroidal, and hollow-cylindrical configurations, respectively. The toroidal configuration is only considered in the cylindrical limit, while the hollow-cylindrical configuration is only considered in the planar limit. The spherical configuration has spherical symmetry. The plasma in the spherical configuration has a radius $r_p$, while a conducting wall is considered to exist having an inner radius $r_w$. The cylindrical, planar, toroidal, and hollow-cylindrical configurations each have cylindrical symmetry about an axis labeled SA in Fig. 1. For the cylindrical configuration, the plasma has radius $r_p$, while a conducting wall is considered to exist having an inner radius $r_w$. For the planar configuration, the plasma has a half width $r_p$, and conducting walls to either side of the plasma are separated by a distance $2r_w$. The same notation, $r_p$ and $r_w$, is used for the first three configurations so that equations can be written that apply simultaneously to all three. For the cylindrical configuration, the axial plasma length is considered to be much larger than twice the wall radius, while for the planar configuration, the radial plasma width is considered to be much larger than the axial separation between the two walls. Gauss' law is used to obtain the radial electric field strength in the spherical configuration and the radial (axial) electric field strength at an axial (radial) position that is far from the axial (radial) edges of the plasma in the cylindrical (planar) configuration. For simplicity, the plasma density is considered to be uniform for each configuration. The electric field strength at the radial (radial) [axial] plasma edge for the spherical (cylindrical) [planar] configuration is

$$E_p = \frac{Zenr_p}{\alpha\varepsilon_0},\tag{1}$$

where $Z$ is the average ion charge state, $e$ is the unit charge, $n$ is the ion density, $\varepsilon_0$ is the permittivity of free space (SI units are used throughout), $\alpha = 3$ for the spherical configuration, $\alpha = 2$ for the cylindrical configuration, and $\alpha = 1$ for the planar configuration. The corresponding electric field strength at each wall illustrated in Fig. 1 is given by

$$E_w = \frac{E_p}{R^{(\alpha-1)}},\tag{2}$$

where $R = r_w/r_p$. Note that $R \geq 1$ is required. The difference in electric potential between the geometric center of the plasma and the wall is

$$V_w = \beta r_p E_p,\tag{3}$$

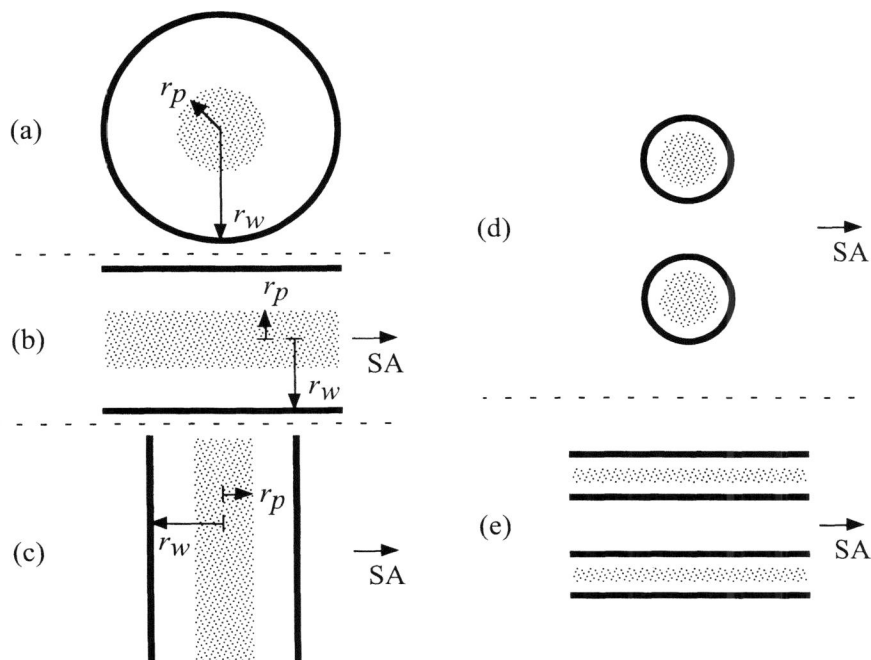

**FIGURE 1.** Spherical (a), cylindrical (b), planar (c), toroidal (d), and hollow-cylindrical (e) plasma confinement geometries. The symmetry axes for (b) - (e) are labeled SA.

where $\beta = (3/2) - (1/R)$ for the spherical configuration, $\beta = (1/2) + \ln(R)$ for the cylindrical configuration, and $\beta = R - (1/2)$ for the planar configuration.

The average fusion reactivity is quantified in terms of an average fusion power density given by

$$P_{avg} = \frac{\zeta E_f n^2 \langle \sigma v \rangle}{R^\alpha}, \tag{4}$$

where $E_f$ is the average energy released per fusion reaction, $\langle \sigma v \rangle$ is the fusion reaction rate coefficient, and $R^{-\alpha}$ approximately equals the ratio of the volume occupied by the plasma to the volume enclosed by the surrounding wall. The volume enclosed by the surrounding wall is considered to represent a lower limit for the volume of the plasma confinement apparatus. Thus, Eq. (4) may be considered an upper limit for the average fusion power density. The fusion reaction rate coefficient depends on the velocity distribution of the plasma. Use of non-Maxwellian plasma components has been identified as an important disciplinary issue for fusion energy production [6, 7]. For the present analysis, a single plasma species of density $n$ is considered that has an isotropic Maxwellian velocity distribution in the local rest frame of the plasma. However, an additional factor $\zeta$ is incorporated in Eq. (4) that can be used to approximate the effects of the presence of: multiple ion species, a non-Maxwellian velocity distribution, a spatially nonuniform density, a temporally nonuniform density, polarized nuclei, etc. The following three equations are obtained by combining Eq. (4) with Eqs. (2), (1), and (3), and eliminating the plasma density:

$$r_p = \frac{\gamma \alpha E_w}{R^{1-\alpha/2}}, \tag{5}$$

$$r_p = \frac{\gamma \alpha E_p}{R^{\alpha/2}}, \tag{6}$$

and

$$r_p = \sqrt{\frac{\gamma \alpha V_w}{\beta R^{\alpha/2}}}. \tag{7}$$

Here, $\gamma = (\zeta E_f \langle \sigma v \rangle / P_{avg})^{1/2} [\varepsilon_0/(Ze)]$. Equations (5) - (7), and some other expressions for $r_p$ below, limit how large $r_p$ can be. The limits on $r_p$ can raise two issues. One is how to achieve a desired total fusion reaction rate or power. A large number of individual subsystems, each containing a separate plasma confinement region, may have to be integrated together. Another issue is how to achieve the fabrication and assembly accuracy needed for the subsystems. It is also noted in passing that the question of what value for $P_{avg}$ would constitute a plasma in which nuclear fusion reactions are taking place at a useful rate is an issue in itself.

Equation (5) indicates that if the electric field strength produced by the plasma at the radial (radial) [axial] wall boundary is limited for the spherical (cylindrical) [planar] configuration, then $r_p$ is limited. An associated issue is how large of an electric field can exist at a wall surface without electric breakdown occurring.

A fully non-neutral plasma produces an electric field that tends to expand the plasma. For the plasma to remain confined, a force balance equilibrium must occur. The outward force due to the plasma's self-electric field must be balanced by an inward force. Equation (6) indicates that if the electric field strength produced by the plasma at the radial (radial) [axial] plasma edge is limited for the spherical (cylindrical) [planar] configuration, then $r_p$ is limited. Suppose an externally applied electric field is used to provide a radial (radial) [axial] force balance equilibrium for a region of the spherical (cylindrical) [planar] configuration using the periodic potential force. Then, the electric field strength produced by the plasma at the radial (radial) [axial] plasma edge cannot be larger than the electric field that is applied to effectively counteract it. An associated issue is how large of an electric field can be applied at the radial (radial) [axial] plasma edge, without electric breakdown occurring at an electrode wall. The issue is similar to the previously identified issue.

Suppose an externally applied electric field is used to provide a radial (axial) [radial] force balance equilibrium for a region of the spherical (cylindrical) [planar] configuration using the electric force. Then, $V_w$ in Eq. (7) must be less than the externally applied difference in potential used to confine the plasma. It should also be noted that if an electrostatic potential energy well is used to confine the plasma, the self-consistently determined well depth must be much deeper than the plasma temperature (in energy units) for good plasma confinement to occur. An associated issue is how large of a difference in potential can be applied to provide plasma confinement.

Suppose an externally applied magnetic field is used to provide a radial (radial) [axial] force balance equilibrium for a region of the spherical (cylindrical) [planar] configuration using the magnetic force. It should be emphasized that the effect of other forces that may be present (e.g., the centrifugal force) is not considered. Such forces must be taken into account when considering specific ways in which forces could be combined to achieve a force balance equilibrium. To simplify matters here, the electric field strength produced by the plasma at the radial (radial) [axial] plasma edge is set equal to the magnetic force that is applied to counteract it. Substituting $E_p = V_d B$ in Eq. (6), where $V_d$ is the speed associated with an ion's motion across the magnetic field, leads to

$$r_p = \frac{\gamma \alpha B}{R^{\alpha/2}} \sqrt{\frac{2K_d}{m}}, \tag{8}$$

where $K_d = (1/2)mV_d^2$ is the ion's kinetic energy associated with its cross-field speed, $m$ is the ion mass, and $B$ is the magnetic field strength. Equation (8) raises two issues. One issue is how large the average kinetic energy per plasma ion can be. For fusion energy production, the average kinetic energy per plasma ion may be limited by a limit on the recirculation power. Another issue is how large of a magnetic field can be applied. If the magnetic force is used in combination with the electric force, then the magnetic field must be produced in the presence of the externally applied difference in potential.

Suppose an externally applied magnetic field is used to provide a radial (radial) [axial] force balance equilibrium for a region of the spherical (cylindrical) [planar] configuration using the magnetic gradient force. The electric field strength produced by the plasma at the radial (radial) [axial] plasma edge is set equal to the magnetic gradient force that is applied to counteract it. Thus, $ZeE_p = \mu|\nabla B| = \delta T/r_p$, where $\delta = r_p|\nabla B|/B$, the magnetic moment is approximated as $\mu = T/B$, and $T$ is the ion temperature in energy units. Substituting for $E_p$ in Eq. (6) and solving for $r_p$ gives

$$r_p = \sqrt{\frac{\gamma \alpha \delta T}{ZeR^{\alpha/2}}}. \tag{9}$$

An associated issue is how to maximize $\delta$.

Suppose centripetal acceleration is used to provide a radial (radial) [axial] force balance equilibrium for a region of the spherical (cylindrical) [planar] configuration. The electric field strength produced by the plasma at the radial (radial) [axial] plasma edge is set equal to the centrifugal force that counteracts it. Thus, $ZeE_p = mV_d^2/(\eta r_p)$, where $V_d$ is now the speed associated with an ion's motion that results in the centrifugal force and $\eta$ is the ratio of the radius of the circular trajectory an ion follows to $r_p$. Although there is not a physical limit on the value of $\eta$, the condition $\eta \gg 1$ is necessary to consider the toroidal configuration in the cylindrical limit or the hollow-cylindrical configuration in the planar limit. (The spherical configuration is only considered for completeness.) Substituting for $E_p$ in Eq. (6) and solving for $r_p$ gives

$$r_p = \sqrt{\frac{2\gamma \alpha K_d}{Ze\eta R^{\alpha/2}}}. \tag{10}$$

An associated issue is how large the average kinetic energy per plasma ion can be. The same issue came up when considering use of the magnetic force.

Suppose an externally applied magnetic field is used, and it is necessary for the guiding center approximation to be valid. Then, the cyclotron radius $r_c$ must be small compared to $r_p$. Defining $\omega = r_p/r_c$,

$$r_p = \frac{\omega\sqrt{mT}}{ZeB}.$$

(11)

Equation (11) indicates how large $r_p$ has to be for the guiding center approximation to be valid, which depends on how large of a magnetic field can be applied. In contrast, the five equations, Eqs. (6) - (10), indicate how small $r_p$ has to be in consideration of using five forces to obtain a force balance equilibrium. It should be noted in passing that, according to Eqs. (6) - (10), $r_p$ decreases with $R$ for all three configurations and all five forces. Thus, $R = 1$ maximizes $r_p$ in Eqs. (6) - (10).

In this paper, an attempt has been made to systematically identify interdisciplinary issues that are associated with the prospect of producing an electron-free fusion plasma. Many other (e.g., disciplinary) issues would no doubt come up by considering: (1) specific ways in which more than one force would be combined to achieve a force balance equilibrium, (2) plasma instabilities (e.g., the two-stream instability) and how to avoid them, and (3) plasma transport processes.

This material is based upon work supported by the Texas Advanced Research Program under Grant No. 3594-0003-2001.

## REFERENCES

1. D. C. Barnes, R. A. Nebel, L. Turner, and T. N. Tiouririne, Plasma Phys. Controlled Fusion **35**, 929 (1993).
2. S. F. Paul, E. H. Chao, R. C. Davidson, and C. K. Phillips, in *Non-Neutral Plasma Physics III*, edited by J. J. Bollinger, R. L. Spencer, and R. C. Davidson, AIP Conf. Proc. No. 498 (AIP, Melville, 1999), p. 435.
3. As background information, the reader is referred to the sections on nonneutral plasma density limits in: C. A. Ordonez and D. D. Dolliver, in *Non-Neutral Plasma Physics IV*, edited by F. Anderegg, L. Schweikhard, and C. F. Driscoll, AIP Conf. Proc. No. 606 (AIP, Melville, 2002), p. 590; and C. A. Ordonez, J. Appl. Phys. (to be published in Sept. 2003).
4. G. Schmidt, *Physics of High Temperature Plasmas,* Second Edition (Academic Press, New York, 1979).
5. See, for example, J. R. Correa and C. A. Ordonez, "Electrostatic Confinement of a Reflecting Ion Beam," these proceedings.
6. T. H. Rider, Phys. Plasmas **2**, 1853 (1995).
7. W. M. Nevins, Phys. Plasmas **2**, 3804 (1995).

# The CiaO Code: Contour Dynamics, Image-Charge Method for the Analysis of O-boundary Systems

Gianni G.M. Coppa, Fabio Peano, Federico Peinetti

*INFM and Politecnico di Torino, Dipartimento di Energetica*
*Corso Duca degli Abruzzi 24, 10129 Torino (Italy)*

**Abstract:** A rigorous extension of the contour dynamics technique to systems with a cylindrical boundary is presented. The new technique makes use of the image-charge method in order to obtain an analytic expression for the velocity field in terms of line integrals on the contours. Selected results are presented to show the high accuracy attainable with the method.

## INTRODUCTION

The contour dynamics (CD) is a method originally developed to study inviscid, two-dimensional flow in an infinite domain [1]. The CD technique is suitable for very accurate studies of the contour evolution for regions of constant vorticity. By resorting to the fact that the Green function of the Poisson equation in an infinite domain, $G(\mathbf{r'} \to \mathbf{r})$, is a function only of $|\mathbf{r'} \to \mathbf{r}|$, the classic CD method provides an analytical expression for the velocity on each vortex contour in terms of line integrals on the contours themselves. In this way, no computational grid is required and, consequently, high precision can be attained in the simulations. The contour dynamics approach cannot be immediately extended to bounded systems, and, in particular, to systems with a cylindrical boundary (the case of interest to study non-neutral plasmas confined in a Penning trap), in which the corresponding Green function satisfies no longer the above-mentioned property. Fajans *et Al.* proposed a solution [2], by using a Fourier expansion to evaluate numerically the electric potential due to the circular electrode. An alternative, fully analytical technique has been proposed by the Authors [3]: the presence of the circular electrode is taken into account by introducing a suitable image-charge distribution in an infinite domain. In this way, the velocity field on the contour of the real vortex is evaluated analytically as a sum of line integrals on both the real and the image contour. The new methodology has been included in the CiaO (Contour dynamics, Image-charge method for the Analysis *of* O-boundary systems) simulation code. Within the CiaO code, the contour of each vortex is approximated with a polygonal line. A proper redistribution routine operates during the simulation, in order to maintain the desired accuracy by varying locally the number of contour points when needed. As the contour becomes increasingly complex during the evolution, new points on the contour are added. To show the effectiveness of the methods, simulations concerning classic phenomena (in particular, the diocotron instability and the interaction of vortices) are presented.

## ANALYTIC DETERMINATION OF THE VELOCITY FIELD

The charge density, $\rho$, confined by a Penning trap evolves in time according to the following mathematical model:

$$
\begin{cases}
\dfrac{\partial \rho(r,\vartheta,t)}{\partial t} + \mathbf{v}(r,\vartheta,t) \cdot \dfrac{\partial \rho(r,\vartheta,t)}{\partial \mathbf{r}} = 0 \\[2mm]
\mathbf{v}(r,\vartheta,t) = \dfrac{1}{B_0}\hat{\mathbf{e}}_z \times \dfrac{\partial \Phi(r,\vartheta,t)}{\partial \mathbf{r}} \\[2mm]
\nabla^2 \Phi(r,\vartheta,t) = -\dfrac{1}{\varepsilon_0}\rho(r,\vartheta,t) \\[2mm]
\Phi(R,\vartheta,t) = 0 \\[2mm]
\rho(r,\vartheta,0) = \rho_0(r,\vartheta)
\end{cases}
\tag{1}
$$

being $\Phi$ the electrostatic potential, $\mathbf{v}$ the velocity field, $R$ the radius of the trap, $B_0$ the external magnetic field and $\varepsilon_0$ the vacuum dielectric constant. The cylindrical boundary conditions can be taken into account by using the image-charge method, i.e., by assuming the system to be unbounded and associating the real charge distribution with its image, by means of the following mapping law:

$$
\begin{cases}
\hat{\rho}(\hat{r},\hat{\vartheta})\hat{r}\,d\hat{r}\,d\hat{\vartheta} = -\rho(r,\vartheta)r\,dr\,d\vartheta \\[2mm]
\hat{r} = \dfrac{R^2}{r} \\[2mm]
\hat{\vartheta} = \vartheta
\end{cases}
\tag{2}
$$

By using Eqs. (2), the image of the charge distribution of a vortex of constant density,

$$
\rho(\mathbf{r}) = \begin{cases} \rho_0, & \mathbf{r} \in D \\ 0, & \mathbf{r} \notin D \end{cases},
\tag{3}
$$

can be expressed as

$$
\hat{\rho}(\mathbf{r}) = \begin{cases} -\rho_0\left(\dfrac{R}{r}\right)^4, & \mathbf{r} \in \hat{D} \\[3mm] 0, & \mathbf{r} \notin \hat{D} \end{cases}
\tag{4}
$$

The electrostatic potential can be written as the sum of two terms, the first due to the real charge distribution, the second to its image:

$$
\begin{aligned}
\Phi(\mathbf{r}) &= \Phi_1(\mathbf{r}) + \Phi_2(\mathbf{r}) \\
\Phi_1(\mathbf{r}) &= \int_D G(\mathbf{r}' \to \mathbf{r})\rho_0\,d\mathbf{r}' \\
\Phi_2(\mathbf{r}) &= \int_{\hat{D}} G(\mathbf{r}' \to \mathbf{r})\hat{\rho}(\mathbf{r}')\,d\mathbf{r}'
\end{aligned}
\tag{5}
$$

According to the classic CD approach, the gradient of $\Phi_1$ can be written as

$$\frac{\partial \Phi_1}{\partial \mathbf{r}} = \left(-\rho_0 \oint_D G(\mathbf{r'} \to \mathbf{r})n_x' dl'\right)\hat{\mathbf{e}}_x + \left(-\rho_0 \oint_{\partial D} G(\mathbf{r'} \to \mathbf{r})n_y' dl'\right)\hat{\mathbf{e}}_y \qquad (6)$$

by resorting to the fact that the Green function $G$ satisfies the property

$$\frac{\partial}{\partial \mathbf{r'}}G(\mathbf{r'} \to \mathbf{r}) = -\frac{\partial}{\partial \mathbf{r}}G(\mathbf{r'} \to \mathbf{r}) \qquad (7)$$

as it depends only on $|\mathbf{r'} \to \mathbf{r}|$. The gradient of $\Phi_2$ cannot be manipulated immediately in the same way, due to the presence of the factor $\hat{\rho}(\mathbf{r'})$. To reduce this term to a contour integral, it turns out convenient to split the gradient of $\Phi_2$ in two terms:

$$\frac{\partial \Phi_2}{\partial \mathbf{r}} = \frac{\partial \Phi_2^I}{\partial \mathbf{r}} + \frac{\partial \Phi_2^{II}}{\partial \mathbf{r}}$$

$$\frac{\partial \Phi_2^I}{\partial \mathbf{r}} = \int_{\hat{D}} -\frac{\partial}{\partial \mathbf{r'}}[G(\mathbf{r'} \to \mathbf{r})\hat{\rho}(\mathbf{r'})]d\mathbf{r'} \qquad (8)$$

$$\frac{\partial \Phi_2^{II}}{\partial \mathbf{r}} = \int_{\hat{D}} G(\mathbf{r'} \to \mathbf{r})\frac{\partial}{\partial \mathbf{r'}}[\hat{\rho}(\mathbf{r'})]d\mathbf{r'}$$

The first term leads readily to a contour integral:

$$\frac{\partial \Phi_2^I}{\partial \mathbf{r}} = \left(-\oint_{\partial \hat{D}} G(\mathbf{r'} \to \mathbf{r})\hat{\rho}(\mathbf{r'})n_x' dl'\right)\hat{\mathbf{e}}_x + \left(-\oint_{\partial \hat{D}} G(\mathbf{r'} \to \mathbf{r})\hat{\rho}(\mathbf{r'})n_y' dl'\right)\hat{\mathbf{e}}_y \qquad (9)$$

The second term needs a different approach: its polar components can be written as

$$\begin{cases} \dfrac{\partial \Phi_2^{II}}{\partial \mathbf{r}} \cdot \hat{\mathbf{e}}_r = \displaystyle\int_{\hat{D}} G(\mathbf{r'} \to \mathbf{r})\frac{d\hat{\rho}}{dr'}\hat{\mathbf{e}}_r \cdot \hat{\mathbf{e}}_{r'} d\mathbf{r'} \\[2mm] \dfrac{\partial \Phi_2^{II}}{\partial \mathbf{r}} \cdot \hat{\mathbf{e}}_{\vartheta} = \displaystyle\int_{\hat{D}} G(\mathbf{r'} \to \mathbf{r})\frac{d\hat{\rho}}{dr'}\hat{\mathbf{e}}_{\vartheta} \cdot \hat{\mathbf{e}}_{r'} d\mathbf{r'} \end{cases} \qquad (10)$$

being

$$\hat{\mathbf{e}}_r \cdot \hat{\mathbf{e}}_{r'} = \cos(\vartheta' - \vartheta), \quad \hat{\mathbf{e}}_r \cdot \hat{\mathbf{e}}_{r'} = \sin(\vartheta' - \vartheta), \quad \frac{d\hat{\rho}}{dr'} = 4\rho_0 \frac{R^4}{r'^5} \qquad (11)$$

The integrals appearing in Eqs. (10) can be written as contour integrals if a new function $\psi(\mathbf{r},\mathbf{r'})$, satisfying the equation

$$G(\mathbf{r'} \to \mathbf{r})\frac{d\hat{\rho}}{dr'} = \frac{1}{r'}\frac{\partial}{\partial r'}[r'\psi(\mathbf{r},\mathbf{r'})] \qquad (12)$$

is introduced. In fact, once $\psi$ is determined, one can write

$$\frac{\partial \Phi_2^{II}}{\partial \mathbf{r}} \cdot \hat{\mathbf{e}}_r = \oint_{\partial \hat{D}} \psi(\mathbf{r},\mathbf{r'})\cos(\vartheta' - \vartheta)n_r' dl'$$

$$\frac{\partial \Phi_2^{II}}{\partial \mathbf{r}} \cdot \hat{\mathbf{e}}_r = \oint_{\partial \hat{D}} \psi(\mathbf{r},\mathbf{r'})\sin(\vartheta' - \vartheta)n_r' dl' \qquad (13)$$

so that $\dfrac{\partial \Phi}{\partial \mathbf{r}}$ is can be expressed as a sum of contour integrals as

$$\frac{\partial \Phi}{\partial \mathbf{r}} = \frac{\partial \Phi_1}{\partial \mathbf{r}} + \frac{\partial \Phi_2^I}{\partial \mathbf{r}} + \frac{\partial \Phi_2^{II}}{\partial \mathbf{r}} \qquad (14)$$

and the velocity field is readily evaluated from the second of Eqs. (1). The analytic expression for $\psi$ is reported in [3].

275

# RESULTS

The following results show the extremely precise contour description provided by the CD algorithm even when the shape of the vortices reaches a high level of complexity. The reported results refer to three typical phenomena: the instability for hollow density profiles (Fig. 1, see [3] for details), the evolution of two-ring patterns of vortices (Figs. 2-3, see [3, 4] for details) and the interaction between a finite size vortex and a point vortex (Fig. 4, see [5] for details). This last case has been studied both experimentally and analytically by Fajans, Pozzoli et al. [6].

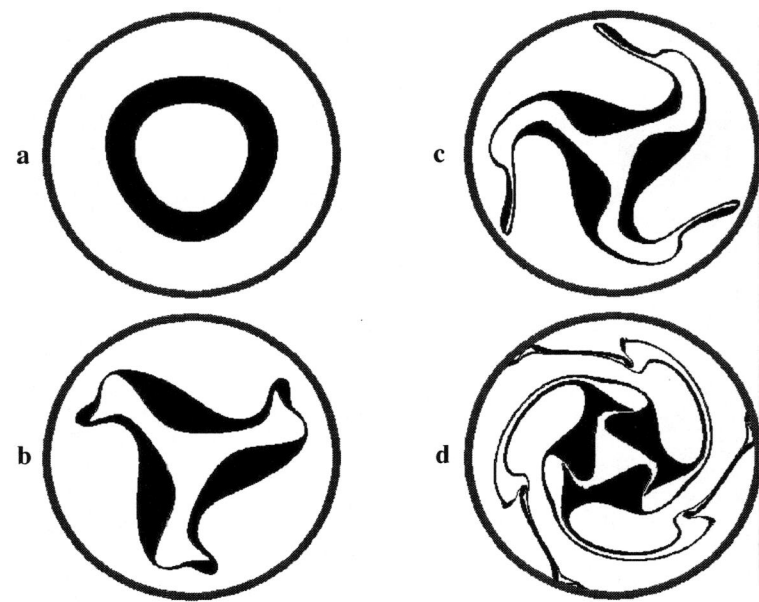

**FIGURE 1.** Evolution of a plasma ring perturbed by exciting the $m = 3$ azimuthal mode [3].

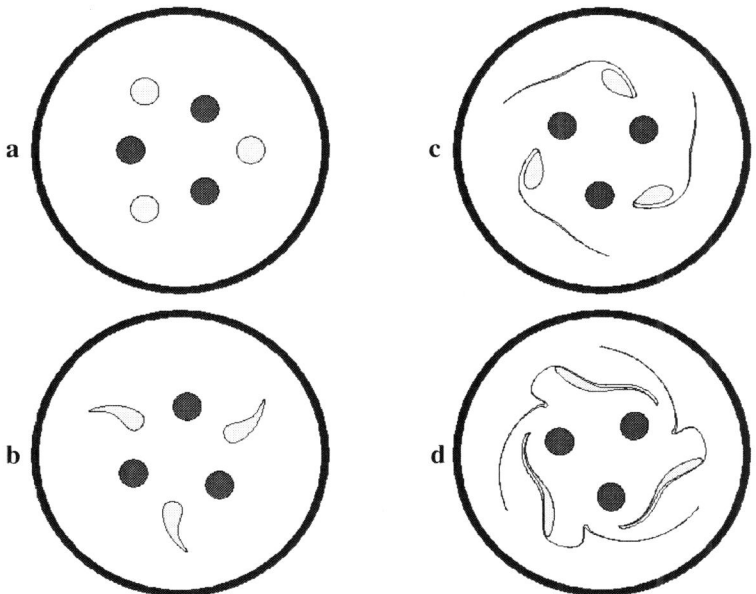

**FIGURE 2.** Evolution of a two-ring pattern of circular vortices with radius $r_0 = 0.1R$. The initial pattern would be an equilibrium configuration in the framework of a point-vortex theory [3, 4].

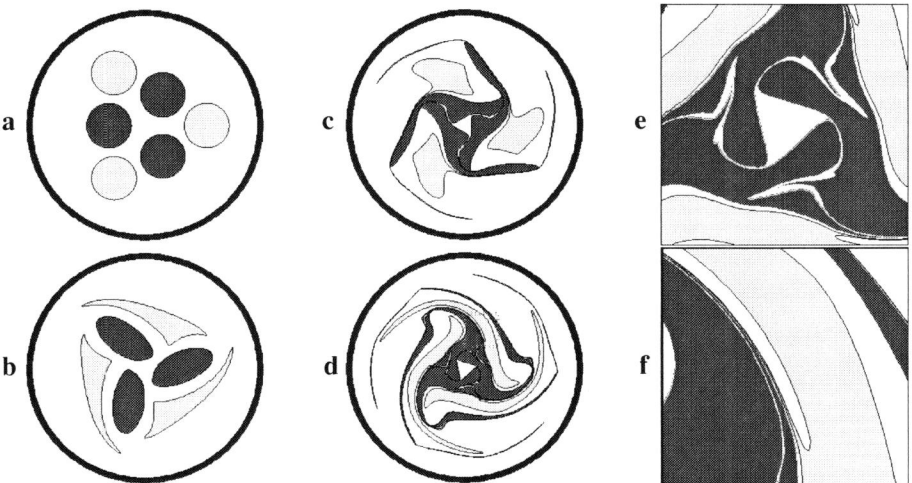

**FIGURE 3.** Evolution of a two-ring pattern of circular vortices with radius $r_0 = 0.2R$. The initial pattern would be a stable configuration in the framework of a point-vortex theory [3, 4]. Figs. 3e and 3f report details of the central part and of the region marked by a dotted line in Fig. 3d.

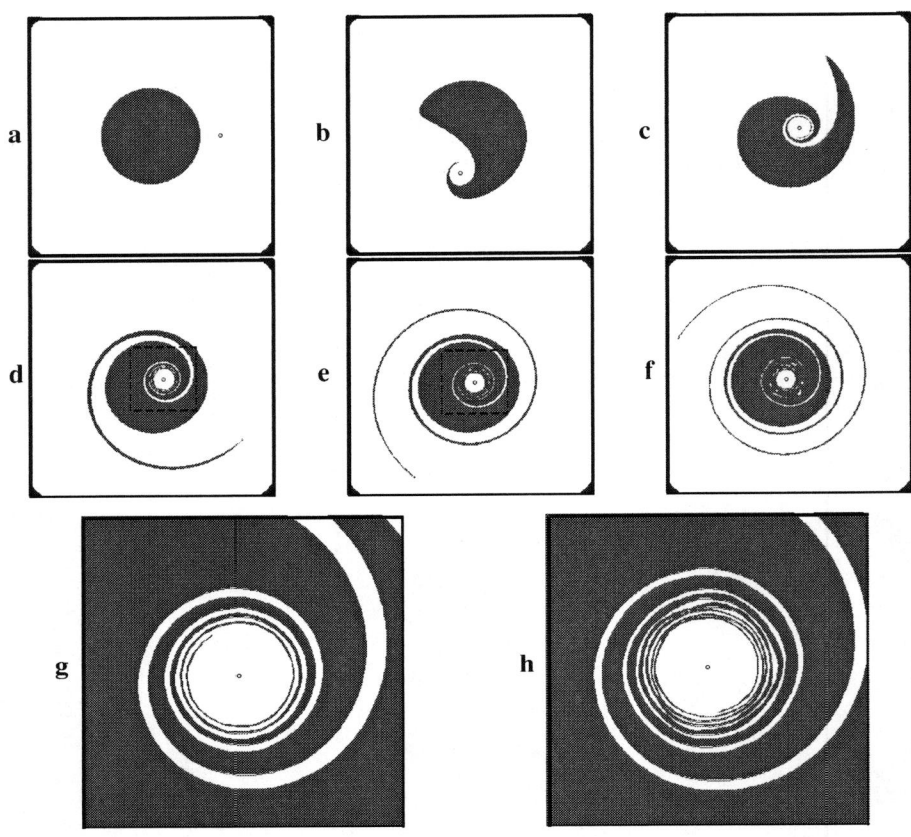

**FIGURE 4.** Merger between a finite-size vortex and a strongly-interacting pointlike vortex [5, 6]. Figs. 4g and 4h report details of the region marked by a dotted line in Figs. 4d and 4e, respectively.

# REFERENCES

1. N.J. Zabusky, M.H. Hughes and K.V. Roberts, *Contour dynamics for the Euler equations in two dimensions*, J. Comput. Phys. **30**, 96 (1979).
2. E. Yu. Backhaus, J. Fajans and J.S. Wurtele, *Application of countour dynamics to systems with cylindrical boundaries*, J. Comput. Phys. **145**, 462 (1998).
3. Gianni G.M. Coppa, F. Peano and F. Peinetti, *Image-charge Method for Contour Dynamics in Systems with Cylindrical Boundaries*, J. Comput. Phys **182**, 392-417 (2002).
4. G.G.M. Coppa, *Analytic studies of Two-Ring Patterns of Vortices in a Penning Trap*, in Workshop on Non-Neutral Plasmas, Princeton, New Jersey USA, (1999).
5. Gianni G.M. Coppa, Antonio D'Angola, F. Peano and F. Peinetti, *Analysis of the merger between plasma vortices in a Penning trap*, to be published in European Physical Journal D.
6. M. Amoretti, D. Durkin, J. Fajans, R. Pozzoli, M. Romé, *Asymmetric vortex merger: Experiments and simulations*, Phys. Plasmas 8, 3865 (2001).

# A new 3D PIC Code for the Simulation of the Dynamics of a Non-Neutral Plasma

Yu. Tsidulko*, R. Pozzoli[†] and M. Romé[†]

*Budker Institute of Nuclear Physics, Novosibirsk, Russian Federation
[†]I.N.F.M., I.N.F.N., Dipartimento di Fisica, Università degli Studi di Milano, Milano, Italy

**Abstract.** The three-dimensional evolution of a pure electron plasma in a Penning-Malmberg trap is studied by means of a newly developed particle-in-cell code in the frame of a cold fluid guiding center electrostatic approximation. Results obtained both in the trapped plasma case and in the beam propagation regime are shown.

The three-dimensional (3D) evolution of a pure electron plasma trapped in a Penning-Malmberg trap [1] or of a non-relativistic electron beam traveling in a drift tube (see Fig. 1) is studied by means of a newly developed particle-in-cell code ("MEP", acronym of Milano Electron Plasma).

The dynamics is analyzed in the frame of a cold fluid guiding center electrostatic approximation. The evolution of the system is followed within a perfectly conducting cylindrical surface of radius $R$ and length $L$, on which the boundary conditions for the electrostatic potential (possibly time-dependent) are imposed. A system of cylindrical coordinates $(r, \theta, z)$ is introduced, where $r$ is the distance from the axis of the trap, $\theta$ the azimuthal coordinate and $z$ the coordinate along the axis, respectively ($z$ is counted from the center of the trap). The magnetic field is assumed to be uniform along $z$, $\mathbf{B} = B\mathbf{e}_z$, where $\mathbf{e}_z$ denotes the unit vector in the $z$ direction. The continuity equation for the electron density is coupled to the Poisson's equation for the electrostatic potential and the equation of momentum balance along the axial coordinate. The perpendicular dynamics is described by the $\mathbf{E} \times \mathbf{B}$-drift, where $E$ is the self-consistent (internal + externally applied) electric field. The equations describing the electron plasma evolution

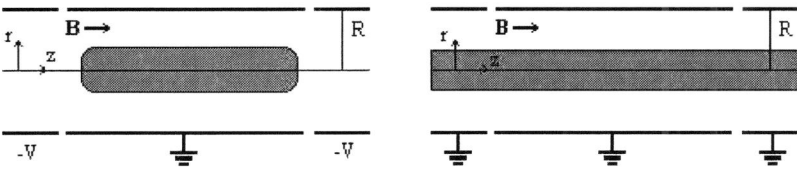

**FIGURE 1.** Scheme of a trapped plasma confi guration (**left**) and a traveling beam confi guration (**right**). $-V$ is the trap potential.

CP692, Non-Neutral Plasma Physics V, edited by M. Schauer et al.
© 2003 American Institute of Physics 0-7354-0165-9/03/$20.00

are then written explicitly as

$$\frac{\partial n}{\partial t} + \nabla \cdot (n\mathbf{v}) = 0, \quad \nabla^2 \varphi = 4\pi e n,$$

$$\mathbf{v} = \frac{c}{B} \mathbf{e}_z \times \nabla \varphi + v_\| \mathbf{e}_z, \quad \frac{dv_\|}{dt} = \frac{e}{m} \mathbf{e}_z \cdot \nabla \varphi, \tag{1}$$

where $n$ is the electron density, $\varphi$ the electrostatic potential, $\mathbf{v}$ the fluid velocity ($v_\|$ denotes its component parallel to the magnetic field), and $-e$, $m$ and $c$ are electron charge, electron mass and speed of light, respectively.

In the following, adimensional quantities are used. Length, time, density and potential are normalized over $R$, $\omega_c/2\omega_p^2$, $n_0$ and $4\pi e n_0 R^2$, respectively, where $\omega_c \equiv eB/mc$ is the non-relativistic electron cyclotron frequency and $\omega_p \equiv (4\pi e^2 n_0/m)^{1/2}$ is the electron plasma frequency, computed for a specified electron density $n_0$. Keeping the same notation for the normalized quantities, Eqs. (1) are rewritten as

$$\frac{\partial n}{\partial t} + \nabla \cdot (n\mathbf{v}) = 0, \quad \nabla^2 \varphi = n,$$

$$\mathbf{v} = \frac{1}{2} \mathbf{e}_z \times \nabla \varphi + v_\| \mathbf{e}_z, \quad \frac{dv_\|}{dt} = \frac{1}{M_{eff}} \mathbf{e}_z \cdot \nabla \varphi. \tag{2}$$

The behavior of the system is therefore characterized by a single parameter, $M_{eff}$, which plays the role of an effective mass, $M_{eff} \equiv 4\omega_p^2/\omega_c^2$.

Using the variable $s \equiv r^2$, the equations of motion for a fluid element, corresponding to Eqs. (2), are written as

$$\frac{ds}{dt} = \frac{\partial \varphi}{\partial \theta}, \quad \frac{d\theta}{dt} = -\frac{\partial \varphi}{\partial s}, \quad \frac{d^2 z}{dt^2} = \frac{1}{M_{eff}} \frac{\partial \varphi}{\partial z}. \tag{3}$$

In the "MEP" code, Eqs. (2) are discretized on an equispaced grid (with the only exception of the radial cell around the cylindrical axis) in the coordinates $s$, $\theta$ and $z$. The number of grid cells is denoted as $N_s$, $N_\theta$ and $N_z$, respectively. The grid for $s$ is defined as $\underline{s}_0 = 0$, $\underline{s}_1 = 1/(N_s N_\theta + 1)$, $\underline{s}_{j+1} = \underline{s}_j + N_\theta/(N_s N_\theta + 1)$, $s_0 = 0$, $s_j = (\underline{s}_j + \underline{s}_{j+1})/2$, $j = 1 \dots N_s$ ($s$ is the position of the center, while $\underline{s}$ denotes the lower boundary of the "radial" cell). The grid for $\theta$ is simply $\theta_l = 2\pi(l-1)/N_\theta$, $l = 1 \dots N_\theta$, with the periodicity relation, $\theta_{N_\theta+1} = \theta_1$ (the central radial cell has no subdivisions in $\theta$), while the grid for $z$ is defined as $z_k = (k - 1/2)L/N_z - L/2$, $k = 1 \dots N_z$. Each cell has the same volume $\Delta V = \Delta s \Delta \theta \Delta z/2$, with $\Delta s = 1/(N_s + 1/N_\theta)$, $\Delta \theta = 2\pi/N_\theta$ and $\Delta z = L/N_z$, respectively.

Using the coordinates $(s, \theta, z)$, the Poisson equation is written as

$$4\frac{\partial}{\partial s} s \frac{\partial \varphi}{\partial s} + \frac{1}{s} \frac{\partial^2 \varphi}{\partial \theta^2} + \frac{\partial^2 \varphi}{\partial z^2} = n. \tag{4}$$

$\varphi|_{s=1} = \breve{\phi}(\theta, z)$ and $\varphi|_{z=\pm L/2} = \breve{\phi}^\pm(s, \theta)$ represent the boundary conditions (possibly time-dependent). A three-point finite differencing is applied to Eq. (4) on the above defined grid. A matrix equation, $\breve{\mathbf{n}} = -\hat{A}\boldsymbol{\varphi}$ (by components $\breve{n}_{j,l,k} = -A_{j,l,k;\bar{j},\bar{l},\bar{k}}\varphi_{\bar{j},\bar{l},\bar{k}}$)

280

can be written, where $\check{\mathbf{n}}$ differs from $\mathbf{n}$ only in the boundary cells: Here the definition of the density is modified to take explicitly into account the boundary conditions for $\varphi$. The operator $\hat{A}$ is self-adjoint, i.e. $A_{j,l,k;\,\bar{j},\bar{l},\bar{k}} = A_{\bar{j},\bar{l},\bar{k};\,j,l,k}$, positively defined and commutative with the rotation operator, $(\hat{R}\varphi)_{j,l,k} \equiv (1-\delta\theta)\varphi_{j,l,k} + \delta\theta\varphi_{j,l+1,k}$. Formally, the solution of the Poisson equation is $\boldsymbol{\varphi} = -\hat{A}^{-1}\check{\mathbf{n}}$, where the inverse matrix $\hat{A}^{-1}$ have to be calculated only once. But the multiplication by $\hat{A}^{-1}$ requires a number of operations proportional to $N_c^2$, where $N_c \equiv N_s N_\theta N_z$ is the total number of grid cells. A more efficient algorithm is introduced in the code. The Poisson equation is transformed both in $\theta$ and in $z$ with a standard Fast Fourier Transform (FFT)

$$\varphi_{j,l,k} = \sum_{m_\theta=1-N_\theta/2}^{N_\theta/2} \sum_{m_z=1-N_z}^{N_z} \tilde{\varphi}_{j,m_\theta,m_z} \exp\left\{ 2\pi i \left[ \frac{m_\theta(l-1)}{N_\theta} + \frac{m_z(k-1)}{2N_z} \right] \right\},$$

$$\varphi_{0,k} = \sum_{m_z=1-N_z}^{N_z} \tilde{\varphi}_{0,0,m_z} \exp\left[ 2\pi i \frac{m_z(k-1)}{2N_z} \right]$$

(5)

(and analogously for the modified density $\check{n}$). In this representation, the number of Fourier complex amplitudes is larger than the number of points in the grid, and additional conditions have to be imposed. Namely, the Fourier amplitudes satisfy the following relations

$$\tilde{\varphi}_{j,m_\theta,m_z} = \tilde{\varphi}_{j,-m_\theta,-m_z}^*, \quad \tilde{\varphi}_{j,m_\theta,m_z} = -\tilde{\varphi}_{j,m_\theta,-m_z} \exp(\pi i m_z/N_z)$$

(and analogously for $\check{n}$). The former relation is simply the reality condition for $\varphi$, while the latter assures that the same relation is obtained between every Fourier amplitudes $\tilde{\varphi}_{j,m_\theta,m_z}$ and $\check{n}_{j,l,k}$, including the boundary components with $k=1$ and $k=N_z$. A tridiagonal system of equations in the variable $s$ is obtained for every Fourier component $\tilde{\varphi}_{j,m_\theta,m_z}$. Each system is solved by means of a standard Gauss elimination procedure. Inverse FFT is then applied to get the solution for the potential on the grid. With this algorithm there is no need of matrix inversion, and the Poisson equation is solved with a number of operations proportional to $N_c \ln(N_\theta N_z)$.

The system governed by Eqs. (2) is simulated numerically as an ensemble of macro-particles with fixed sizes $\Delta s$, $\Delta\theta$ and $\Delta z$, using a PIC method. Let the total number of macro-particles be $N_p$ and let the $\alpha$-th macro-particle be characterized by the position $s_\alpha, \theta_\alpha, z_\alpha$ and the $z$-momentum $p_\alpha$. It is assumed that the macro-particle gives a contribution $w_{j,l,k;\alpha} N_c/N_p$ to the electron density $n_{j,l,k}$ in the cell $\{j,l,k\}$. The weight functions $w_{j,l,k;\alpha}(s_\alpha, \theta_\alpha, z_\alpha)$ satisfy the condition $\sum_{j,l,k} w_{j,l,k;\alpha} = 1$, which provides the normalization of the particle density $n$ and the charge conservation. Piecewise quadratic weight functions (defined on the cell containing the particle and the relevant neighboring cells) continuous with their first derivative are used in the "MEP" code. The use of smooth weights accelerates the time integration procedure in comparison with the usual "area weighting" method. The same weight functions are used for the computation of the potential corresponding to the particle position, $\varphi_\alpha(s_\alpha, \theta_\alpha, z_\alpha) = \sum_{j,l,k} w_{j,l,k;\alpha}(\varphi_{j,l,k}^0 + \varphi_{j,l,k}^{vac})$, where $\varphi_{j,l,k}^0$ is the solution of the Poisson equation with zero boundary conditions and

$\varphi_{j,l,k}^{vac}$ is the vacuum potential in the cell corresponding to the solution of the Laplace equation, $\nabla^2 \varphi^{vac} = 0$ with the boundary conditions $\breve{\phi}$ and $\breve{\phi}^{\pm}$.

The equations of motion corresponding to Eqs. (3) can be written in Hamiltonian form, using the "momenta" $s_\alpha, p_\alpha$ and the corresponding conjugated variables $\theta_\alpha, z_\alpha$. The Hamiltonian function is written as

$$H(\mathbf{s}, \mathbf{p}; \theta, \mathbf{z}) = \sum_\alpha \frac{p_\alpha^2}{2M_{eff}} + U(\mathbf{s}, \theta, \mathbf{z}), \tag{6}$$

$$U = \frac{N_c}{2N_p} \sum_{\alpha,\beta,j,l,k,j',l',k'} w_{j,l,k;\alpha} A_{j,l,k;j',l',k'}^{-1} w_{j',l',k';\beta} - \sum_{\alpha,j,l,k} w_{j,l,k;\alpha} \varphi_{j,l,k}^{vac}. \tag{7}$$

Since $\hat{A}^{-1}$ is self-adjoint, the derivatives of the Hamiltonian appearing in the equations of motion can be computed as $\partial H/\partial \zeta_\alpha = -\partial \varphi_\alpha/\partial \zeta_\alpha$, where $\zeta_\alpha$ is either $s_\alpha$, $\theta_\alpha$ or $z_\alpha$. The equations of motion for the computational particles are solved in the code by means of the Runge-Kutta-Fehlberg predictor-corrector scheme [2]. Note that in this Hamiltonian formulation of the physical problem and of the numerical scheme, the symmetry of $\hat{A}$ plays a fundamental role: It assures the conservation of phase integrals (Liouville theorem) and the energy conservation in a time-independent problem.

Examples of results obtained with the 3D PIC "code" MEP are reported here both in the trapped plasma case and in the beam propagation regime. Fig. 2 refers to a trapped electron plasma in low rigidity condition, i.e., to a case in which the axial bounce frequency and the azimuthal rotation frequency of the electron plasma are of the same order. The plasma is initially set off-axis. Its cross-section is strongly distorted as it rotates around the axis of the trap, and big oscillations are evident along the axial direction.

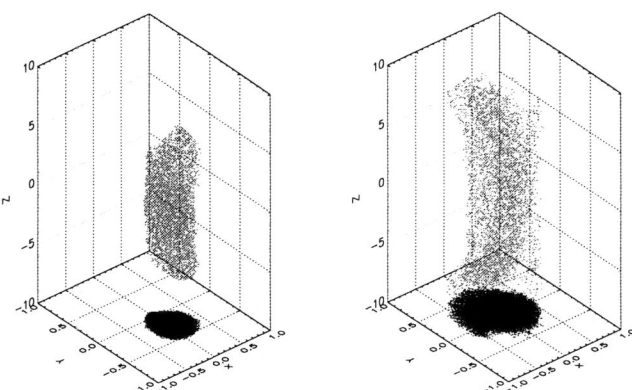

**FIGURE 2.** 3D plot of the particle configuration, in the case of a trapped plasma, at $t = 1.0$ (**left**) and $t = 4.0$ (**right**). The projection on a plane perpendicular to the axis of the trap is also shown. The parameters of the run are: $M_{eff} = 0.325$, $N_s = 64$, $N_\theta = 64$, $N_z = 64$, $N_p = 10^5$.

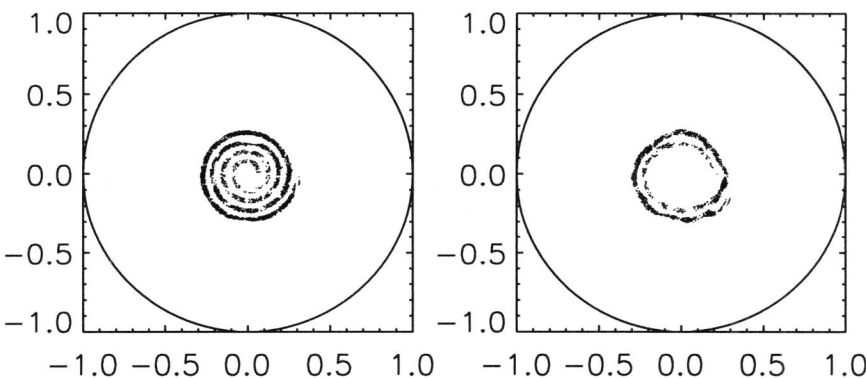

**FIGURE 3.** From left to right: projection on a plane perpendicular to the trap axis of the particles in the interval $-10 \le z \le -5$ and $5 \le z \le 10$, respectively. The beam is injected at $z = -10$. The data refer to $t = 1.0$. The parameters of the run are: $M_{eff} = 5 \cdot 10^{-4}$, $N_s = 64$, $N_\theta = 64$, $N_z = 64$, $N_p = 10^5$, and $p_z = 20.0$ at $t = 0$.

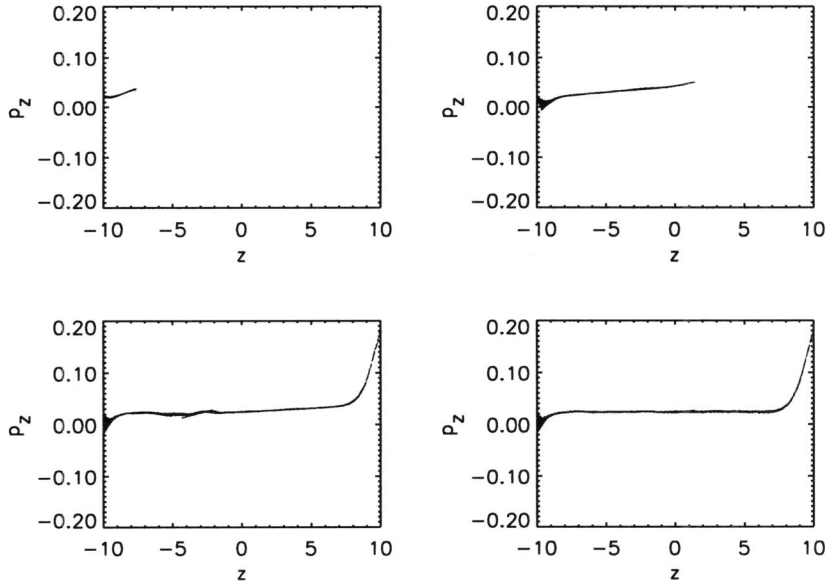

**FIGURE 4.** Plot of the parallel momentum versus the axial coordinate, for the same case as in Fig. 3. From left to right, top to bottom, the data refer to $t = 0.05$, $t = 0.15$, $t = 0.35$, and $t = 1.00$, respectively.

In Figs. 3-4 the case of a beam injected from a spiral cathode is considered. The formation of a "virtual cathode" close to the injection boundary is evidenced. The central part of the beam is reflected back to the cathode by the space-charge of the beam itself, and only the outer part of the beam is trasmitted to the opposite end of the trap, held at positive potential. In almost stationary conditions, three regions can be distinguished in the reported case: a reflection region close to the injection boundary, a central part with almost constant parallel velocity, and an acceleration region close to the exit boundary.

# REFERENCES

1.  R. C. Davidson, *"An Introduction to the Physics of Nonneutral Plasmas"* Addison-Wesley, Redwood City, USA (1990).
2.  G. E. Forsythe, M. A. Malcolm and C.B. Moler, *"Computer methods for mathematical computations"*, Prentice-Hall inc., Englewood Cliffs, N.J. 0763 (1977).

# Magnetorotational Instability in a Couette Flow of Plasma

Koichi Noguchi* and Vladimir I. Pariev[†]

*Applied Physics Division, Los Alamos National Laboratory, Los Alamos, NM 87545, USA
[†]Department of Physics and Astronomy, University of Rochester, Rochester, NY 14627, USA and
Lebedev Physical Institute, Leninsky Prospect 53, Moscow 119991, Russia

**Abstract.** All experiments, which have been proposed so far to model the magnetorotational instability (MRI) in the laboratory, involve a Couette flow of liquid metals in a rotating annulus. All liquid metals have small magnetic Prandtl numbers, $Pm \sim 10^{-6}$, the ratio of kinematic viscosity to magnetic diffusivity. With plasmas both large and small $Pm$ are achievable by varying the temperature and the density of plasma. Compressibility and fast rotation of the plasma result in radial stratification of the equilibrium plasma density. Evolution of perturbations in radially stratified viscous and resistive plasma permeated by an axial uniform magnetic field is considered. The differential rotation of the plasma is induced by the $\mathbf{E} \times \mathbf{B}$ drift in applied radial electric field. Global unstable eigenmodes are calculated by our newly developed matrix code. The plasma is shown to be MRI unstable for parameters easily achievable in experimental setup.

## INTRODUCTION

A central problem in the theory of accretion disks in astrophysics is understanding the fundamental mechanism of angular momentum transfer. A robust anomalous outward angular momentum transport must operate in order for accretion to occur [1]. A phenomenological theory of turbulent angular momentum transport ($\alpha$-disks) proposed by Shakura [2] was in the basis of our understanding of how accretion occurs and still remains viable to date. The puzzle of the origin of turbulence in hydrodynamically stable disks was resolved when magnetorotational instability (MRI), originally discovered in Refs. [3] and [4], was applied to accretion disks by Balbus and Hawley [5]. MRI causes MHD turbulence to develop in an initially weakly magnetized fluid.

Despite its importance, MRI has never been observed in the laboratory. Recently, two experiments have been proposed to test MRI in a differentially rotating flow of liquid metal (Couette flow) between two rotating cylinders [6, 7]. A great deal of theoretical work on investigating MRI in Couette flow of liquid metals has been done confirming that MRI can be excited for magnetic Reynolds number, Rm, exceeding a few [8, 9, 10, 11]. A particular attention was given to why MRI was not observed in previous experiments with hydromagnetic Couette flows of liquid metals [8, 10]. Because of the very small magnetic Prantdl number of metals, $Pm \sim 10^{-6}$, the rotation needs to be very fast to achieve $Rm > 1$. Indeed, kinetic Reynolds number needs to exceed $Re = Rm/Pm \sim 10^6$. In previous experiments the rotation speed was not high enough to achieve $Re \sim 10^6$. For such high Reynolds numbers the flow in the experiment is likely to become turbulent even without a magnetic field for Rayleigh stable rotation profiles. The instability may have a nonlinear hydrodynamic nature [12] or may be due to the Ekman circulation induced by the end plates [7]. The presence of such turbulence can affect the conditions for the excitation of MRI [7].

Here we consider plasma as alternative to liquid metals to use in MRI experiment. By changing temperature and density of the plasma by a few times around values $T \sim 5\,\mathrm{eV}$ and $n \sim 10^{14}\,\mathrm{cm}^{-3}$ one can vary Pm in the range of about $10^{-2}$ to $10^2$. For velocities of plasma of the order of the thermal speed of ions and the typical size of the apparatus of about 50 cm Rm is in the range from $10^2$ to $10^3$, while Re is in the range from 5 to $10^4$. These plasma parameters are readily achievable in the laboratory [13]. Therefore, in plasma experiment it can be relatively easy to have high enough Rm allowing for MRI to grow, while keeping Re modest and the flow laminar. Laminar character of the flow can allow a detailed study of the structure of the unstable mode and the secondary flow without intervening noise from turbulence. For higher Re $\sim 10^4$, the transition from laminar to turbulent flows in hydromagnetics can be investigated. Moreover, the effects of wide variations of Pm can be studied with a plasma MRI experiment.

## DESCRIPTION OF THE EXPERIMENT

The basic setup of possible plasma MRI experiment is illustrated in Fig. 1. This experiment is now under construction at Los Alamos National Laboratory. It can be also used for observing laminar plasma dynamos [13]. Tentative set of specifications and typical parameters of plasma for that experiment is shown in Table 1. The plasma is produced by the electric discharge in hydrogen and is injected into the space between two coaxial conducting cylinders. Mean free path for Coulomb collisions is much smaller than the radii of the cylinders, and MHD description of plasma is appropriate. Cylindrical coordinate system $r$, $\phi$, $z$ is used with the axis of symmetry coinciding with the central axis of the cylinders. Axial uniform magnetic field $B_{0z}$ is produced by the coils around the cylinders and is preexistent of the discharge. Therefore, plasma is coupled to the axial magnetic field lines already at the formation and can slide along the axial magnetic field lines filling the space between two cylinders. The outer cylinder is grounded and the voltage $\Phi_0$ is applied to the inner cylinder during and after the plasma injection, in order to maintain the rotation. Neglecting the boundary and sheath modification of any applied electric field, plasma rotates with $\mathbf{E} \times \mathbf{B}$ drift,

$$\mathbf{E}_0 = \frac{\Phi_0}{\ln(R_2/R_1)} \frac{1}{r} \mathbf{e}_r = -\frac{\mathbf{V}_0 \times \mathbf{B}_0}{c}, \qquad (1)$$

where $\mathbf{E}_0$ is the equilibrium electric field, $\mathbf{B}_0 = B_{0z}\mathbf{e}_z$ is the equilibrium axial magnetic field, $R_1$ is the radius of the inner cylinder, $R_2$ is the radius of the outer cylinder, and $c$ is the speed of light.

Since plasma is highly collisional and is thermalized before it reaches to the experimental region of the chamber, we further make an assumption that the stationary state is isothermal:

$$\Gamma P_0/\rho_0 = C_s^2 = \text{constant}, \qquad (2)$$

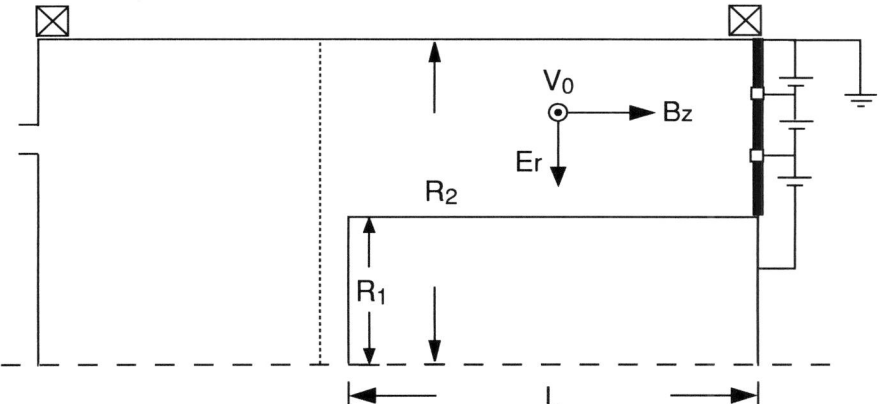

**FIGURE 1.** The meridional cross section of the device. Center line of the chamber is shown as a dashed line. Dotted line indicates the boundary of the working space between two cylinders (to the right). Plasma is injected from the left and rotates due to $\mathbf{E} \times \mathbf{B}$ drift. Axial magnetic field is produced by coils. Coaxial metal rings at the right end plate are charged to fractions of $\Phi_0$ ($\Phi_0 < 0$ as illustrated) to support differential rotation of plasma against the formation of Ekman layer.

where $P_0(r)$ is the equilibrium pressure, $\rho_0(r)$ is the equilibrium density, $C_s$ is the sound speed, and $\Gamma$ is the ratio of specific heats ($\Gamma = 5/3$ for hydrogen).

The Larmor radius of electrons in the experiment is also much smaller than the radii of the cylinders, but is comparable to the Coulomb mean free path. Despite this, here we assume for simplicity that the conductivity of the plasma is isotropic and is given by non-magnetized expression. The coefficient of magnetic diffusivity, $\eta$, and dynamic viscosity coefficient $\rho v$ are independent of density and depends only on temperature. Then, $\eta$ and $\rho v$ remain constant throughout the volume of the plasma in the isothermal approximation considered here.

In the equilibrium state of the steady rotation the magnetic field is $B_{0z} = $ constant and the velocity is $\mathbf{V}_0 = (0, r\Omega(r), 0)$, with the angular velocity profile

$$\Omega(r) = -\frac{\Phi_0}{B_{0z} \ln(R_2/R_1)} \frac{c}{r^2}. \tag{3}$$

Balance between the centrifugal force and the gradient of pressure in the equilibrium state is

$$-r\Omega^2 \rho_0 = -\frac{dP_0}{dr}. \tag{4}$$

Using isothermal condition, Eq. (4) can be solved to give the radial dependence of $\rho_0$

$$\rho_0 = \rho_0(R_1) \exp\left[\int_{R_1}^r \frac{r\Gamma\Omega^2}{C_s^2} dr\right]. \tag{5}$$

When the rotation speed $\Omega(R_1)R_1$ is comparable to the sound speed, the centrifugal force is significant enough to cause the compression of the plasma toward the outer cylinder.

**TABLE 1.** Parameters of the plasma.

| | |
|---|---|
| Inner Radius $r_1$ (cm) | 15 |
| Outer Radius $r_2$ (cm) | 52 |
| Length $L$ (cm) | 100 |
| Density n (cm$^{-3}$) | $1 \times 10^{14}$ |
| Electron Temperature $T_e$ (eV) | 5 |
| Kinematic Viscosity, $v$ (cm$^2$ s$^{-1}$) | $3 \times 10^6$ |
| Magnetic Diffusivity, $\eta$ (cm$^2$ s$^{-1}$) | $2.7 \times 10^5$ |
| Prandtl number $P_m$ | 11 |
| Sound Speed $C_s$ (cm s$^{-1}$) | $4 \times 10^6$ |
| Maximum Frequency $\Omega_{1\max}$ (s$^{-1}$) | $1.9 \times 10^5$ |

## THE INSTABILITY IN A RADIALLY STRATIFIED PLASMA

We consider perturbations of the equilibrium state described above: $\mathbf{V} = \mathbf{V}_0 + \mathbf{v}$, $\mathbf{B} = \mathbf{B}_0 + \mathbf{b}$, $\rho = \rho_0 + \rho_1$, $P = P_0 + P_1$, and linearize MHD equations in small perturbations. We idealize the problem by considering cylinders of infinite length and take the dependence of perturbations on $t$, $\theta$, and $z$ as $\propto \exp[i(-\omega t + m\theta + k_z z)]$. If the equilibrium density $\rho_0(r)$ varies significantly between $R_1$ and $R_2$, general perturbations of the plasma are compressible. However, the phenomenon of MRI is due to the stretching of the magnetic field lines by the differential rotation coupled with the action of centrifugal force. Boussinesq approximation neglects the changes of the volume of the displaced parcel of plasma as it gets quickly adjusted to the new pressure equilibrium at a new radial location in the stratified plasma. It allows to capture centrifugal force acting on a displaced parcel of plasma but excludes compressible modes, which are not essential for the development of MRI. This simplifies calculations substantially. Here we adopt Boussinesq approximation. Linearized continuity equation becomes equivalent to two equations:

$$\frac{\partial \rho_1}{\partial t} + (\mathbf{V}_0 \nabla)\rho_1 + \mathbf{v} \cdot \nabla \rho_0 = 0, \tag{6}$$

$$\nabla \cdot \mathbf{v} = 0. \tag{7}$$

Other linearized MHD equations are:

$$\nabla \cdot \mathbf{b} = 0, \tag{8}$$

$$\frac{\partial \mathbf{b}}{\partial t} = \nabla \times (\mathbf{v} \times \mathbf{B}_0) + \nabla \times (\mathbf{V}_0 \times \mathbf{b}) + \eta \nabla^2 \mathbf{b}, \tag{9}$$

$$\rho_0 \frac{\partial \mathbf{v}}{\partial t} + \rho_0 (\mathbf{v} \nabla)\mathbf{V}_0 + \rho_0 (\mathbf{V}_0 \nabla)\mathbf{v} + \rho_1 (\mathbf{V}_0 \nabla)\mathbf{V}_0 = -\nabla P_1 + \rho_0 v_0 \nabla^2 \mathbf{v}$$

$$-\frac{1}{4\pi} \nabla (B_{0z} b_z) + \frac{1}{4\pi}(\mathbf{b}\nabla)\mathbf{B}_0 + \frac{1}{4\pi}(\mathbf{B}_0 \nabla)\mathbf{b}, \tag{10}$$

where $v_0$ is the unperturbed value of kinematic viscosity. Note, that $\rho v = \rho_0 v_0$, so that the viscosity coefficient $\rho v$ remains unperturbed. The following reductions are done. First, we note that any solution of Eq. (9) satisfies Eq. (8) automatically. This means that out of four equations provided by (8) and (9) together, one should be omitted. We choose to omit $z$-component of Eq. (9) and use Eq. (8) to express $b_z$ via $b_r$ and $b_\theta$ in

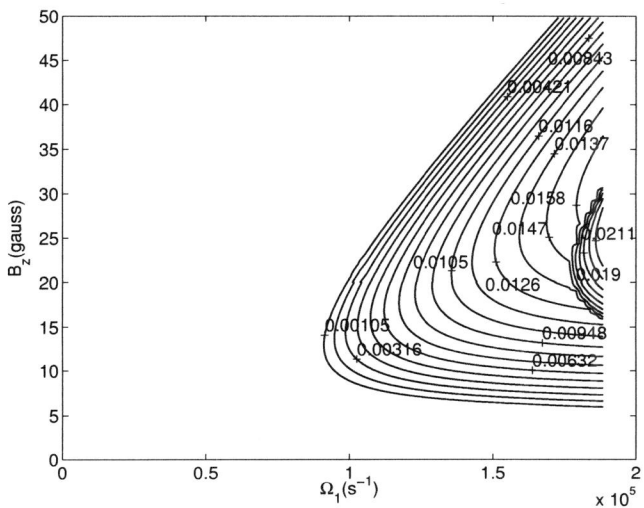

**FIGURE 2.** Growth rate, $\mathrm{Im}\,\omega/\Omega_1$, of the unstable $m = 0$, $k_z = \pi/L$ mode as a function of $\Omega_1$ and $B_{0z}$

the $r$ and $\theta$ components of Eq. (9) as well as in all components of Eq. (10). We also use Eq. (7) to express $v_z$ via $v_r$ and $v_\theta$. Eq. (6) is used to express $\rho_1$ via $v_r$ and substitute into momentum Eq. (10). As for Eq. (10) we replace its $z$-component by the equation obtained by taking the divergence of Eq. (10). Using Eqs. (8) and (7) the divergence of Eq. (10) can be reduced to the equation with the only second order radial derivative being the $\partial^2 \Pi/\partial r^2$, where $\Pi = P_1 + B_{0z} b_z/(4\pi)$ is the perturbation of the total pressure. In summary, we obtain five second order linear differential equations in $r$ for five variables, $\Pi, b_r, b_\theta, v_r$, and $v_\theta$: the divergence of Eq. (10), $r$ and $\theta$ components of induction Eq. (9), and $r$ and $\theta$ components of momentum Eq. (10)

An ideal conductor boundary is a good approximation, since $\eta$ for metallic walls is $< 800\,\mathrm{cm}^2\,\mathrm{s}^{-1}$, which is always much smaller than that of plasma, so even a thin metallic wall can be considered as a good conductor. Boundary conditions for conducting walls are given by $b_r = 0$, $db_\theta/dr + b_\theta/r = 0$, $v_r = 0$, $v_\theta = 0$, $dv_r/dr = 0$ at both $r = R_1$ and $R_2$.

We discretize the equations and boundary conditions on a uniform grid in rescaled variable $x = \ln(r/R_1)$ with $N = 200$ grid points. The problem is reduced to finding values of $\omega$ such that the determinant of the matrix $5N \times 5N$ is zero. We calculate the determinant by using LU-decomposition and search for zeros by using Newton iteration method. After an eigenvalue of $\omega$ is found, the corresponding eigenfunction is calculated by using back substitution.

We find $m = 0$ axisymmetric unstable mode in a wide range of parameters achievable in the experiment. Let us focus on one case with fixed temperature and density given in Table 1. Then, $C_s(T)$ and $\mathrm{Pm}(\rho, T)$ are also fixed. The remaining parameters, which can be adjusted in the experiment, are $\Phi_0$ and $B_{0z}$. For angular velocity profile (3) the value of $\Phi_0$ is uniquely related to the value of $\Omega_1 = \Omega(R_1)$. The velocity of the plasma should be subsonic everywhere for Boussinesq approximation to be a reasonable approximation. At supersonic speeds, plasma is pushed close to the outer wall leaving very rarefied region in the bulk between cylinders. In this case, it is more appropriate to discuss buoyant or Parker instabilities than MRI. The velocity of rotation is maximum near the inner cylinder, therefore, our calculations are valid for $\Omega_1 R_1 < C_s$. The maximum value $\Omega_{1\max} = C_s/R_1$ is given in Table 1.

We plotted the MRI growth rate $\mathrm{Im}(\omega)/\Omega_1$ for $m = 0$, $k_z = \pi/L$ mode in Fig. 2. We also searched for the unstable non-axisymmetric modes, but did not find any. In a small region to the right in Fig. 2 with small contour separation the mode is purely growing $(\mathrm{Re}(\omega) = 0)$, in the rest of the unstable region the mode is growing and oscillating $(\mathrm{Re}(\omega) > 0)$. The MRI is absent for weak magnetic fields $(v_{Az}/C_s \ll 10^{-2})$ and small wave numbers $(k_z L < 2\pi)$ because driving force from magnetic field is too weak to overcome the viscosity and diffusivity. The MRI is also suppressed in high magnetic field region due to magnetic tension. For a given magnetic field, only finite number of modes in the $k_z$ direction will be excited.

The drawback of using plasmas is limited time available before plasma recombines enough to significantly reduce its coupling with the magnetic field. During the confinement time of the order of $10^{-3}$ s initial perturbations grow by $\sim 10$ times, and can be registered experimentally. This confinment time is within the range of planned experiment.

## ACKNOWLEDGMENTS

The authors are grateful to Zhehui Wang of Los Alamos National Laboratory for encouraging the present work and support. Discussions with Hantao Ji and Jeremy Goodman were very useful. KN acknowledges support from DOE through the LDRD-ER program at Los Alamos National Laboratory. VP acknowledges support from DOE grant DE-FG02-00ER54600.

## REFERENCES

1. Frank, J., King, A., and Raine, D.J., *Accretion Power in Astrophysics: Third Edition*, Cambridge University Press, Cambridge, 2002.
2. Shakura, N.I., *Astron. Zhurnal* **49**, 921-929 (1972).
3. Velikhov, E.P., *J. Exptl. Theoret. Phys.* **36**, 1398-1404 (1959).
4. Chandrasekhar, S., *Proc. Natl. Acad. Sci.* **46**, 253-257 (1960).
5. Balbus, S.A., and Hawley, J.F., *Astrophys. J.* **376**, 214-233 (1991).
6. Ji, H., Goodman, J., and Kageyama, A., *Mon. Not. Roy. Astron. Soc.* **325**, L1-L5 (2001).
7. Noguchi, K., Pariev, V.I., Colgate S.A., Nordhaus J., and Beckley H.F., *Astrophys. J.* **575**, 1151-1162 (2002).
8. Goodman, J., and Ji, H., *J. Fluid Mech.* **462**, 365-382 (2002).
9. Rüdiger, G., and Zhang, Y., *Astron. Astrophys.* **378**, 302-308 (2001).
10. Rüdiger, G., Schultz, M., and Shalybkov, D., *Phys. Rev. E* **67**, 046312 (2003).
11. Willis, A.P., and Barenghi, C.F., *Astron. Astroph.* **388**, 688-691 (2002).
12. Richard, D., and Zahn, J.-P., *Astron. Astroph.* **347**, 734-738 (1999).
13. Wang, Z., Pariev, V.I., Barnes, C.W., and Barnes, D.C., *Phys. Plasmas* **9**, 1491-1494 (2002).

# SECTION 4

# TOROIDAL SYSTEMS

# Experimental Investigation of Helical Non-neutral Plasmas

H. Himura, H. Wakabayashi, M. Fukao, and the CHS group*

*Department of Advanced Energy, Graduate School of Frontier Sciences,
University of Tokyo, Hongo, Bunkyo Ward, Tokyo 113-0033, Japan*

*\*National Institute for Fusion Science, Toki, Gifu 509-5292, Japan*

**Abstract.** For the first time, an experimental study on helical nonneutral plasmas is performed on the Compact Helical System (CHS) device. (1) Remarkably, despite being launched from the outside of closed magnetic surface, the injected electrons travel across the magnetic field and penetrate deeply inside the magnetic surfaces. This penetration of electrons is caused by a collisionless mechanism that has never been observed in past toroidal nonneutral plasmas confined in axisymmetric geometry. (2) The penetrated electrons form a helical non-neutral plasma inside the magnetic surface. The electron density is about $10^{11\text{-}13}$ $m^{-3}$ much smaller than the Brillouin density limit. (3) The stable phase of the helical non-neutral plasma continues only for $1 - 4$ ms and then starts to disrupt. About 50 kHz of disruptive instability is observed. This frequency and the other parameters related to the onset time of the instability suggest an ion-related instability as the possible mechanism.

## INTRODUCTION AND SUMMARY

Recently, many intensive theoretical [1] and experimental [2] studies on plasma flow have been performed. In particular, work on the generation of a strong electric field, **E,** in the boundary layer of a closed magnetic surface in toroidal plasmas has been important to research aimed at improving confinement properties by means of fast shear flows [3] and also to the production of high-$\beta$ plasmas with a fast ion perpendicular flow [4] as an alternative fusion confinement concept. One of the methods proposed for producing **E** inside toroidal plasmas is a toroidal, non-neutralized, two-fluid plasma [5]. For this reason, pure electron plasmas confined in a toroidal geometry have intensively been studied [6] in order to address the basic physics of toroidal non-neutral plasmas.

One of the key issues of these experiments is the injection of electrons across closed magnetic surfaces. Usually, in axisymmetric toroidal plasmas, electrons cannot be injected across the magnetic surface. This calls for some new method to cause the electrons to move across the magnetic surface. As one possible method, we propose to inject electrons from a stochastic magnetic layer. In this case, all adiabatic invariants of the electrons are no longer conserved. This may result in operating electrons in a completely non-ordered motion in the stochastic layer, and consequently, some of the electrons might move across the magnetic surface.

CP692, *Non-Neutral Plasma Physics V*, edited by M. Schauer et al.

In order to test this idea, we have conducted experiments on the Compact Helical System (CHS) device [7]. In this machine, a static helical magnetic surface can be produced with an asymmetric configuration. In addition, a stochastic magnetic layer exists between the last closed flux surface (LCFS) and the separatrix. In the stochastic magnetic region no (global) adiabatic invariant is inherently conserved, raising the possibility of penetration of electrons across the stochastic magnetic region.

Electrons are injected into the CHS vacuum magnetic field, $\mathbf{B}$, from a $LaB_6$ cathode which is located in the stochastic magnetic region. As expected, despite launching from the outside of the LCFS, the electrons travel across the LCFS and penetrate deeply into the closed magnetic surface. The length scale of the penetration reaches the average minor radius $r$ of the helical magnetic surface which is about 20 cm, while the Larmour radius of electrons, $\rho_e$, is estimated to be about 2 mm in this experiment.

Simultaneously, the electron flux measured inside the separatrix increases significantly within 100 µs. This is shorter than the electron-neutral collision time ($\sim$ 350 µs). Thus, the observed penetration is caused by a collisionless phenomenon. In fact, the measured probe current, $I_p$, shows apparent fluctuations with 40-100 kHz frequency. No dependence of the penetration on the initial direction of electron injection is observed in the experiments; the electrons penetrate even when being launched quasi-parallel to $\mathbf{B}$.

No penetration of electrons occurs when the electron gun is placed about 5 cm outside the LCFS, a distance which approximately corresponds to the width of the stochastic magnetic region, nor is there penetration when the stochastic magnetic region is short-circuited. Finally, significant penetration of electrons is only observed when the beam current, $I_b$, is larger than roughly 1 mA (the beam density $n_b$ is about $10^{12}$ m$^{-3}$). These results show that the key to the observed penetration is the stochastic region in which some electrostatic collective motion of electrons is induced with increasing electron density, $n_e$. The collective motion causes the inward motion of some electrons across the helical magnetic surfaces.

In consequence of the electron penetration, a broad profile of space potential, $\phi_s$, constituted by the injected electrons is formed inside the helical magnetic surface. The maximum of $\phi_s$ (equivalently $n_e$) is achieved at the magnetic axis $R_{ax}$ and is less than the initial beam energy $eV_{acc}$. This result can be understood by conservation of energy. The value of $\phi_s$ is almost constant inside the magnetic surface of $\sqrt{\Psi} \sim 0.6$. Thus, the corresponding $\mathbf{E}$ ($= -\nabla\phi_p$) is localized to the boundary region. The value of $E_r$ is about 3 kV/m for the present configuration which implies $n_e \sim 10^{11-12}$ m$^{-3}$, a value substantially lower than the Brillouin density limit. The confinement time of the plasmas, $\tau_N$, seems to be in the range between 0.5 and 1.5 ms, and a preliminary analysis shows that the equipotential surface appears to deviate from the magnetic surface. The observed deviation is larger than the banana width.

About 1-4 ms after the injection is complete, a disruption is observed. The unstable mode has a frequency of about 50 kHz and a growth time of about 50 µs. The onset time of the instability depends on the cross section for background gas ionization. This indicates an ion-resonance instability (IRI) produced via electron impact ionization of the background gas[8]. The instability appears throughout the plasma, implying a global mode. A simple calculation of the m = 1 IRI indicates that only a few percent of the ions need participate to drive the instability. The corresponding frequency and growth time are comparable to the observed values. However, more data is necessary to confirm the IRI as the culprit.

## APPARATUS

Experiments are conducted on the CHS machine. CHS is a medium size device whose major radius, $R$, and averaged minor radius, $r$, are 1.0 and 0.2 m, respectively. Here, $r$ is defined as $r \equiv \sqrt{a \cdot b}$ where $2a$ and $2b$ are the major and minor axes of an elliptic shape of the separatrix of the helical magnetic surface.

On the CHS experiments, there are two parameters which are essential to form the helical configuration. One of these is the radial position of $R_{ax}$ which is usually set around 92.1 cm. However, for this setting, the inside edge of magnetic surface intersects the vacuum chamber. This results in a short-circuiting of the stochastic magnetic region. Therefore, for the present experiments, $R_{ax}$ is shifted outward and fixed at 101.6 cm. In this case, the stochastic region is electrically isolated from the chamber.

The second essential parameter is the strength of **B**. The experiments presented here were performed under DC operation to create a well-controlled helical field and avoid experimental difficulties. The field strength can vary between 0.2 and 0.9 kG with the typical strength of **B** for the experiments presented here being approximately 0.9 kG at $R_{ax}$ for $R_{ax}$ = 101.6 cm. This yields an electron gyroradius of approximately 2 mm when the beam velocity, $v_b$, is about $2 \times 10^7$ m/s ($eV_{acc} \sim 1.2$ keV).

Electrons are injected from a typical diode-type electron gun (henceforth called e-gun) which comprises a LaB$_6$ emitter as the cathode. The LaB$_6$ cathode is quadrate (1.5 cm each) and the emitter has also a quadrille shape with tungsten wires. The beam current ($j_b$) of the electrons can be varied. However, for the present research we have kept $j_b$ at a constant value of ~10 mA for $V_{acc}$ = 1.2 kV. The beam energy is also variable in the range from 0.2 to 1.2 keV. The e-gun is installed in the equatorial plane ($z = 0$), and the cathode is usually placed in the stochastic magnetic region, however the cathode is movable. The injection angle, $\alpha$, of the electrons can be varied by rotating the barrel of e-gun. The angle of $\alpha$ is defined as $0°$ when the electrons are injected in the counterclockwise toroidal direction, $90°$ for the +z-direction, $180°$ for the clockwise toroidal direction, and $270°$ for the –z-direction. A detailed description of the e-gun is given in a companion paper [9].

For diagnostics, we have employed two probes, each of which works as not only an electrostatic but also as an emissive probe [10] in the same way as the Proto-RT experiments [6]. With the probes, both space potential $\phi_p$ and probe current $I_p$ are measured. One of the probes is installed along the z axis and the other is inserted along the r axis. Both are movable, which provides spatial distribution of $\phi_p$ and $I_p$ in the magnetic surface.

# RESULTS AND DISCUSSION

In this section, we briefly explain the experimental observation of (A) injection of electrons via stochastic magnetic region, (B) confinement of helical non-neutral plasmas, and (C) disruption of electron plasmas. Further details of the experiments will be reported in future publications.

## A. Injection of Electrons via Stochastic Magnetic Region

Figure 1 shows the spatial evolution of $\phi_s$ for $\alpha = 300°$ . The $\sqrt{\Psi} = 0$ and 1 of the horizontal axis correspond to $R_{ax}$ and LCFS, respectively. The trigger is activated at $t = 0$ μs. Just after the triggering, $\phi_s$ increases significantly. Since the electrons reach the neighborhood of $R_{ax}$, the depth, $\delta$, of penetration is about 20 cm. This value is about 100 times larger than $\rho_e$ (~ 2 mm as explained earlier).

The measured $\phi_s$ takes its maximum value of ~ 1.2 kV at $R_{ax}$. This value is almost equal to the beam energy ($eV_{acc}$). In fact, the value of $\phi_s$ decreases linearly with decreasing $eV_{acc}$. These results can be understood by invoking energy conservation for the injected electrons. As seen in the data, $\phi_s$ is almost constant inside $\sqrt{\Psi} \sim 0.6$, which suggests a hollow profile for the electrons in the magnetic surface. In fact, little electric field, $\mathbf{E}$ (= $-\nabla\phi_s$), is built in the neighbor of $R_{ax}$. On the other hand, around $\sqrt{\Psi} \sim 1$, a gradient of $\phi_s$ is clearly formed. From the profile of $\phi_s$, the maximum of $E_r$ (about 3 kV/m for these parameters) is attained at the LCFS.

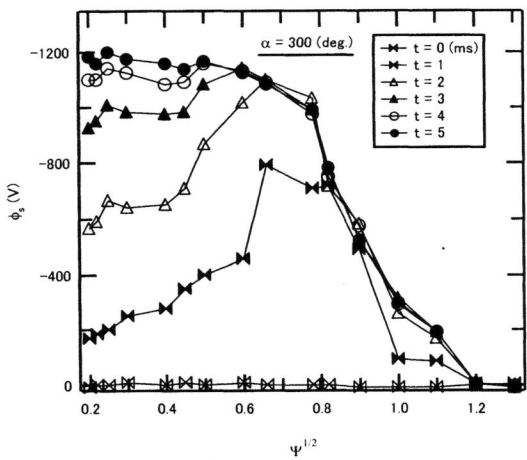

**FIGURE 1.** Time evolution of space potential $\phi_s(z)$ constituted by the penetrated electrons for $\alpha = 300$ deg. case. Data are measured along the z axis. Significant penetration of the injected electrons is recognized. Because of the high-impedance emissive method, the time constant is relatively long.

The data of $\phi_s$ are measured by the high impedance emissive technique so that $t_c$ is very long (~ 50 ms). The time scale of penetration can more exactly be recognized from the $I_p$ data having $t_c \sim 0.1$ μs. In the experiment, $I_p$ rises in ~ 100 μs inside LCFS.

This shows the penetration of electrons happens within 100 μs which is much faster than the electron-neutral collision time (~ 350 μs) for the current experimental parameters: $v_e \sim 10^7$ m/s and $P_0 \sim 1 \times 10^{-7}$ Torr. No orbital motion of the injected electrons extends inside the LCFS. The observed penetration of electrons is thus caused by a collisionless process.

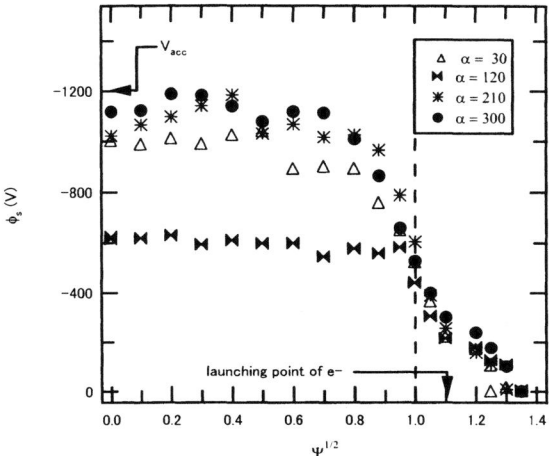

**FIGURE 2.** Radial profiles of $\phi_s(r)$ measured by an emissive probe [10]. Data are shown for four different angles of injection, $\alpha$. These data show that as long as the electron gun is placed in the stochastic magnetic region, the electron penetration always occurs, regardless of $\alpha$.

Because both $\phi_s$ and $I_p$ have directly been measured, we can compare these measured values with the Poisson's equation as $\nabla^2 \phi \sim \phi / (\Delta r)^2 = en_e / \varepsilon_0$ to verify the measured data. From values of $\phi_s \sim 1.2$ kV at $\sqrt{\Psi} \sim 0.8$ and the distance of $\sim 10$ cm between $\sqrt{\Psi} \sim 0.8$ and $\sqrt{\Psi} \sim 1$, $n_e$ is calculated to be $\sim 10^{13}$ m$^{-3}$ from the equation. On the other hand, because the electron temperature $T_e$ just inside LCFS can be calculated as $eV_{acc} - \phi_p \sim 600$ eV, the mean velocity, $<v_e>$, of the injected electrons is roughly estimated to be $\sim 5 \times 10^6$ m/s, by assuming that the total energy of the injected electron is conserved. This yields the value of $n_e \sim 2 \times 10^{12}$ m$^{-3}$ from the probe measurement because $n_e <v_e> \sim 1 \times 10^{19}$ /m$^2$s. These values of $n_e$ agree with each other within experimental error.

Since the injection angle, $\alpha$, of the electrons can be varied, it is possible to evaluate the dependence of the penetration on $\alpha$. Figure 2 shows profiles of $\phi_s(r)$ for different values of $\alpha$. For each value of $\alpha$, $\phi_s$ increases. In other words, regardless of $\alpha$, the collisionless penetration occurs. We also shifted the magnetic surface so as to bring it into contact with the vacuum vessel. In this case, the stochastic magnetic layer was electrically short-circuited and no penetration of electrons occurred, regardless of the value of $\alpha$. Presumably, this is due to loss of electrons to the vacuum vessel.

Finally, the e-gun was pulled back in 0.5 cm increments in order to change the injection point. No penetration was observed when the injection point was outside of the stochastic region. These results indicate that the key to penetration of the electrons is the stochastic region.

To investigate the penetration mechanism, FFT analyses of $I_p$ measured around LCFS have been performed. Fluctuations in the range of 50 – 100 kHz have clearly been observed during the injection. This is in the range of m = 1 diocotron mode frequency, and since the injected electrons initially form a hollow profile, it may be that this instability is driving the observed penetration of electrons.

## B. Confinement of Helical Non-neutral Plasmas

The measured $I_p$ saturates after about 100 μs. As seen in Fig. 2, the profile of $\phi_s$ achieves its maximum in the neighbor of $R_{ax}$. Figure 3 shows $\phi_s$ at $R_{ax}$ measured against $eV_{acc}$ for $\mathbf{B} = 0.9$ kG. As seen from the data, $\phi_s$ increases significantly with $V_{acc}$. In addition, higher $\mathbf{B}$ resulted in the larger $\phi_s$. This indicates that the maximum $\phi_s$ (equivalently $n_e$) is limited by $eV_{acc}$ and the strength of $\mathbf{B}$ inside the magnetic surface. As already mentioned, the dependence on $eV_{acc}$ can be understood by conservation of energy of the electrons.

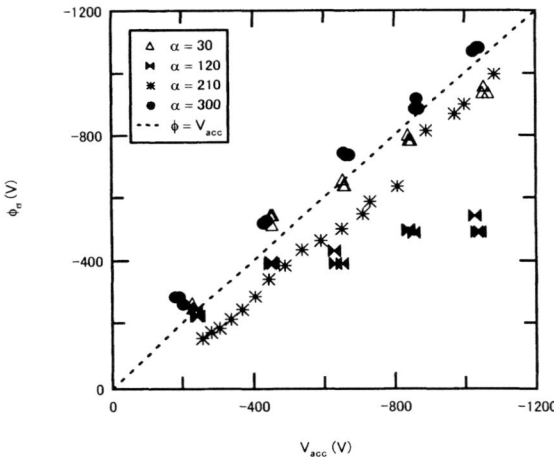

**FIGURE 3.** The maximum $\phi_s$, which is equivalently electron density $n_e$, at the magnetic axis $R_{ax}$ against the beam energy $eV_{acc}$ for $\mathbf{B} = 0.9$ kG case. The dashed line shows $\phi_s = V_{acc}$. The data indicate that the $\phi_s$ is limited by $eV_{acc}$.

On the other hand, the dependence on $\mathbf{B}$ is probably due to better confinement of electrons in the stronger $\mathbf{B}$, not to an increased Brillouin density limit, $n_B$. We estimate $n_e$ in our experiment to be $10^{11\text{-}13}$ m$^{-3}$ for $\phi_s \sim 1$ kV, but $n_B$ is calculated to be $\sim 10^{16}$ m$^{-3}$ for $\mathbf{B} = 0.9$ kG, too large to explain the experimental value.

No change in saturation of $\phi_s$ (density limitation) was observed when the beam current $I_b$ was increased. For $I_b < 10$ mA, $\phi_s$ increased almost linearly, however for $I_b > 10$ mA, $\phi_s$ saturated. This indicates that only a few more electrons drift inward across the magnetic surface with increased $I_b$ and additional injected electrons are lost.

The reason $\phi_s$ saturated at $I_b \sim 10$ mA can be roughly understood by a simple calculation of the electron number confined in the stochastic region.

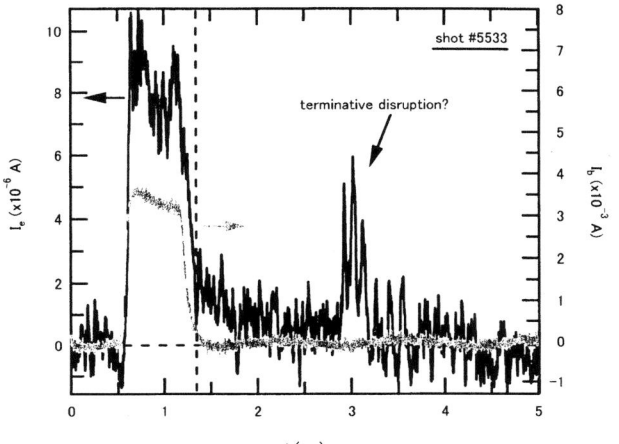

**FIGURE 4.** Typical signals of probe current, $I_p$, measured at the last closed flux surface (LCFS) and beam current, $I_b$. At $t \sim 1.4$ ms, the electron beam is completely turned off. The output from $I_p$ continues for approximately 1 ms more, which implies that the confinement time, $\tau_N$, is about 1 ms for this shot.

An estimate of the confinement time, $\tau_N$, of electrons in the magnetic surface can be obtained from the signal of the electrostatic probe located at LCFS. Typical data are shown in Fig. 4. It can be seen that $I_p$ persists for $0.5 - 1$ ms after the e-gun is turned off at t $\sim$ 1.4 ms. This suggests that $\tau_N$ of the helical electron plasmas is about $0.5 - 1$ ms for the $R_{ax} = 101.6$ cm case. It should be noted that, in most shots, a spike appears at the end of the duration of $I_p$, as seen at t $\sim$ 3 ms in Fig. 4. This might be the signal of a terminative disruption of the plasma inside the magnetic surface.

Since $\tau_N$ is on the order of 0.1 ms, the transport coefficient is calculated to be $D_{exp} \sim r/\tau_N \sim 10^2$ m²/s in the experiment. In most cases of neutral helical plasmas, the dominant transport mechanism is helically trapped particle (HTP) loss [11]. However, this does not account for the observed transport in non-neutral helical plasmas. The speed of HTP is about $v_h \approx m_e v_\perp^2 / qBR_h \sim 3\times10^4$ m/s. Thus, the rotation frequency of HTP in poloidal direction $f_{pol}$ is estimated to be 100 kHz. Therefore, the coefficient due to the HTP loss is approximately $D_h \approx (\varepsilon_t / \varepsilon_h)^2 r^2 v_{en} / \sqrt{\varepsilon_h} \sim 10^6$ m²/s, which is too large to explain $D_{exp}$. Therefore, some other mechanism appears to govern the transport.

In the experiment, $\phi_s$ has been measured at two different cross sections along the $r$ and $z$ axes. Thus, we can compare the equipotential surface with the magnetic surface. The equipotential surface seems to deviate from the magnetic surface by $\sim$ 2 cm for the case of $R_{ax} = 101.6$ cm on CHS. The banana width of a toroidally trapped particle does not explain the observed deviation. The value of the width is calculated as $\Delta r \sim 2qp_e(R \quad r)^{1/2} \sim 5$mm. Further measurements of $\phi_s$ at other cross sections are required.

The **E** formed by the penetrating electrons is about 3 kV/m, as is also indicated in Fig. 1 and 2. This magnitude of **E** would produce an ion rotation velocity of $\sim 10^5$ m/s in a neutral plasma. But **E** is formed only near the LCFS. This suggests the presence of few electrons around $R_{ax}$, even in the possible equilibrium phase.

## C. Disruption of Electron Plasmas

The stable phase described above lasts for 1 - 4ms and then starts to disrupt. This event can be recognized from typical $I_p$ data shown in Fig. 5. Multiple disruptions are observed when electrons are continuously injected into the **B** field. The observed frequency of the disruption is about 50 kHz and seems to be independent of the strength of **B**. The growth time of the disruption is about 50 μs. Evidence of the disruptions was also visible in the $\phi_s$ signal.

We found the time interval between disruptions, $t_{int}$, to depend on the cross section for ionization of the background gas in the apparatus. This strongly suggests that the electron beam ionizes the background gas and produces residual ions. Although the details are still unknown, some ion-related instability (IRI) may occur inside the magnetic surface, driving both electrons and ions out of the closed magnetic surface.

In fact, in the experiment, such a signal of multiple disruptions was observed over the whole region inside the magnetic surface, even around $R_{ax}$. In fact, some shots seem to indicate that the disruption originates from the neighborhood of $R_{ax}$. In the present configuration of the magnetic surface, the iota profile approaches 0.5 around $R_{ax}$, producing the rational surface of m/n = 1/2 there.

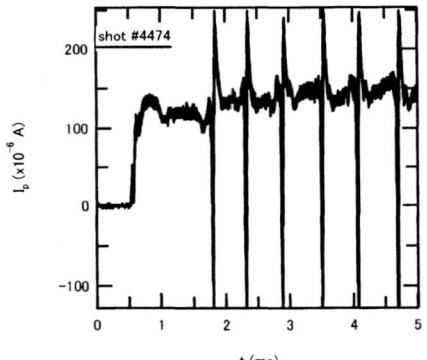 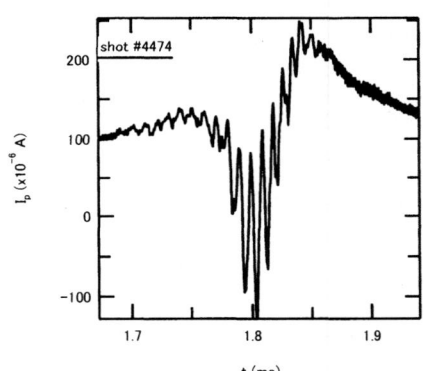

**FIGURE 5.** Typical waveform of $I_p$ in the graph on the left showing a disruption of helical electron plasma. Multiple disruptions occur when electrons are continuously injected into the B-field. In the graph on the right one of the disruptions is magnified. The observed frequency and growth rate are about 50 kHz and 50 μs, respectively.

Assuming the conventional IRI found in cylindrical systems occurs in the helical geometry, a calculation of the m = 1 mode agrees with the observed frequency and growth rates. Thus, this may account for the observed instability. Actually, the penetrated electrons must be ordered in drifting by $\mathbf{E} \times \mathbf{B}$ in the magnetic surface. This implies that, even in the helically twisted magnetic field, a diocotron wave due to the drift motion could be generated because of the inhomogeneous $\mathbf{B}$ and $\mathbf{E}$.

If the residual ions start the slow rotation, and the ion angular velocity is different from the angular velocity of the electrons, a two-stream instability would be possible due to the differential rotation of electrons and ions. Once such instability is initiated, it would immediately grow in the neighborhood of $R_{ax}$ because of the rational surface. Also, due to the small value of $\mathbf{E}$ there, little $\mathbf{E} \times \mathbf{B}$ drift occurs, possibly resulting in little short-circuiting of perturbations in $\mathbf{E}$ due to the diocotron oscillation. Thus, little suppression of the instability may occur around $R_{ax}$ for this configuration. This speculation will be experimentally tested in the next series of experiment.

## ACKNOWLEDGMENTS

The authors are grateful to T. Kurihara, H. Saito, and J. Morikawa for their help in preparing parts used for the experiment, and to Profs. Z. Yoshida and Y. Ogawa for discussions and comments. This work is performed with the support and under the auspices of the NIFS LHD Project Research Collaboration.

## REFERENCES

1. *for example*, Steinhauer, L. C. and Ishida, A., Phys. Rev. Lett., **79**, 3423-3426 (1997); Mahajan, S. M. and Yoshida, Z., Phys. Rev. Lett., **81**, 4863-4866 (1998).
2. *for example*, Kaneko, T., Tsunoyama, H., Hatakeyama, R., Phys. Rev. Lett., **90**, 125001 (2003).
3. Burrell, K. H., Phys. Plasmas, **6**, 4418-4435 (1999).
4. Ogawa, Y., Himura, H. et al., J. Plasma Fusion Res. SERIES, **5**, 100-105 (2002).
5. Himura, H. et al., Fusion Energy 2000, IAEA-CN-77-ICP/14 (2000).
6. Himura, H. et al., Phys. Plasmas, **8**, 4651-4658 (2001); Nakashima, C., Yoshida, Z., Himura, H. et al., Phys. Rev. E, **65**, 036409 (2002).
7. Nishimura, K. et al., Fusion Technol., **17**, 86-100 (1990).
8. Levy, R.H. et al, Phys. Fluids, **12**, 2612 (1969).
9. Wakabayashi, H., Himura, H. et al., *"Parameter dependence of inward diffusion on injected electrons in helical nonneutral plasmas"* in this proceedings.
10. Himura, H. et al., *submitted to Review of Scientific Instruments.*
11. Wakatani, M., *Stellarator and Heliotron Devices*, (Oxford Univ. Press).

# Confinement of non-neutral plasmas in the Columbia Non-neutral Torus Experiment

Thomas Sunn Pedersen*, Jason P. Kremer* and Allen H. Boozer*

*Columbia University, New York, NY 10027

**Abstract.** The physics of non-neutral plasmas confined on magnetic surfaces is discussed. The Columbia Non-neutral Torus (CNT), a table-top ultrahigh vacuum stellarator being constructed at Columbia University, is being built to systematically study non-neutral plasmas confined on magnetic surfaces. The experimental design is discussed in the context of relevant physics parameters, such as the number of Debye lengths in the device, and the parallel versus perpendicular time scales.

## INTRODUCTION

Confinement systems that use magnetic field lines alone have several advantages over those that use magnetic and electric fields, such as the Penning trap, including the ability to confine positive and negative particles simultaneously, and the ability to confine light, energetic particles. Closed toroidal field line systems have been used to confine pure electron plasmas [1, 2, 3, 4], and more recently, magnetic surface configurations have become of interest as confinement devices for non-neutral plasmas [5, 6]. The physics of pure electron plasmas confined on magnetic surfaces is fundamentally different from previously studied configurations. A magnetic surface configuration is characterized by magnetic field lines that lie on toroidal surfaces (the magnetic surfaces), with each field line coming arbitrarily close to any point on the surface that it lies on. The parallel dynamics of the plasma then determines not just what happens on an isolated field line, but an entire surface. Magnetic surface configurations are well-known in fusion science, in particular the tokamak and the stellarator. A stellarator is a magnetic surface configuration created entirely from magnets external to the plasma, so it has the advantage that it does not require any plasma currents, and can be steady state. A stellarator, the Columbia Non-neutral Torus (CNT) is currently being constructed specifically to investigate the physics of non-neutral plasmas confined on magnetic surfaces. This paper reviews the theory of non-neutral plasmas confined on magnetic surfaces, and discusses some of the experimental parameters that are of importance to a non-neutral stellarator experiment. The CNT experiment is discussed in the context of these parameters.

# CONFINEMENT OF PURE ELECTRON PLASMAS

## Equilibrium

The equilibrium of a pure electron plasma in a magnetic surface configuration is described by a self-consistent equation for the electrostatic potential [6]:

$$\varepsilon_0 \nabla^2 \phi = eN(\psi) \exp\left(e\phi/T_e(\psi)\right) \tag{1}$$

Here $\psi$ is the magnetic surface coordinate, that is, each magnetic surface is described by $\psi = constant$. The temperature is taken to be constant on a magnetic surface due to rapid thermalization along field lines, $T_e = T_e(\psi)$. The function $N(\psi)$ indirectly specifies the density profile. The equilibrium plasma flow is:

$$\mathbf{v}_e = \frac{(\nabla p/en_e - \nabla \phi) \times \mathbf{B}}{B^2} + v_\parallel \frac{\mathbf{B}}{B} \tag{2}$$

It can be shown that this flow cannot cross the magnetic surfaces [6]. The parallel flow adjusts itself to make the total particle flux divergence free, even if the perpendicular particle flux is not. With closed toroidal field lines, or in a Penning trap, the parallel flow cannot do this, and hence, contours of constant density and electrostatic potential must coincide in order to keep the perpendicular particle flux divergence free.

## Stability

The equilibrium electrostatic potential given by Equation 1 minimizes an energy-like quantity [7]:

$$U = \int \left(\frac{1}{2}\varepsilon_0 (\nabla \phi)^2 + N(\psi)T_e(\psi) \exp\left(e\phi/T_e(\psi)\right)\right) dV = \int \left(\frac{1}{2}\varepsilon_0 E^2 + p\right) dV \tag{3}$$

The equilibrium electron density increases near positive image charges. This is in contrast to the Penning trap [8] and the pure toroidal field trap [9], which have electrostatic potentials that maximize the potential energy, and the electron plasma tends to move away from positive image charges. However, the energy-like quantity that is minimized in equilibrium in a magnetic surface configuration is not the free energy, so this does not guarantee stability of all possible configurations. Further work is needed to develop proper stability criteria of pure electron plasmas confined on magnetic surfaces.

## Confinement

Confinement in a non-neutral stellarator is limited by neoclassical diffusion. The guiding center drifts that cause the particles to drift away from the magnetic surfaces are the $E \times B$ drift as well as curvature and $\nabla B$ drifts:

$$\mathbf{v}_D = \frac{\mathbf{E} \times \mathbf{B}}{B^2} + \left(\frac{1}{2}mv_\perp^2 + mv_\parallel^2\right)\frac{\nabla B \times \mathbf{B}}{eB^3} \tag{4}$$

The $E \times B$ drift causes drift excursions away from the magnetic surfaces because the electrostatic potential contours do not coincide exactly with the magnetic surfaces. The electrostatic potential contours do match very closely to the magnetic surfaces in any region with appreciable plasma density, unless the Debye length is large. A simple scaling estimate yields the following particle confinement time.

$$\tau_p \approx \tau_e \frac{a^4}{\lambda_D^4} \tag{5}$$

Here $\tau_e$ is the electron collision time, and the estimate above is valid for small Debye lengths.

## Initial plasma formation in a stellarator

Equation 5 suggests that single particles injected into an empty stellarator will not be confined. However, the expression above ignores the confinement due to the magnetic surfaces themselves, which provide confinement for $a/\lambda_D \to 0$. In a classical stellarator, such as CNT, there will be two kinds of particles, well-confined and poorly confined particles (particles on bad orbits). Generally, passing particles are well-confined with very long confinement times. The mirror-trapped poorly confined particles will be confined for at least $\tau \approx a/v_D$, where $a$ is the minor radius and $v_D$ is the guiding center drift velocity, Equation 4. Even though this confinement time is low, it is sufficient to start the accumulation of electrons and establish a finite Debye length. In addition, the well-confined particles will be confined much longer. Hence, the confinement of the magnetic surfaces will allow initial accumulation of electrons even in the zero density limit, and confinement will improve as the number of Debye lengths increases. It should be mentioned that quasi-symmetric stellarators, which have specially optimized magnetic surfaces, the fraction of poorly confined particles is essentially zero, and neoclassical confinement times are very long in the complete absence of space charge. Such stellarators are generally characterized by having rather complicated coils.

## CONFINEMENT OF PARTLY NEUTRALIZED AND ELECTRON-POSITRON PLASMAS

A stellarator confines both positive and negative particles simultaneously whether space charge and internal currents are present or not. This allows the study of plasmas of arbitrary neutralization, a field of plasma physics that is currently largely unexplored. Positive particles, ions or positrons, will be very well confined in an electron-rich plasma, by the overall negative space charge as well as by the magnetic surfaces. As one slowly neutralizes the plasma, the electric field weakens, as the density rises. The electron confinement time will then be [10]:

$$\tau_p \approx \tau_e \frac{a^4}{\lambda_C^4} \tag{6}$$

This is a similar scaling to that of a pure electron plasma, however, the Debye length $\lambda_D$ is replaced by the Coulomb length,

$$\lambda_C^2 = \frac{|n_e - n_p|}{n_e + n_p} \lambda_D^2 \qquad (7)$$

Here, $n_e$ is the electron density and $n_p$ is the density of the positive species, assumed to be a proton or a positron. In a quasi-neutral plasma $a/\lambda_C$ is on the order of 1 or less, whereas in a pure electron plasma, $a/\lambda_C = a/\lambda_D \gg 1$. A partly neutralized plasma may be characterized by $a/\lambda_D > a/\lambda_C \gg 1$. The confinement time given by Equation 6 is long in this limit, and that may allow significant accumulation of positrons injected into a stellarator containing an initially pure electron plasma, even with the relatively weak positron sources available today [10]. Hence, this may be an attractive way to create the first laboratory electron-positron plasma.

## DESIGN OF THE COLUMBIA NON-NEUTRAL TORUS

The Columbia Non-neutral Torus (CNT), a small stellarator currently being constructed at Columbia University, is the first experiment specifically designed to study the physics of non-neutral and electron-positron plasmas confined in a stellarator. The coil configuration is very simple, consisting only of four circular, planar coils. Two of these coils are interlocked and will be inside the vacuum chamber. The stellarator magnetic field is characterized by B=0.2 T on the magnetic axis, (average) minor radius $a = 0.1$ m, (average) major radius $R = 0.3$ m, rotational transform $\iota = 0.2 - 0.6$, and the device is designed to have a neutral base pressure of $< 3 \times 10^{-10}$ Torr. $\iota$ is the rotational transform, which is the number of times a magnetic field line winds around poloidally divided by the number of times it winds around toroidally. The magnets will be driven by a 200 kW DC power supply, and the pulse length will be at least 15 seconds at full field, limited by the allowable temperature rise of the copper conductors.

### Important physics parameters

In order to guide the design of CNT, we have identified important physics parameters for a non-neutral stellarator. Specifically, we have focused on parameters of importance to confining pure electron plasmas. The most fundamental physics parameter of any plasma physics experiment is $a/\lambda_D$, where $a$ is the smallest characteristic size of the plasma, in the case of a stellarator, the minor radius. In order for the electron cloud to be a plasma, $a/\lambda_D \gg 1$. In a non-neutral plasma experiment, including CNT, this is a non-trivial constraint that requires careful matching of the injected electron energy to the plasma potential, or some method of cooling the plasma after it has been injected. $a/\lambda_D \gg 1$ is particularly important in a non-neutral stellarator, given the predicted strong scaling of the confinement time with $a/\lambda_D$, Equation 5. Another important parameter is the time scale for ion accumulation due to ionization of background neutrals, $\tau_i$. When $\tau_i \gg \tau_p$, electron plasmas will decay before being significantly contaminated. When

$\tau_i \ll \tau_e$, ions will significantly neutralize an initially pure electron plasma before it decays away. It is desirable to maximize both time scales, since either one can trivially be decreased. A large $\tau_i$ will be achieved through the ultrahigh vacuum design and operation at low plasma temperatures. A large $\tau_p$ can be achieved by making the Debye length short compared to the system size, although there is a tradeoff involved which will be addressed in the following.

A key issue for a non-neutral plasma on magnetic surfaces is whether the plasma truly equilibrates on a magnetic surface through parallel dynamics faster than the $E \times B$ drift can take the plasma away from the magnetic surfaces. In a quasi-neutral plasma, this is basically always true, but the $E \times B$ drift can be very large in non-neutral plasmas. The time scale for perpendicular distortions is the $E \times B$ rotation time $\tau_\perp = 2\pi a/(E/B)$. Any breaking of the parallel force balance will lead to plasma oscillations which are subsequently Landau damped in a finite temperature plasma. We approximate the parallel relaxation time by the time it takes a thermal particle to move along the magnetic field to fully explore the magnetic surface, $\tau_\parallel = 2\pi R/(\iota\sqrt{(T_e/m_e)})$. Then in order to ensure that the parallel force balance dominates over diocotron-type perpendicular dynamics, we must have $\tau_\perp/\tau_\parallel \gg 1$. The ratio of the two can be expressed as:

$$\frac{\tau_\perp}{\tau_\parallel} \approx 2\sqrt{2}\frac{a}{R}\iota\sqrt{\frac{n_B}{n_e}}\frac{\lambda_D}{a} \tag{8}$$

Here, $n_B = \varepsilon_0 B^2/2m_e$ is the Brillouin density. Since $\iota < 1$, $a/R < 1$, the conditions that $a/\lambda_D \gg 1$ and $\tau_\perp/\tau_\parallel \gg 1$ can only be satisfied simultaneously if $\sqrt{n_B/n_e} \gg 1$, which is well satisfied in most non-neutral plasma experiments. CNT is designed to operate in this regime as well.

## Engineering design optimization

One of the main design considerations for CNT is the ability to reach ultrahigh vacuum levels. It is particularly important to avoid ion contamination in toroidal pure electron plasma experiments, because the ions will accumulate due to the electrostatic attraction to the electrons. CNT has been designed to reach ultrahigh vacuum levels ($< 3 \times 10^{-10}$ Torr) in order to allow studies of pure electron plasmas for long times.

CNT was also designed to maximize the ratio $\tau_\perp/\tau_\parallel$, Equation 8. It can be re-expressed as:

$$\frac{\tau_\perp}{\tau_\parallel} \approx 2\sqrt{2}\,(\varepsilon_a \iota B)\left(\frac{\lambda_D}{a}\sqrt{\frac{\varepsilon_0}{2m_e n_e}}\right) \tag{9}$$

Here, we have separated plasma parameters from experimental design parameters, and introduced the inverse aspect ratio, $\varepsilon_a = a/R$. It is clear that in order to achieve $a/\lambda_D \gg 1$ and $\tau_\perp/\tau_\parallel \gg 1$, one should maximize the product $\varepsilon_a \iota B$ of the stellarator magnetic field. CNT was specifically optimized for this ratio. The maximization of $\varepsilon_a \iota B$ led to a design in which the two interlocking coils were placed inside the vacuum chamber, despite the extra complexity of having in-vessel, interlocking coils. This allowed us to use a large, cylindrical vacuum chamber, instead of a tight-fitting toroidal chamber. The elimination of the tight-fitting toroidal chamber has the following advantages:

- An increased copper cross section, allowing a stronger magnetic field and a longer pulse length with the same power supply.
- Increased plasma minor radius $a$ for the same major radius. The original vacuum chamber actually cut off good magnetic surfaces.
- Relatively large tilt angles became much more easily accommodated, allowing a large increase in $\iota$.
- The ability to change the angle between the two interlocking coils

The ability to change the tilt angle in the experiment gives tremendous flexibility in changing the magnetic topology. For example, $\iota$ is decreased by a factor of 3 from 0.6 to 0.2 by decreasing the angle between the interlocking coils from 88 to 64 degrees. Details about the magnetic configuration and field error sensitivity are presented elsewhere [11]. The redesign and reoptimization of the experiment led to an improvement of a factor of 8 in $\varepsilon_a \iota B$ over the previous design which had the interlocking coils external to the plasma. The improvement came in all three variables, $\varepsilon_a$ was increased by a factor of 1.3, $\iota$ by a factor of 3, and $B$ by a factor of 2. As shown in the next section, this improves the operating space for CNT significantly.

## Physics parameter regimes of CNT

There are numerous important physics parameters in a non-neutral stellarator. For clarity, we will primarily focus on three key parameters, the number of Debye lengths in the device, $a/\lambda_D$, the ratio of perpendicular to parallel dynamical time scale $\tau_\perp/\tau_\parallel$ (Eq. 8), and the characteristic time of ion contamination, $\tau_i$. Larger values are more desirable for all of these parameters, but as discussed earlier, there is a tradeoff between maximizing $a/\lambda_D$ and maximizing $\tau_\perp/\tau_\parallel$. In Figure 1, we show contours of constant values of each of these three parameters for the optimized CNT design. It is seen that there is a reasonably large range of densities and temperatures where all three parameters have acceptable values. A desirable operational point could be $n_e = 10^{12}$ m$^{-3}$ and $T_e = 1$ eV. At this point, $a/\lambda_D = 13.4$, $\tau_\perp/\tau_\parallel = 18.5$, $\tau_i = 2.7 \times 10^5$ s. The calculations here assume for simplicity that the neutrals are hydrogen atoms, which is not an unreasonable assumption as hydrogen often dominates in the ultrahigh vacuum range. The electron confinement time is more than one hour as predicted by Equation 5, dominated by electron-electron collisional transport. The ion contamination time is a very strong function of temperature though. At $T_e = 5$ eV, $\tau_i = 2.4$ s, still rather long, but now significantly shorter than the electron confinement time as well as the experimentally achievable pulse length. Finally, in Figure 2, we show the optimized CNT but with a poorer vacuum, $10^{-8}$ Torr. Interesting experiments can still be performed, but unless one operates at very low temperatures ( $T_e < 2$ eV), ion contamination will be a limiting factor.

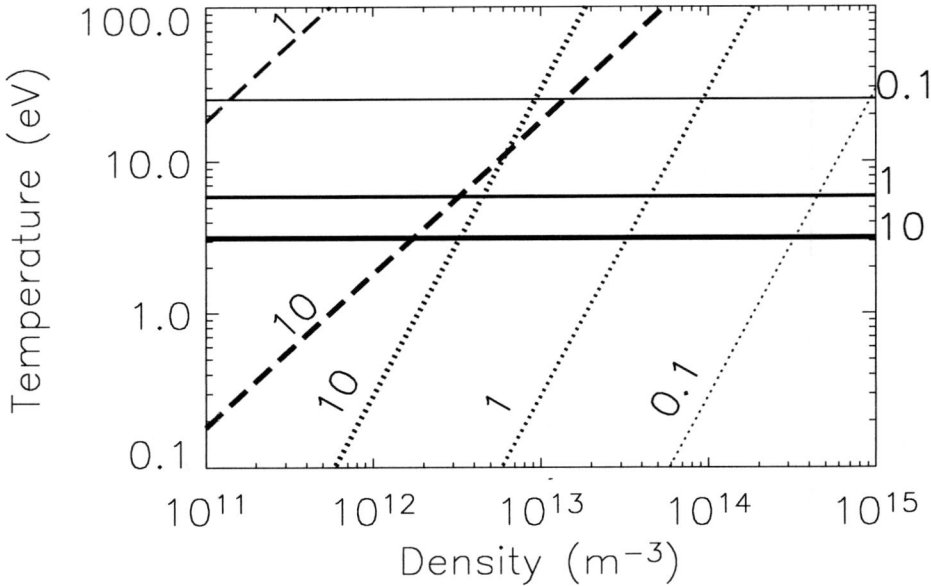

**FIGURE 1.** Some physics parameters for the optimized design of CNT. Contours of constant $a/\lambda_D$ (dashed lines), $\tau_\perp/\tau_\parallel$ (dotted lines) and $\tau_i$ (in seconds, solid lines). The thickest lines indicate a value of 10, the intermediate lines a value of 1, and the thinnest lines a value of 0.1

## CONCLUSION

CNT is being constructed specifically to study non-neutral and electron-positron plasmas confined in a stellarator. The design is a compromise between the need to build the device as easily and economically as possible, and the desire to access the most interesting parameter regimes of such a device. We arrived at a design with a simple coil configuration, two adjustable in-vessel coils, a significant magnetic field strength, and an ultrahigh vacuum. The experiment is currently under construction.

## ACKNOWLEDGMENTS

The authors would like to acknowledge the contributions of F. Dahlgren, W. Reiersen and N. Pomphrey on the stellarator design, and W. Dorland on gyrokinetic simulations of electron-positron plasmas. This work was supported by the United States Department of Energy Grant DE-FG02-02ER54690.

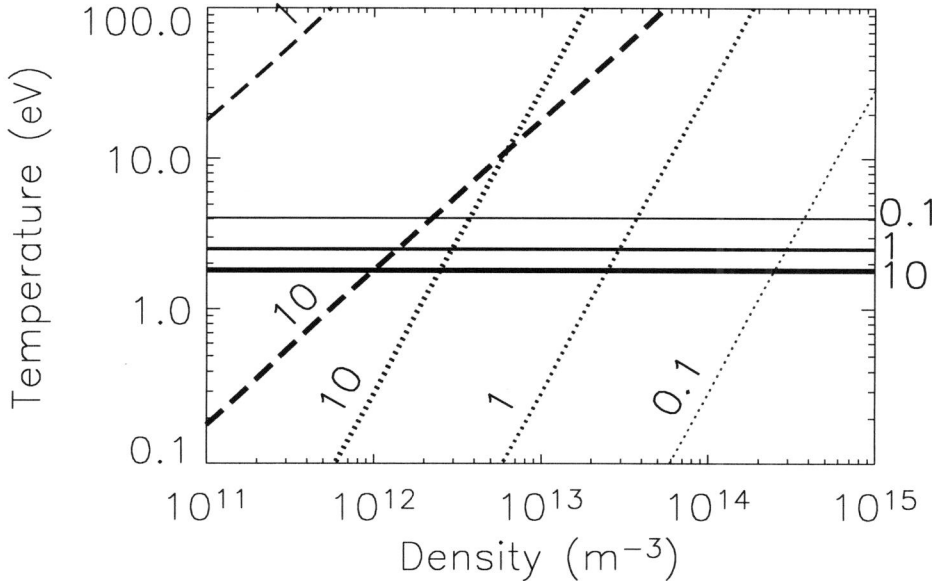

**FIGURE 2.** Some physics parameters for CNT if the neutral pressure were to be $1 \times 10^{-8}$ Torr. Contours of constant $a/\lambda_D$ (dashed lines), $\tau_\perp/\tau_\parallel$ (dotted lines) and $\tau_i$ (in seconds, solid lines). The thickest lines indicate a value of 10, the intermediate lines a value of 1, and the thinnest lines a value of 0.1

# REFERENCES

1.  Daugherty, J., Eninger, J., and Janes, G., *Phys. Fluids*, **12**, 2677 (1969).
2.  Clark, W., Korn, P., Mondelli, A., and Rostoker, N., *Phys. Rev. Letters*, **37**, 592 (1976).
3.  Zaveri, P., John, P., Avinash, K., and Kaw, P., *Phys. Rev. Letters*, **68**, 3295 (1992).
4.  Stoneking, M. R., Fontana, P., and Sampson, R., *Phys. Plasmas*, **9**, 766 (2002).
5.  Yoshida, Z., Ogawa, Y., Morikawa, J., et al., *AIP Conf. Proceedings*, **498**, 397 (1999).
6.  Pedersen, T. S., and Boozer, A. H., *Phys. Rev. Letters*, **88**, 205002 (2002).
7.  Pedersen, T. S., *Phys. Plasmas*, **10**, 334 (2003).
8.  Notte, J., Peurrung, A. J., Fajans, J., Chu, R., and Wurtele, J. S., *Phys. Rev. Letters*, **69**, 3056 (1992).
9.  O'Neil, T., and Smith, R., *Phys. Plasmas*, **1**, 2430 (1994).
10. Pedersen, T. S., Boozer, A. H., Dorland, W., Kremer, J. P., and Schmitt, R., *J. Phys. B*, **36**, 1029 (2003).
11. Kremer, J. P., Pedersen, T. S., Pomphrey, N., Reiersen, W., and Dahlgren, F., *Non-neutral Plasma Workshop, AIP Proceedings* (2003).

# Millisecond Confinement and Observation of the *m*=1 Diocotron Mode in a Toroidal Electron Plasma

M.R. Stoneking, M. A. Growdon, M. L. Milne, and R. T. Peterson

*Department of Physics, Lawrence University, Appleton, WI 54911*

**Abstract.** Electron plasmas with density in excess of $10^6$ cm$^{-3}$ are trapped for about 1 ms in a purely toroidal magnetic field ($B_o = 200$G, $R_o = 43$cm, $a = 4.5$cm). The trap is a partial torus, or curved Penning-Malmberg trap. The measured confinement time is an order of magnitude longer than previously published results in the same device, and is substantially longer than the confinement measured in all previous electron plasma experiments with purely toroidal magnetic fields. The confinement time exceeds all characteristic single particle drift timescales and implies the existence of an equilibrium in which the space charge generated $\mathbf{E} \times \mathbf{B}$ drift acts as an effective rotational transform. Confinement appears to be limited by the growth of a *m*=1 diocotron-like mode, observed with wall patch probes and a phosphor screen imaging detector. The mode grows to large amplitude whereupon electrons are lost to material surfaces.

## INTRODUCTION: POLOIDAL E × B ROTATION REQUIRED FOR EQUILIBRIUM

Toroidal confinement of nonneutral plasmas has received relatively little experimental attention compared to either cylindrical confinement of nonneutral plasmas or toroidal confinement of (quasi-) neutral plasmas. While the first toroidal nonneutral plasma experiment dates back thirty-four years [1] only five experimental papers on the topic, of which we are aware, were published between 1969 and 1997 [2, 3, 4, 5, 6]. In the last few years, there has been renewed interest in such experiments. Four toroidal electron plasma experiments are either operating [7, 8, 9] or are under construction [10] (also see the following papers in these proceedings: H. Himura, *et. al.*, T. Sunn Pedersen, *et. al.*, J. P. Kremer, *et. al.*, H. Saitoh, *et. al.*, and H. Wakabayashi, *et. al.*). New results from the Lawrence Nonneutral Torus (LNT) [8] are presented in this paper.

The LNT is the only contemporary toroidal nonneutral plasma device that employs a purely toroidal magnetic field. We rely on poloidal $\mathbf{E} \times \mathbf{B}$ rotation of the electron plasma to compensate for curvature and $\nabla B$ drifts, rather than employing a rotational transform. Theory supports the expectation that poloidal rotation due to the space charge electric field can provide a stable equilibrium for a toroidal electron plasma [11, 12, 13, 14, 15, 16]. Substantial space charge is required in order to generate sufficient poloidal rotation to access the theoretically predicted equilibrium. This requirement presents a challenge to the experimentalist: to inject charge into an initially empty trap rapidly enough to initiate poloidal rotation. The work described here makes use of a partially toroidal confinement strategy to meet this challenge.

CP692, *Non-Neutral Plasma Physics V*, edited by M. Schauer et al.

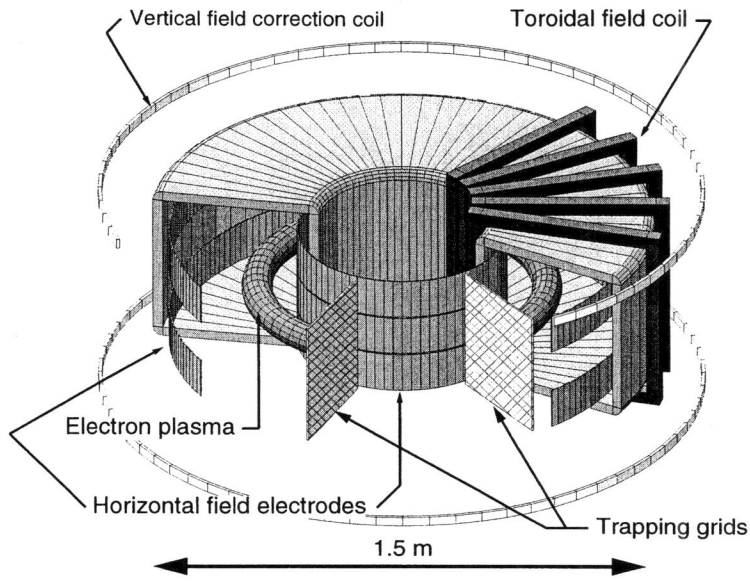

Vertical field correction coil     Toroidal field coil

$r_0$

Electron plasma

Horizontal field electrodes

Trapping grids

1.5 m

**FIGURE 1.** Cutaway view of the toroidal vacuum chamber showing the electron plasma, the trapping grids, the horizontal electric field electrodes, a representative set of toroidal field coils, and the vertical magnetic field correction coil. The electron source and phosphor screen detector reside back-to-back between the grids but are not shown.

## APPARATUS: PARTIAL TORUS AND DIAGNOSTICS

The partially toroidal design of the LNT is its novel feature. By restricting the confinement region to a 300° toroidal arc, electrons are injected onto toroidal magnetic field lines that carry them into the trapping region. Likewise, the plasma is dumped along magnetic field lines onto a phosphor screen diagnostic for measurement. Figure 1 is a cutaway schematic diagram of the device showing some of its components: trapping grids, electrodes for applying a horizontal electric field, a representative number of toroidal magnetic field windings, and the vertical magnetic field correction coil. Not shown in the diagram are the electron source (a pancake spiral consisting of 12 turns of tungsten wire) and the phosphor screen imaging detector, both of which are located in the 60° toroidal sector that is excluded from the trapping region. The major radius of the plasma is 43 cm and the minor radius is nominally 5 cm. The magnetic field is generated by a toroidal inductor and is therefore purely toroidal (except for small ripple, error fields, the earth's field and the vertical field correction), falling off as $\frac{1}{R}$, with a maximum strength of 196 G at the major radius. The base vacuum pressure during operation is $\approx 5 \times 10^{-7}$ Torr. A more complete description of the design and operation of the LNT is provided in the references [8].

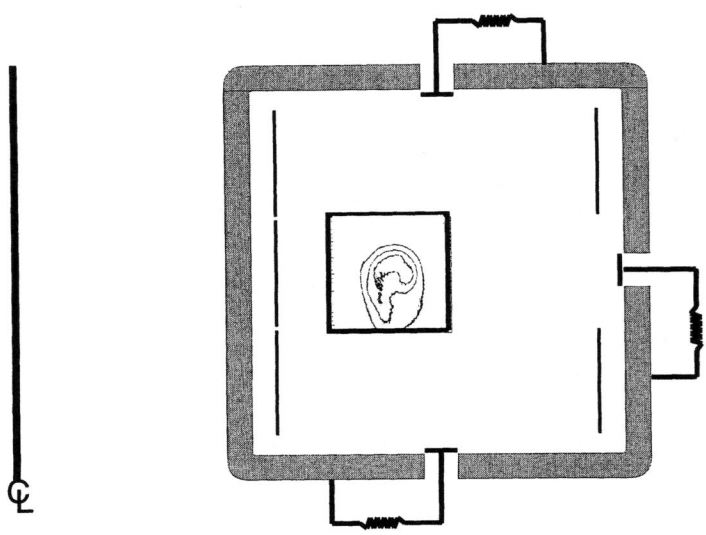

**FIGURE 2.**  A schematic of the poloidal cross-section of the machine, showing the projected locations of the wall patch probes and the phosphor screen with representative intensity contour plot.

Recently, a number of hardware alterations were made to improve the symmetry of the applied fields and reduce the base pressure in the LNT. The trapping grids were enlarged to nearly fill the vertical plane and present more nearly planar equipotential surfaces to the plasma at its reflection points. The tungsten filament was redesigned to reduce the magnetic perturbation caused by the D.C. heating current. A vertical field correction coil was added to null out large scale vertical magnetic field errors. The base vacuum pressure was reduced by a factor of about five with the addition of a cryopump. We believe the dramatic improvement in confinement reported in this paper is due to the upgrades discussed here, particularly the improvements in trap symmetry.

The diagnostics used to acquire data for this paper are a phosphor screen and CCD camera imaging system and four wall patch probes. The phosphor screen is located behind one of the trapping grids in the 60° toroidal sector excluded from the trapping region. The screen is positively biased (2.0-3.5 kV) to accelerate the dumped electron plasma to sufficient energy to illuminate the phosphor. The image is captured by a CCD camera viewing through a window in the outboard midplane via a 45° mirror. When the plasma is dumped, a fraction of the current collected by the phosphor screen (with a conductive coating) is detected by a capacitively coupled circuit. The four wall patch probes are each approximately 1.5" in diameter, located nearly flush with the vacuum chamber wall and also nearly fill the port hole in the vacuum chamber. Each probe is connected to the grounded vacuum chamber across 12 kΩ so that the flow of image charge to and from the probe in response to plasma motion, can be detected. Three of the four probes are located at the same toroidal angle, approximately 113° from the grid nearest the filament, one on the top, one on the bottom, and one on the outboard midplane. The fourth probe is also on the outboard midplane and is separated from the other probes by 60°. Fig. 2 shows a poloidal cross-section of the device, indicating the locations of the probes and a representative intensity contour plot derived from the phosphor screen image of the dumped plasma. In this view, the magnetic field is directed out of the page. The reader should keep in mind that the phosphor screen and the wall patch probes are located at different toroidal angles.

## RESULTS: MILLISECOND CONFINEMENT AND THE DIOCOTRON MODE

Confinement in the LNT is measured by the same method used in cylindrical Penning-Malmberg traps [17]. Suitably timed gate pulses are applied to the trapping grids to execute a load, trap, dump sequence. During the "load" phase of the cycle the potential on the grid furthest from the tungsten filament is maintained at negative potential to reflect electrons, while the grid nearest the filament is grounded to permit electrons to enter the trapping region. During the "trap" phase of the cycle, both grids are maintained at negative potential to seal the trap. After a specified time, the potential on the far grid is removed (or made positive) to "dump" the remaining contents of the trap onto the phosphor screen. The confinement time is determined by varying the duration of the "trap" phase of the cycle and measuring the charge that is dumped versus the trapping time. For the measurements presented here, the effects of stray capacitance were not adequately accounted for in the charge measurement circuit. Therefore, although the absolute magnitude of the dumped charge is not obtained from this measurement, a fixed fraction of the dumped charge is reliably detected. Below we use the frequency of the observed $m=1$ oscillations to estimate the trapped charge.

The timescale for decay of the trapped charge (*i.e.* the confinement time) in LNT is about 1 ms (Fig. 3a). This result, as far as the authors are aware, is a record for electron plasma confinement in a device with a toroidal magnetic field. It also represents an order of magnitude improvement over the previously published data from this device [8]. More significantly, since 1 ms is longer than all other characteristic timescales for single particle motion, an equilibrium has been accessed. This result verifies the theoretical expectation that the poloidal $\mathbf{E} \times \mathbf{B}$ rotation can provide an effective rotational transform for a toroidal electron plasma. We attribute the improvement in confinement over previous results in LNT to the improved field symmetry resulting from the hardware improvements discussed earlier. The present data were also obtained in a different regime and likely represent a different electron velocity distribution for the trapped population compared to our previous work. Previously reported data [8] were obtained by applying a strong horizontal electric field via the electrodes shown in Figs. 1 and 2. We inferred that the electron population had a $\approx 100$ eV beamlike distribution with the horizontal electric field providing centripetal acceleration. The present data were obtained with a very weak horizontal electric field and we infer that the electron distribution is slower/colder.

The technique for measuring confinement discussed above is a destructive technique; the plasma is sacrificed in the process of making the measurement. Wall patch probes

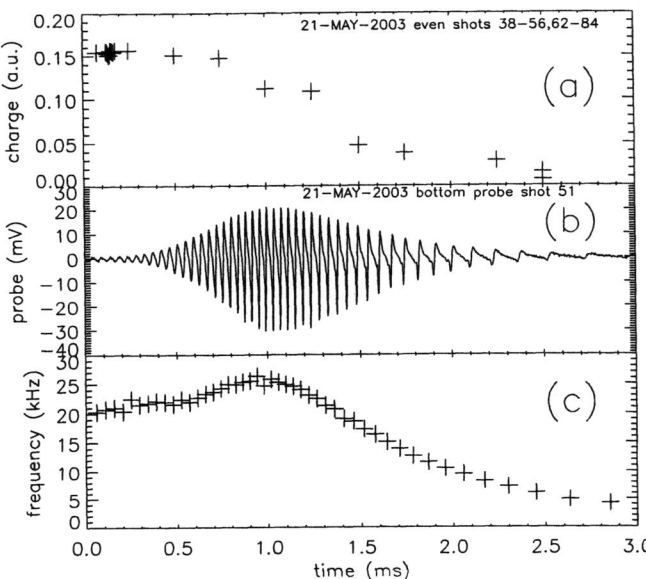

**FIGURE 3.** a) The charge detected by the circuit capacitively coupled to the phosphor screen versus trapping time. b) The raw signal on the bottom wall patch probe for a typical shot when the plasma is dumped later than 3 ms. c) The frequency of the wall probe signal versus time for the same shot shown in panel (b).

provide a means of nondestructively sampling the dynamics of the trapped plasma. As the plasma moves around, the image charge on the inner surface of the grounded vacuum chamber is redistributed to maintain ground potential. For image charge to reach the isolated wall patch probe, it must flow through an external resistance and be detected. Fig. 3b shows the raw signal on the bottom wall probe for a shot where the plasma was dumped sometime after 3 ms. The signal is shown on the same time scale as the dumped charge measurements (Fig.3a) for a set of shots with varied trapping times. Presumably the time evolution shown in the two panels is in correspondence. An oscillation is observed on the wall probe signal. The amplitude grows, saturates, and then decay as the plasma decays. Casual viewing of the raw signal will convince the reader that the frequency of the oscillation drops as the plasma decays, a signature of diocotron oscillations. The measured frequency shown in Fig. 3c confirms that observation and also shows that during the growth phase, the frequency of the oscillation rises slightly.

Information on the spatial structure of the oscillations discussed above is obtained by observing the phase relationship between measurements made with wall probes at different spatial locations. Fig. 4 shows simultaneous data from all four wall probes for a portion of a typical shot. The top two panels show that the top and bottom wall probe signals are nearly 180° out of phase. The bottom two panels show that the signals for two

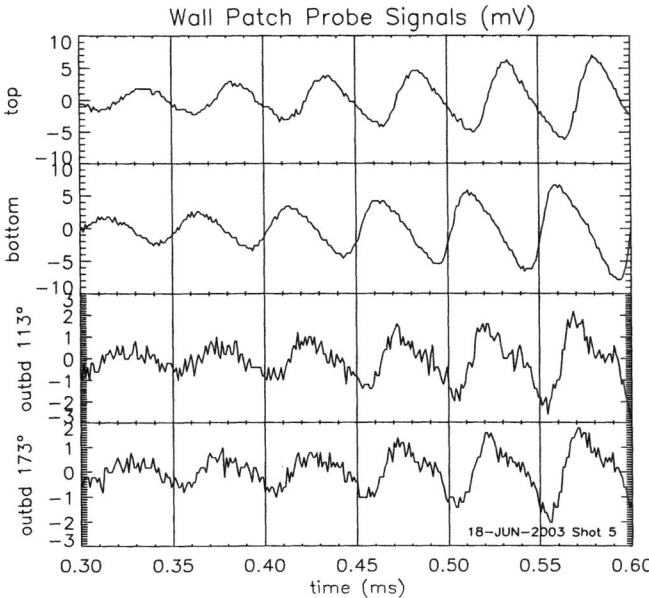

**FIGURE 4.** Simultaneous raw signals from the four wall patch probes for a typical shot indicating the $m = 1, k = 0$ spatial structure of the mode.

probes on the outboard midplane, separated by 60° toroidally, are nearly in-phase. The data in Fig. 4 strongly suggests an $m$=1, $k$=0 spatial structure, where $m$ is the poloidal mode number and $k$ is the parallel (or toroidal) wavenumber.

Images acquired with the phosphor screen support the conclusion that the observed oscillations have poloidal mode number $m$=1 and parallel wavenumber $k$=0. The top panel in Fig. 5 is a 12 shot average of the signal from the bottom wall patch probe early on in the trapping phase of the cycle. The average of twelve shots robustly shows the $\approx$20 kHz oscillation indicating that the initial phase of the oscillations is very reproducible. Contour plots of the phosphorescent intensity obtained by dumping the plasma onto the phosphor screen at times corresponding to the vertical lines in the top panel of Fig. 5 are shown in the bottom panels. The $m$=1 rotation of the plasma in the poloidal plane is evident and the sense (CCW) is consistent with diocotron motion. That the sequence of images on the phosphor screen display the $m$=1 evolution of the plasma column supports the previous conclusions that 1) the phase of the oscillations is reproducible with respect to time when the trap is closed, and 2) the parallel wavenumber of the oscillations is $k$=0.

Diocotron oscillations have a well-known dependence on the trapped charge and the magnetic field strength, $f \propto \frac{Q}{B}$. The measured frequency of the oscillations (early in the trapping phase) has a $\frac{1}{B}$ dependence as shown in Fig. 6. The good fit to an inverse magnetic field dependence seen in Fig. 6 also suggests that the quantity of trapped

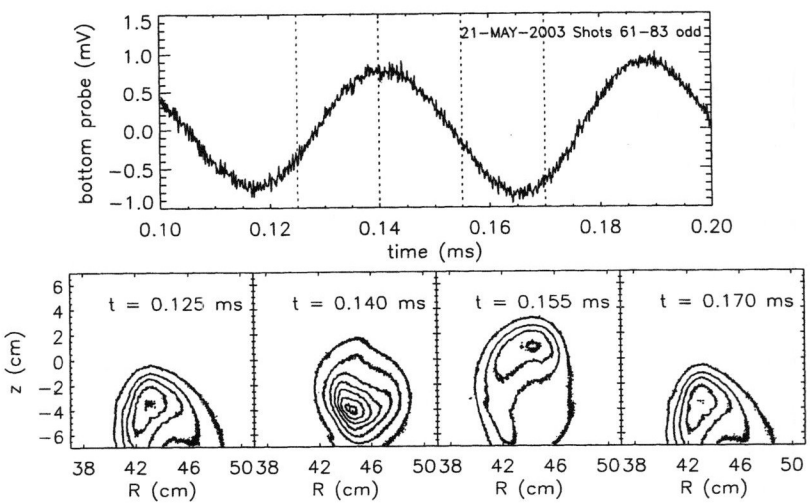

**FIGURE 5.** A 12-shot average of the bottom wall probe signal (top panel) showing the 4 times (vertical dashed lines) at which the plasma was dumped to generate the phosphor intensity contour plots (bottom panel).

charge is independent of the magnetic field strength. An estimate of the trapped charge and the associated mean electron density is obtained by using the relationship between frequency, magnetic field, trapped charge, and geometric dimensions for the *cylindrical* $m=1$ diocotron mode. The frequency can be written as [18]

$$f = \frac{Q}{4\pi^2 \varepsilon_o L b^2} \frac{1}{B} \qquad (1)$$

where $L$ is the length of the plasma (2.25 m here) and $b$ is the wall radius (effectively $\approx$ 0.2 m here). The least squares fit of the observed oscillation frequency versus magnetic field strength, $f = \frac{378 \pm 3}{B}$ Hz T, and Eq. 1 imply a total trapped charge of approximately 12 nC. The corresponding mean density is about $4 \times 10^6 \text{cm}^{-3}$. A uniform density electron column with density equal to $4 \times 10^6 \text{cm}^{-3}$ and a radius of 5 cm generates a space charge potential of 48 V relative to its edge. The radial potential drop across the tungsten filament responsible for injecting the electron plasma is $\approx$ 100 V. The reason for the discrepancy is not understood at this time, however, a 48 V space charge potential is adequate to generate closed, nested equipotential contours given the applied potentials on the horizontal field electrodes (+25 V on the inner middle and outer electrodes and +35 V on the inner top and bottom electrodes) and the location of the grounded vacuum chamber wall.

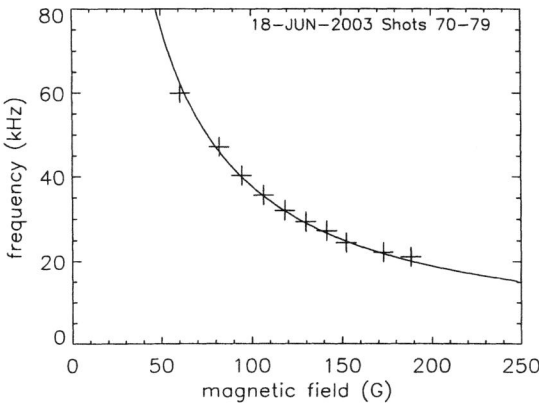

**FIGURE 6.** The frequency of the $m=1$ oscillation at early times ($<0.20$ ms) versus magnetic field strength (crosses) and a fit to $\frac{\kappa}{B}$ where $\kappa = 378 \pm 3$ Hz T (solid line).

The $m=1$ diocotron oscillation is unstable (Fig. 3). The data presented in this paper represent the results of efforts to maximize the confinement time by varying the experimental parameters (potentials on the horizontal field electrodes and the offset bias on the edge of the tungsten filament). Bhattacharyya and Avinash [15] predict a $m=1$ mode in a toroidal nonneutral plasma, the stability of which depends on the value and the spatial variation of the applied horizontal electric field. In addition, they predict the mode to have either a purely real frequency (stable) or a purely imaginary frequency (unstable). We see an oscillation that grows and therefore has both real and imaginary components. Also, although the experimentally observed growth rate depends on the horizontal electric field, the dependence is rather weak, and the frequency is nearly independent of the applied fields within the limits where trapping is achieved. Future work will attempt to determine the extent to which the observed mode matches the theoretical description of Bhattacharyya and Avinash, however at this time, the mode character more nearly resembles the cylindrical $m=1$ diocotron mode. The $m=1$ mode may be unstable for reasons other than the toroidal geometry of the experiment. The ion resonance instability [19] was seen in the earliest toroidal electron plasma experiment [1], where ionization of the residual gas was believed to lead to substantial accumulation of ions in a completely toroidal trap. More recently, Fajans [20] has argued that even transient (untrapped) ions can catalzye this mode, a prediction that is supported by some experimental evidence in cylindrical traps [21, 22] (see also the paper by A. Kabantsev in this volume). Given the relatively high pressure at which the LNT operates ($\approx 5 \times 10^{-7}$ Torr), ionization of neutrals is to be expected and the ion resonance instability is a plausible explanation for the growth of the observed mode.

The intensity contour plots shown in Fig. 5 represent the early, small amplitude phase of the evolution of the mode. Later, the mode grows to large amplitude, passes out of the field of view of the CCD camera and off the edge of the phosphor screen. This is the reason the charge begins to decay (Fig. 3a) before the $m=1$ mode frequency begins to drop (Fig. 3c). Charge is not lost entirely from the trap. It is just no longer detectable with the phosphor screen collector. The principle limitation on confinement in the present experiment appears to be the growth of the $m=1$ mode to large amplitude where electrons begin to scrape off on material surfaces. In the near future we plan to attempt active feedback experiments to stabilize the mode and perhaps extend the confinement period.

One additional feature of the observed $m=1$ oscillations begs for explanation. As the mode grows in amplitude, its frequency also rises (Fig. 3). This result is in qualitative agreement with the upshift in the frequency of the $m=1$ diocotron mode at large amplitude (*i.e.* the nonlinear diocotron mode) seen in cylindrical traps [23]. It is also possible that ionization of the residual gas leads to a noticeable "fueling" of the electron plasma. The ionization of neutrals would add electrons and ions to the trap. The ions would drift out of the ends of the trap on a several hundred microsecond timescale, but the electrons would remain in the trap. This explanation for the early time frequency rise dovetails with the ion resonance instability argument presented above, but appears to contradict the data in Fig. 3a showing a monotonic decrease in the trapped charge with time. However, as mentioned above, at later times the dumped plasma spills off the edge of the screen when the mode grows to large amplitude and therefore the measured charge does not reflect all of the charge remaining in the trap. A definitive explanation for the frequency rise seen in Fig. 3c therefore requires further experiments.

## CONCLUSIONS

Electron plasmas with density in the range of $4 \times 10^6$ cm$^{-3}$ are trapped for 1 ms in a partially toroidal trap ($B = 200$ G). Millisecond confinement is ten times longer than previously published confinement times in the Lawrence Nonneutral Torus and longer than any other purely toroidal nonneutral trap of which we are aware. The confinement measurement confirms the expectation that space charge generated poloidal rotation provides an effective rotational transform to produce toroidal equilibrium in this system. Data obtained with a phosphor screen imaging detector and a set of four wall patch probes suggest that confinement is limited by the growth of a poloidal mode number $m=1$, parallel wavenumber $k=0$ mode that has many features of the $m=1$ diocotron mode observed ubiquitously in cylindrical traps.

## ACKNOWLEDGMENTS

The project is funded by Lawrence University and by grants from Research Corporation (Award No. CC4593) and the U.S. Department of Energy (Grant No. DE-FG02-98ER54503).

# REFERENCES

1.  Daugherty, J.D., Eninger, J.E., and Janes, G.S., Phys. Fluids **12**, 2677 (1969).
2.  Mohri, M., Masuzaki, M., Tsuzuki, T., and Ikuta, K., Phys. Rev. Lett. **34**, 574 (1975).
3.  Clark, W., Korn, P., Mondelli, A., and Rostoker, N., Phys. Rev. Lett. **37**, 592 (1976).
4.  Fisher, A., Gilad, P., Goldin, F., and Rostoker, N., Appl. Phys. Lett. **37**, 531 (1980).
5.  Puravi Zaveri, John, P.I., Avinash, K., and Kaw, P.K., Phys. Rev. Lett. **68**, 3295 (1992).
6.  Khirwadkar, S.S., John, P.I., Avinash, K., and Kaw, P.K., Phys. Rev. Lett. **71**, 4334 (1993).
7.  Himura, H., Nakashima, C., Saito, H., and Yoshida, Z., Phys. Plasmas **8**, 4651 (2001).
8.  Stoneking, M. R., Fontana, P. W., Sampson, R. L., and Thuecks, D. J., Phys. Plasmas **9**, 766 (2002).
9.  Nakashima, C., Yoshida, Z., Himura, H., Fukao, M., Morikawa, J., and Saitoh, H., Phys. Rev. E **65**, 036409 (2002).
10. Pedersen, Thomas Sunn and Boozer, Allen H., Phys. Rev. Lett. **88**, 205002 (2002).
11. Daugherty, J.D. and Levy, R.H., Phys. Fluids **10**, 155 (1967).
12. Elsässer, K., Yu, M.Y., and Shukla, P.K., Phys. Lett. A **152**, 59 (1991).
13. Avinash, K., Phys. Fluids B **3**, 3226 (1991).
14. Hurricane, O.A., Phys. Plasmas **5**, 2197 (1998).
15. Bhattacharyya, S.N. and Avinash, K., Phys. Fluids B **4**, 1702 (1992).
16. O'Neil, T.M. and Smith, R.A., Phys. Plasmas **1**, 2430 (1994).
17. Malmberg, J. H., and Driscoll, C. F. Phys. Rev. Lett. **44**, 654 (1980).
18. Ronald C. Davidson *Physics of Nonneutral Plasmas* (Addison-Wesley, Redwood City, CA, 1990), p. 306.
19. Levy, R.H., Daugherty, J.D., and Buneman, O., Phys. Fluids **12**, 2616 (1969).
20. Fajans, J., Phys. Fluids B **5**, 3127 (1993).
21. Peurrung, A.J., Notte, J., and Fajans, J., Phys. Rev. Lett. **70**, 295 (1993).
22. Pasquini, Thomas and Fajans, Joel, Bull. Am. Phys. Soc. **47**, 127 (2002).
23. Fine, K. S., Driscoll, C. F., and Malmberg, J. H. Phys. Rev. Lett. **63**, 2232 (1989).

# The Status of the Design and Construction of the Columbia Non-neutral Torus

J. P. Kremer*, T. S. Pedersen*, N. Pomphrey[†], W. Reiersen[†] and F. Dahlgren[†]

*Dept. of Applied Physics and Applied Mathematics, Columbia University, New York, NY 10027*
*[†]Princeton Plasma Physics Laboratory, Princeton, NJ 08543*

**Abstract.** The Columbia Non-neutral Torus (CNT) is a tabletop (R=0.3 m, a=0.1 m, B=0.2 T) stellarator now being constructed at Columbia University. The goal of CNT is to study the equilibrium, stability, and transport of non-neutral plasmas confined on magnetic surfaces. CNT will use four circular, planar coils: two interlocking coils with a variable tilt angle, plus two additional poloidal field coils. By varying the angle between the interlocking coils, the rotational transform can be varied from 0.2 to 0.6 and the magnetic shear from essentially zero to 20%. The results of a numerical study of how error fields affect the quality of the magnetic surfaces will be presented. The plasma will be diagnosed by numerous Langmuir and sector probes, connected to a computer data acquisition and control system.

## INTRODUCTION

Magnetic surface configurations have long been used to confine fusion plasma but have only recently become of interest as confinement devices for non-neutral plasmas [1], [2]. The physics of non-neutral plasmas on magnetic surfaces has been shown to be fundamentally different from previous configurations [3], [4], including the Penning trap. The goal of the Columbia Non-neutral Torus (CNT) is to study the equilibrium, stability, and transport of non-neutral plasmas which are confined on magnetic surfaces.

Our definition of a magnetic surface is a surface where a single field line comes arbitrarily close to every point on that surface. Stellarators and tokamaks are magnetic surface configurations while Penning traps and pure toroidal traps are not. An important parameter when discussing magnetic surfaces is the rotational transform, $\iota$, which is the ratio of the number of times a field line winds around poloidally to the number of times it wind around toroidally. A great deal of care has been taken in choosing parameters such that the bad surfaces, also known as stochastic regions and magnetic islands, exist only outside a large confining region of magnetic surfaces. Error fields can reduce the confining volume by introducing bad surfaces. The effects of a full range of expected error fields were studied computationally and will be discussed.

CP692, *Non-Neutral Plasma Physics V*, edited by M. Schauer et al.

**FIGURE 1.** A cutaway drawing of CNT showing the vacuum vessel and the four circular coils.

## PHYSICS TO BE STUDIED ON CNT

**Fundamental plasma physics:** CNT will study the largely unexplored area of confinement of non-neutral plasmas on magnetic surfaces. The physics of such plasmas is fundamentally different from non-neutral plasmas in other devices. Magnetic surface configurations can also explore the whole range of plasma neutrality from single component to quasi-neutral. CNT is a rather unique concept. The device is simple, small, elegant, and, as will be described later, can explore a wide range of rotational transform profiles.

**Antimatter physics:** Stellarators can confine particles of opposite charge and can confine light, energetic particles. CNT may therefore be an excellent confinement device for exotic antimatter plasmas such as positron/electron or positron/anti-proton plasmas[5].

**Physics relevant to fusion and astrophysics:** CNT could be used to study the physics of strongly rotating toroidal plasmas which are of great interest to both the fusion and astrophysics communities. Partially neutralized electron rich helium plasmas are of particular interest to fusion science and will be studied in CNT.

## THE PRESENT DESIGN SPECIFICATIONS

### Basic Parameters

CNT is a small ultrahigh vacuum stellarator now being constructed at Columbia University. CNT consists of only four circular, planar coils. Two of the coils are interlocking, have a radius of 0.405 m, and will be placed inside the vacuum vessel. These two coils will be copper, water cooled, and driven by a 200 kW power supply. The pulse length and repetition rate will be limited by the ability to cool the copper conductors. The other

| Parameter | Value |
|-----------|-------|
| $n_e$ | $10^{12} - 10^{14}$ m$^{-3}$ |
| $T_e$ | 1 - 100 eV |
| B | 0.05 - 0.2 T |
| R | 0.3 m |
| a | 0.1 m |
| p | $10^{-10}$ Torr |

two coils are called the poloidal field coils or PF coils. They will have a radius of 1.08 m and will reside outside the vacuum vessel. The angle between the two interlocking coils (the tilt angle) can be varied between 63° and 90°. This allows us to explore rotational transforms of 0.2 to 0.6 and shear from essentially zero to 20%.

## Plasma Diagnostics

Methods have been developed to measure plasma temperature and density profiles accurately in Penning traps and other open field line configurations [6]. Toroidal configurations, including those which have magnetic surfaces, do not have open field lines so the same diagnostics cannot be used. Like many other toroidal devices, the temperature and density of plasmas in CNT will therefore be measured using an array of Langmuir-type probes. Current-voltage characteristics of such probes can be determined and used to estimate the density and temperature of plasmas. [7]. Fluctuations in the image charge on a set of capacitive probes surrounding the plasma will be used to measure large scale fluctuations in plasma density and location. We expect to have between 15 and 20 such probes surrounding the plasma so that fluctuations can be localized. These probes can also be biased to alter the plasma equilibrium or excite waves.

## COMPUTATIONAL STUDIES

The confinement of plasmas in CNT or any other magnetic surface configuration is limited by stochastic regions and magnetic islands. Parameters are chosen such that these so-called bad surfaces are kept to the outside of a large region of good surfaces. The confining magnetic field of a stellarator is created wholly by the magnetic coils. It is therefore possible to computationally study the magnetic topology. In the case of CNT, which has only planar, circular coils, this requires relatively little computational might. The real magnetic coils have been approximated by sets of 100 circular thin wire coils which occupy a volume similar to the windings of the real coils. In the case of the interlocking coils, the structural mounts that the coils are to be wound on, were approximated as 1 cm thick regions which are void of any windings . Poincare plots of numerous field lines are made at three different toroidal cross sections. Stochastic regions can be seen as thick or undefined surfaces and magnetic islands can be seen as holes either in the surfaces or between them.

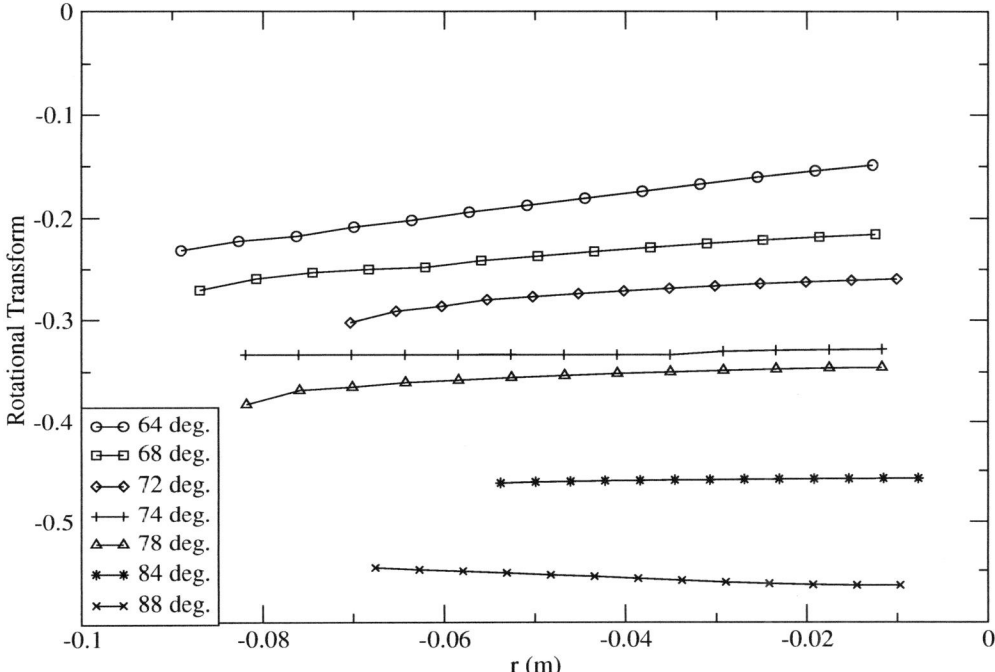

**FIGURE 2.** The rotational transform profiles for various tilt angles. The bottom axis represents the radial distance away from the magnetic axis.

## Parameter Optimization

We have chosen to maximize the size of the confining volume by varying two parameters: the angle between the two interlocking coils (the tilt angle), and the current flowing through the poloidal field (PF) coils. Different rotational transform profiles can be accessed through a change in tilt angle. Engineering constraints dictate that tilt angles must be chosen before the coils are placed in the vacuum vessel and that there can at most be three working tilt angles. The PF current was optimized for a range of tilt angles from 63° to 90°. Tilt angles were chosen such that each had a unique $\iota$ profile and had a large volume of good surfaces. The three tilt angles chosen were 64°, 78°, and 88°. Poincare plots of the 64° tilt angle are shown in Fig. 3.

## Error Field Analysis

There are numerous sources of error fields in any magnetic configuration. Sources of such error fields are slight errors in coil placement or alignment or imperfections in coil windings. In stellarators, the error fields can introduce bad surfaces to the topology

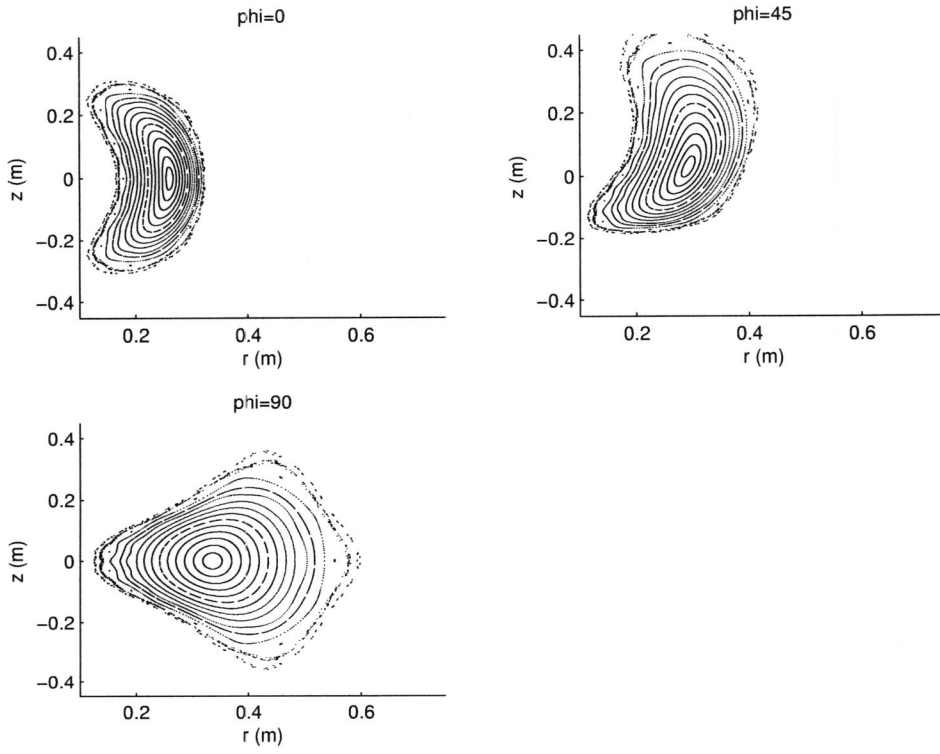

**FIGURE 3.** Poincare plots for the 64° tilt angle. The plots are toroidal cross-sections at 0°, 45°, and 90°.

thereby limiting the confinement. An error field analysis was carried out to understand and quantify the effects of error fields on the magnetic topology of CNT. Poincare plots for each chosen tilt angle were made as various sources of field errors were added. The tolerance to error fields was determined by the resulting loss of confining volume.

Error field sources that were primarily investigated were those resulting from slight perturbations to coil positions and/or alignments. First, the effect of perturbing a single coil 1 cm along all of the axes of symmetry and in numerous random directions, presumably off the axes of symmetry, was studied. Perturbations of 1 cm were made to the alignments by tilting the coil such that the edge of the coil would move 1 cm from its proper location. For all tilt angles, all perturbations of 1 cm to the PF coils were found to be tolerable. Next, the effect of introducing similar sorts of perturbations to the interlocking coils was studied. These perturbations had a much more drastic effect. This is because the interlocking coils are closer to the confining region and play a more significant role in the topology than the PF coils do. The effect was found to be more pronounced in the configurations with larger tilt angles. All of the chosen tilt angles were found to be stable to single 0.5 cm perturbations to the interlocking coils.

To estimate a tolerance in coil placement, the collective effects of random perturbations to all of the coils were studied. Poincare plots were produced after random perturbations limited to a given tolerance were made to all coil locations and alignments. The tolerance was determined as the maximum tolerance which resulted in limited loss of confining volume. Perturbations of this type were found to have the most drastic effect on the 88° tilt angle; the working maximum value for the tolerance was limited to that for the 88° tilt angle. By way of this method, a final value for the tolerance in coil placement was determined to be 0.2 cm.

The effects of field errors from current leads and winding transitions have been approximated and appear not to have a noticeable effect on the magnetic surface quality, if concentrated in the region away from the plasma.

# CONCLUSIONS

The goal of CNT is to study the equilibrium, stability, and transport of non-neutral plasmas on magnetic surfaces. CNT will primarily be a basic plasma physics experiment used to confine pure electron and partially neutralized electron-ion plasmas but may also be used to study exotic plasmas such as electron-positron plasmas. Details of the present design of the device were presented. The magnetic topology was studied computationally. Parameters were optimized such that the effects of stochastic regions and magnetic islands were minimized. A thorough error field analysis was conducted. The magnetic surfaces are tolerant to rather large perturbations to single coils. The tolerance decreases as the number of perturbed coils increases. The tolerance was found to be lower for larger tilt angles, and lower for the two interlocking coils. It was found that a general tolerance in coil placement and alignment of 0.2 cm consistently produces a large volume of good magnetic surfaces for all tilt angles.

# ACKNOWLEDGMENTS

This work is supported by U.S. DOE Grant # DE-FG02-02ER54690

# REFERENCES

1. Yoshida, Z., Ogawa, Y., Morikawa, J., and et al., "Toroidal magnetic confinement of non-neutral plasmas," in *Non-Neutral Plasma Physics III*, AIP, New York, 1999, vol. 498 of *AIP Conf. Proceedings*, p. 397.
2. Pedersen, T. S., and Boozer, A. H., *Phys. Rev. Lett.*, **88** (2002).
3. Pedersen, T. S., *Physics of Plasmas*, **10** (2003).
4. Pedersen, T. S., Kremer, J. P., and Boozer, A. H., "Confinement of non-neutral plasmas in the Columbia Non-neutral Torus," in *Non-neutral Plasma Workshop*, AIP Conference Proceedings, American Institute of Physics, New York, 2003.
5. Pedersen, T. S., Boozer, A. H., and et al., *J. Phys. B.*, **36** (2003).
6. Driscoll, C. F., and Malmberg, J. H., *Phys. Rev. Lett.*, **50**, 167 (1983).
7. Himura, H., Nakashima, C., Saito, H., and Yoshida, Z., *Physics of Plasmas*, **8**, 4651–4658 (2001).

# Long-Time Confinement of
# Toroidal Electron Plasmas in Proto-RT

H. Saitoh\*, Z. Yoshida\*, H. Himura\*, J. Morikawa†, M. Fukao† and
H. Wakabayashi\*

\**Graduate School of Frontier Sciences, The University of Tokyo, Bunkyo-ku, Tokyo 113-8656 Japan*
†*High Temperature Plasma Center, The University of Tokyo, Bunkyo-ku, Tokyo 113-8656 Japan*

**Abstract.** Long-term confinement (the upper limit is set by the diffusion due to neutral collisions) of toroidal electron plasmas is achieved in a dipole magnetic field configuration of an internal conductor device by the optimization of internal potential profiles. The application of a negatively biased electrode makes possible the elimination of a potential hall inside the plasma and results in the stabilization of the diocotron instability. In the experimental parameters of the magnetic field $B \sim 100$ G and back pressure $P \sim 10^{-4}$ Pa ($\sim 10^{-6}$ Torr), the obtained maximum decay time of the trapped charge (number density $n_e \sim 10^{12}$ m$^{-3}$) is 200 msec, which is comparable to the classical neutral diffusion time. It is demonstrated that toroidal magnetic surface configurations have excellent confinement properties for non-neutral plasmas, and might be useful as applications for novel traps of charged particles such as anti-matters or other multi-fluid non-neutral plasmas.

## INTRODUCTION

Toroidal traps for non-neutral plasmas [1-3] are attracting renewed interest in spite of the long history of study regarding them. In toroidal geometry, we can simultaneously confine multiple species of charged particles at any degree of non-neutrality, because toroidal devices use no electrostatic potential well along the magnetic field lines. The confinement of anti-matter and its mixtures such as pure positron and electron-positron plasmas [1, 4], or the formation of fast $\mathbf{E} \times \mathbf{B}$ drift flow due to the strong self electric fields and testing for the resultant high $\beta$ equilibrium of two fluid non-neutral plasmas (Double Beltrami state) [2, 5] are the expected applications of toroidal non-neutral traps.

The experiments on toroidal non-neutral plasmas have been carried out in a pure toroidal magnetic field configuration [3] for several decades. In these devices, many interesting properties of toroidal electron plasmas, such as equilibria, confinement, and electrostatic fluctuation modes, etc., have been intensively investigated. Recently, the use of another magnetic field configuration for toroidal non-neutral plasmas has been proposed and these traps are expected to show different confinement properties for non-neutral plasmas. The trapping of non-neutral plasmas in a magnetic surface configuration is currently conducted or under design using stellarator [1] or internal conductor devices [2].

In this study, we report the experimental investigation of the confinement properties of torus electron plasmas in Proto-RT (Prototype-Ring Trap) [2] (Fig. 1). The use of magnetic surface (dipole) configuration and potential optimization has made possible the stabilization of diocotron instability, and long-term (i.e., comparable to the diffusion

CP692, *Non-Neutral Plasma Physics V*, edited by M. Schauer et al.

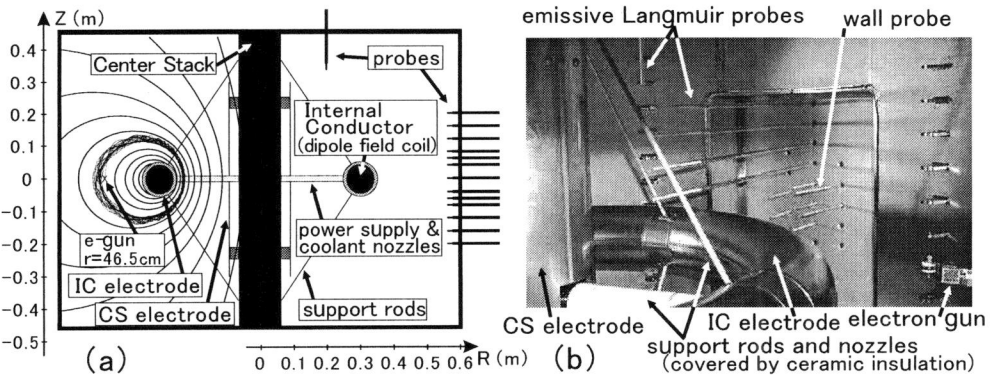

**FIGURE 1.** (a) The poloidal cross-section of Proto-RT, the dipole field magnetic surfaces, and the typical electron orbit. (b) Photographic view of experimental configurations inside Proto-RT. In order to suppress perturbations of the plasma, each probe was inserted independently while the other probes were located outside the confinement region.

time due to the neutral collisions) confinement of toroidal electron plasmas was demonstrated. The device setup, diagnostics, and recent results of pure electron experiments in Proto-RT will be described in the following sections.

## SETUP AND DIAGNOSTICS

Proto-RT is a normal conducting toroidal device constructed for investigating the relaxation states of two fluid flowing plasmas [5], the injection and confinement properties of non-neutral plasmas on magnetic surfaces [2], and the chaos-induced anomalous resistivity at the magnetic null line [6]. The chamber is evacuated to a base pressure of $8 \times 10^{-5}$ Pa ($\sim 6 \times 10^{-7}$ Torr) by a turbomolecular pump. Besides toroidal field coils, Proto-RT also has an internal conductor for a dipole magnetic field and a pair of vertical field coils, and the combination of these coils allows us to use

a variety of magnetic field configurations. As an initial experiment, the trapping of electron plasmas on the magnetic surfaces of dipole magnetic field was carried out in this study. The coil current of the internal conductor is 10.5 kA and the strength of the typical magnetic field is of the order of 0.01 T in the confinement region of the torus. For the optimization of the potential profiles of toroidal electron plasmas, two electrodes are installed in the confinement region of the Proto-RT vessel. In this experiment, we used a ring electrode on the internal conductor, and the effects of potential biasing up to 350V DC relative to the vessel wall were examined, while the electrode on the center stack was shorted to the chamber. Electrons are injected by a LaB$_6$ cathode electron gun located at $r = 46.5$ cm and $z = 0$. At an operating acceleration voltage of 300 V, applied between the cathode and anode grid located 2 mm behind the cathode, the electron beam current obtained was $\sim 10$ mA. The circuit of the electron gun is operated by an FET high-speed semiconductor switch.

Internal potential distribution is measured using an array of emissive Langmuir

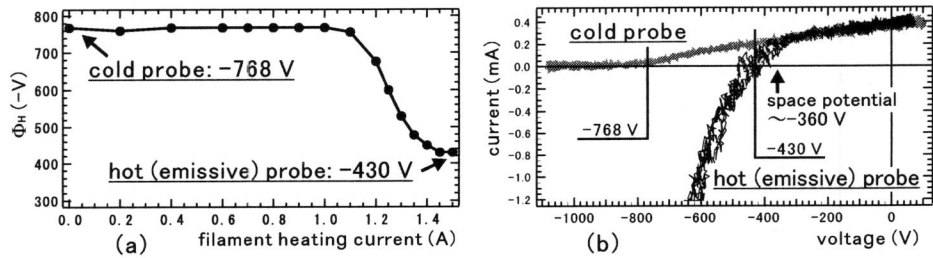

**FIGURE 2.** (a) Measured potentials ($\Phi_H$) when the Langmuir probe tip is terminated by a high impedance (100 M$\Omega$) voltage probe $vs$ the heating current of the emissive Langmuir probe filament $I_{filament}$. When the tip is sufficiently heated, $\Phi_H$ saturates to a certain value due to the space charge limit. (b) I-V curves of emissive ($I_{filament} = 1.5$ A) and cold (non-emitting, $I_{filament} = 0$ A) Langmuir probes. $\Phi_H$ of the emissive probe gives a good approximation of the space potential, while $\Phi_H$ of the cold probe is close to the sum of the potential and kinetic energy of the plasmas.

probes [7]. The probe tip is a thoria-tungsten spiral filament and is heated by the passage of a current. For the measurements of potential profiles, the tip is terminated across a high impedance (100 M$\Omega$) voltage probe, in order to avoid perturbations to the plasma. When compared with the I-V characteristics of hot (emissive) and cold (non-emitting) probes in Fig 2, the measured potential ($\Phi_H$) gives a crossing point of the I-V curve and the load line of the high impedance. The floating potential of a sufficiently heated emissive probe is close to the space potential, and thus $\Phi_H$ of the emissive probe gives a good approximation of the space potential of the electron plasmas. Although the charge of the escaping thermoelectrons of the emissive probes during the measurements of I-V curves (up to 1 mA) might be comparable to the confined electron charge, the I-V curves above the space potential show that the electron emission of the probe tip does not affect the electron density of the plasmas. This is possibly because the density limit is set by the acceleration voltage of the electron gun and further addition of electrons does not contribute to the plasma density.

For the measurements of electrostatic fluctuations and the remaining charge of electron plasmas, we have employed a wall probe [3]. As a wall tip, sensor foil (a copper sheet of $5 \times 15$ mm) is installed in an insulating quartz tube and located just outside the confinement region in the chamber. The foil is grounded to the chamber via a current amplifier, and the detected image current indicates the oscillation in the plasmas. By integrating the escaping image current when the plasma is externally destroyed, the charge confinement time is also measured by wall probes.

## EXPERIMENTAL RESULTS

Non-neutral plasmas in a ring trap have a hollow distribution around the internal conductor. When $V_{IC}$ is not externally controlled (i.e., the electrode is shorted to the chamber), the potential profile is also hollow, as shown in Fig. 3 (a1), resulting in a strong $\mathbf{E} \times \mathbf{B}$ flow shear that may destabilize the diocotron instability. By applying a potential ($V_{IC}$) to the electrode, the potential distribution was successfully modified in the toroidal elec-

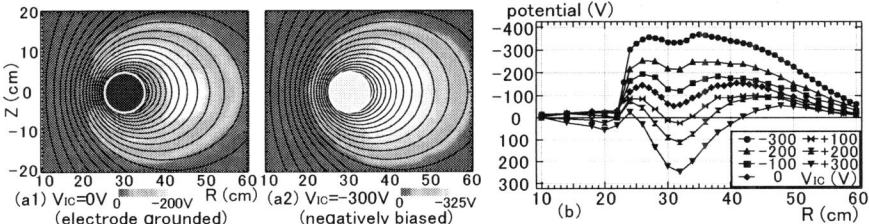

**FIGURE 3.** (a1) 2-d potential profiles of electron plasmas around the IC electrode, in the *r-z* cross-section of the Proto-RT chamber, when the potential is not externally controlled (i.e., the IC electrode is shorted to the chamber) and (a2) when the IC electrode is negatively biased. Thin lines show the magnetic surfaces of the dipole field. The contour images are reconstructed from 284 data points taken shot by shot. (b) Radial potential profiles at $z = +6$ cm while $V_{IC}$ was varied from $-300$ to $+300$ V. The potentials in these figure are $\Phi_H$ of the emissive Langmuir probes during the electron injection phase.

tron plasmas to form a stable equilibrium. The potential hole is eliminated using the negatively biased electrode, and in Fig. 3 (a2), potential contours surround the internal conductor. Fluctuations in the wall probe signal when $V_{IC} = -300$ V are reduced by a factor of 10, in comparison with the fluctuation that occur when the electrode is grounded. After the electron injection was stopped, long-lasting oscillating signals indicating the confinement of electron plasmas were observed by controlling the bias voltage of the IC electrode.

An example of fluctuation and trapped charge, when the electrode is negatively biased, is shown in Fig. 4 (a). During the electron injection ($t = -0.1$ to 0 msec), the observed dominant frequency was 510 kHz. After the electron injection was stopped, both the amplitude and frequency of the oscillation were damped, and a quiescent state followed. The first stable phase lasted for $\sim 0.3$ msec, until the fluctuation grew rapidly. The amplitude decayed again, when the frequency dropped to 43 kHz. As shown in Fig. 4 (b), the fundamental frequencies at this period (i.e., just before the second quiescent phase) were proportional to $E/B$, where $E$ is the strength of the external electric fields and $B$ is the dipole magnetic field strength. These scalings and the observed frequency drop during the charge decay suggest that the observed fluctuations are due to diocotron oscillation. The second stable phase lasted relatively long ($t^* \sim 200$ msec) and ended with a fast growth of fluctuation (a typical time constant was 0.1 msec), possibly caused by the ion resonance instability [9].

As shown in Fig. 5, the life time $t^*$ of the electron cloud scaled as $\propto P^{-1}B^{2.5}$, where $P$ is the background neutral gas pressure and $B$ is the dipole magnetic field strength, and both $t^*$ and the decay time $\tau_2$ of the trapped charge have a strong dependency on the strength of the magnetic field and the degree of vacuum. From the potential profiles in Fig. 3, the number density and the charge of the trapped electron cloud during the injection phase were calculated to be $1 \times 10^{13} \mathrm{m}^{-3}$ (at peak) and $3 \times 10^{-7}$ C, using the Poisson equation. The potential profiles after the electron injection was stopped were not obtained because of the perturbation problem of the Langmuir probes. However, judging from the frequency drop in the diocotron oscillation and also from the change of image charge on the wall probe, the peak number density and trapped charge of the remaining electrons in the quiet phase were estimated to be $\sim 10^{12}$ m$^{-3}$ and $\sim 10^{-8}$ C,

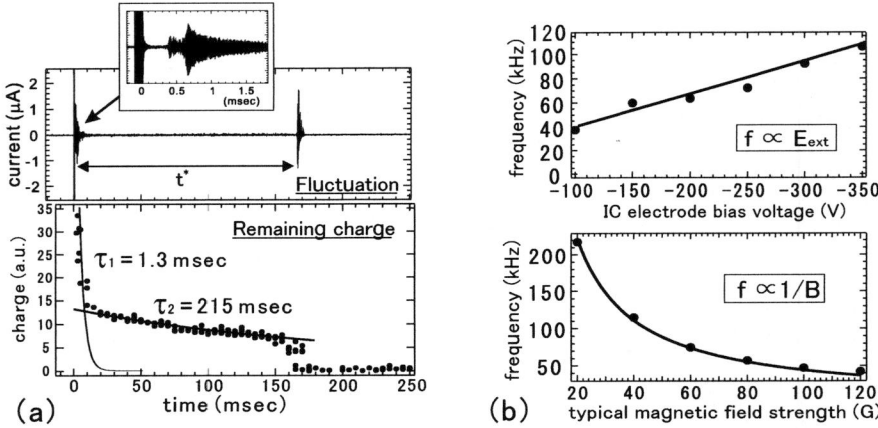

FIGURE 4. (a) Wall probe signal and decay of image charge on the wall foil. Electrons are injected to the chamber from $t = -0.1$ to 0 msec. (b) Fundamental frequencies of electrostatic fluctuations just before the quiescent phase, as functions of the magnetic field strength $B$ and the potential of the IC electrode $V_{IC}$.

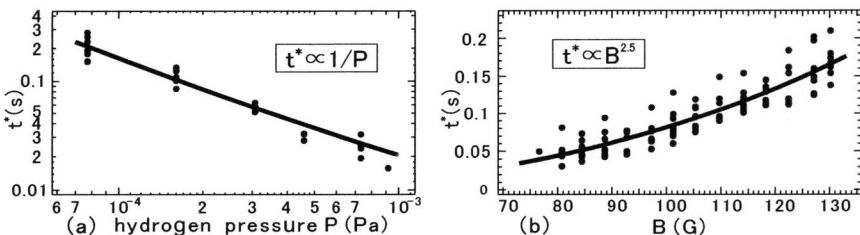

FIGURE 5. The life time $t^*$ (stable confinement time before the sudden growth of instability) of electron plasmas as functions of (a) background pressure $P$ and (b) typical magnetic field strength $B$.

respectively.

In the current relatively high base pressure $\sim 10^{-4}$ Pa ($\sim 10^{-6}$ Torr), neutral collisional effects are dominant in the diffusion process of electron plasmas. Using the experimental parameters of magnetic field strength $B = 0.01$ T, electron number density $n_e \sim 10^{12}$ m$^{-3}$, and estimated electron temperature $T_e \sim 1$ eV (which is close to the $\mathbf{E} \times \mathbf{B}$ drift speed of electron plasmas), the classical diffusion time was $t_D \sim v_{en}^{-1} \lambda_D^2 r_L^{-2} \sim$ 0.1 sec. Both the decay time $\tau_2$ and the life time $t^*$ are comparable to $t_D$, suggesting that the current confinement time is set by the diffusion due to the electron-neutral collisions. Some preliminary experiments have also shown that the electrostatic fluctuations during the trap phase are stabilized by the effects of magnetic shear [8].

In conclusion, long-term confinement (comparable to the diffusion time due to neutral collisions) of toroidal electron plasmas was achieved in the dipole magnetic field configuration of an internal conductor device by controlling the internal potential distribution. It was demonstrated that a magnetic surface configuration has excellent confinement

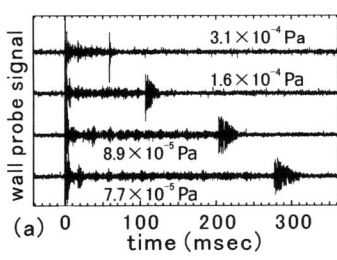
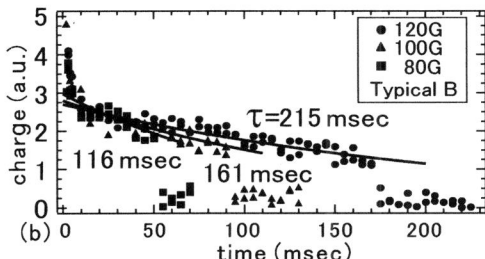

**FIGURE 6.** (a) Wave form of wall probe signals (electrostatic fluctuation of the plasma) in the variation of back pressure (hydrogen) and (b) decay of trapped charge on the wall probe in the variation of magnetic field strength. Electrons are injected from $t = -0.1$ to 0 msec.

properties for non-neutral plasmas, and that it might be usefully applied in novel traps of charged particles such as anti-matter or other multi-fluid non-neutral plasmas.

## ACKNOWLEDGMENTS

This work was supported by a Grant-in-Aid for Scientific Research (14102033) from MEXT Japan. The work of HS was supported in part by a JSPS Research Fellowship for Young Scientists.

## REFERENCES

1. T. S. Pedersen and A. H. Boozer, Phys. Rev. Lett. **88**, 205002 (2002); T. S. Pedersen, Phys. Plasmas **10**, 334 (2003); T. S. Pedersen et al., J. Phys. B: At. Mol. Opt. Phys. **36**, 1029 (2003).
2. Z. Yoshida, Y. Ogawa, H. Himura et al., in *Nonneutral Plasma Physics III* and *IV*, (AIP, New York); S. Kondoh and Z. Yoshida, Nucl. Inst. and Meth. in Phys. Res. A **382**, 561 (1996); C. Nakashima and Z. Yoshida, Nucl. Inst. and Meth. in Phys. Res. A **428**, 284 (1998); C. Nakashima et al., Phys. Rev. E **65**, 036409 (2002); H. Saitoh, Z. Yoshida, and C. Nakashima, Rev. Sci. Instrum. **73**, 87 (2002).
3. J. D. Daugherty and R. H. Levy, Phys. Fluids **10**, 155 (1967); W. Clark et al., Phys. Rev. Lett. **37**, 592 (1976); J. D. Daugherty et al., Phys. Fluids **12**, 2677 (1969); L. Turner, Phys. Fluids B **3**, 1355 (1991); K. Avinash, Phys. Fluids B **3**, 3226 (1991); S. N. Bhattacharyya and K. Avinash, Phys. Fluids B **4**, 1702 (1992); P. Zaveri et al., Phys. Rev. Lett. **68**, 3295 (1992); S. S. Khirwadkar et al., Phys. Rev. Lett. **71**, 4334 (1993); M. R. Stoneking et al., Phys. Plasmas **9**, 766 (2002).
4. C. M. Surko et al., Phys. Rev. Lett. **62**, 901 (1989); M. Hori et al., Phys. Rev. Lett. **89**, 093401 (2002); M. Amoretti et al., Nature **419**, 456 (2002).
5. S. M. Mahajan and Z. Yoshida, Phys. Rev. Lett. **81**, 4863 (1998); Z. Yoshida and S. M. Mahajan, Phys. Rev. Lett. **88**, 095001 (2002).
6. T. Uchida, Jpn. J. Appl. Phys. **33**, L 43 (1994); Z. Yoshida et al., Phys. Rev. Lett. **81**, 2458 (1998).
7. A. Tsushima, Jpn. J. Appl. Phys. **35**, 2820 (1996);
   H. Himura et al., Phys. Plasmas **8**, 4651 (2001).
8. S. Kondoh, T. Tatsuno, and Z. Yoshida, Phys. Plasmas **8**, 2635 (2001).
9. R. H. Levy, J. D. Dagherty, and O. Buneman, Phys, Fluids **12**, 2616 (1969); J. Fajans, Phys. Fluids B **5**, 3127 (1993); A. J. Peurrung, J. Notte, and J. Fajans, Phys. Rev. Lett. **70**, 295 (1993).

# Parameter Dependence of Inward Diffusion on Injected Electrons in Helical Non-Neutral Plasmas

H. Wakabayashi, H. Himura, M. Fukao and Z. Yoshida

*Department of Advanced Energy, Graduate School of Frontier Sciences, The University of Tokyo,*
*7-3-1 Hongo, Bunkyo Ward, Tokyo 113-0033, Japan*

**Abstract.** Experimental studies on an electron injection into a helical magnetic field and character-istics of non-neutral plasmas have been performed. It is found that the space potential $\phi_s$ has a weak dependence on the injection angle except for a narrow 'window' region in which $\phi_s$ significantly drops. A calculation shows that because of the electric field $\mathbf{E}_g$ of the electron gun (e-gun), the emit-ted electrons are launched quasi-parallel to the helical magnetic field $\mathbf{B}$, regardless of $\alpha$. This seems to agree with the observation. The 'window' seen in the data may be attributed to an current-driven instability which might result in the insufficient electron penetration or the degradation of electron confinement in the magnetic surface.

## INTRODUCTION

In recent years, many studies have been performed focused on plasma flow. For example, several studies have been made on the boundary layer of plasmas in order to improve particle confinement by fast flow. To generate the flow inside plasmas, two methods are considered, the one is to insert a pair of electrodes to apply an electric field to the plasmas [1], and the other is to inject excess electrons into plasmas in order to generate a self-$\mathbf{E}$ field [2].

For the latter case, one difficulty is the injection of the electrons into magnetic surfaces. Several methods had been used to inject the electrons in toroidal geometry([3]-[4]). Recently, a new method of injecting electrons through a stochastic magnetic layer has been performed [5] on the Compact Helical System (CHS) device [6]. In this case, the magnetic field $\mathbf{B}$ of the layer is chaotic because of an asymmetric structure. Thus, this suggests a possibility of an enhanced chaotic motion of electrons magnetized in the layer. In fact, the field lines have long connection lengths in the stochastic layer, thus the injected electrons could stay in the region for a long time. This may offer a collective motion of the electrons which may cause the electrons to penetrate across the closed magnetic surface.

Based on the idea mentioned above, we have performed the experiment on CHS. In the experiment, an electron gun (e-gun) with a $LaB_6$ cathode is placed in the stochastic layer. It is observed that the injected electrons successfully penetrate into magnetic surfaces [5], and the penetration depends upon injection parameters such as beam density and injection angle $\alpha$, where $\alpha = 0°$ and $\alpha = 90°$ correspond to toroidal(horizontal) and vertical directions, respectively. Moreover, the experiment strongly indicates that the

CP692, *Non-Neutral Plasma Physics V*, edited by M. Schauer et al.
© 2003 American Institute of Physics 0-7354-0165-9/03/$20.00

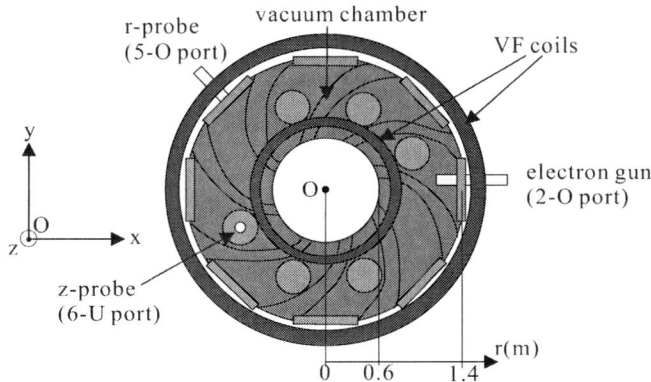

**FIGURE 1.** A top view of CHS. The e-gun is inserted holizontally on the 2-O port. The r-probe is inserted along the *r* axis on the 5-O port, as well as the *z*-probe is inserted along the *z* axis on the 6-U port.

electron penetration is caused by a collisionless mechanism.

In order to study the mechanism, the parameter dependence of the e-gun has been systematically examined. At the first series of the experiment, it is observed that the electron penetration depends strongly on $\alpha$. In this paper, the result is intensively discussed.

## EXPERIMENTAL SETUP

Experiments were carried out on CHS, a stellarator device of the Heliotron/Torsatron type [6]. The major radius of CHS is 1 m, and the poloidal cross section of the vac-uum vessel is elliptic. The major and minor radii of the ellipse are 0.4 m and 0.2 m, respectively. The elliptic cross section rotates eight times around the magnetic axis $R_{ax}$, thus the magnetic configuration repeats every 45°. The device has a pair of four-turn windings of helical coils to generate **B**. And four pairs of poloidal coils are provided to modify the plasma shape and shift $R_{ax}$. The magnetic field is DC and the strength of **B** is variable between 0.2 and 0.9 kG at $R_{ax}$.

A pair of Langmuir probes are inserted into the vacuum chamber to measure space potential $\phi_s$ and probe current $I_p$. One is inserted along the *z* axis, and the other is along *r* axis. They can be moved along the *z* and the *r* axes so that radial profiles of $\phi_s$ and $I_p$ can be measured.

As for the e-gun, the size of the cathode is 1cm (W) × 1cm (H), and the emission current is ∼ 400 mA at the cathode temperature ∼ 2000 K. A plane diode geometry is employed for the cathode and anode. The anode is made of a tungsten mesh, and the gap width between the electrodes is 5 mm. Some electrons emitted from the cathode hits the anode and drains out directly to the ground. The rest of electrons pass through the anode and are launched out into the chamber as an effective beam current $I_b$. The maximum directed energy to the electrons is ∼ 1.2 keV. The e-gun can also be moved along the *r* axis and rotated around its barrel to vary $\alpha$. The definition of $\alpha$ is shown on the right-

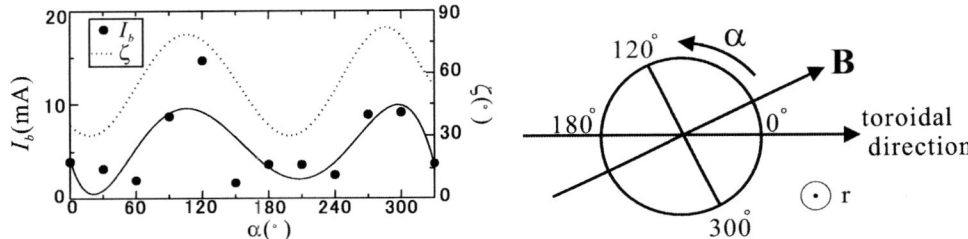

**FIGURE 2.** The definition of $\alpha$ is shown on the right-hand side, the end-on view of the e-gun port. When $\alpha$ is varied, $\zeta$ as well as $I_b$ also vary as shown on the left hand-side. The pitch angle $\zeta$ is in the range of 30° to 80° for the presented setting.

hand side of Fig.2, an end-on view of the e-gun port. Since the field line of **B** is curved at the e-gun port, a radial component of $B_r$ exists. The value of $B_r$ is about $\sqrt{|\mathbf{B}|^2}/2$.

As shown in the left-hand side of Fig.2, the pitch angle $\zeta$ of the e-gun to **B** varies in the range of 30° to 80° when $\alpha$ is varied. The beam current $I_b$ also changes with respect to $\alpha$.

## RESULTS

In the experiment, it is observed [5] that the profile of $\phi_s$ strongly depends on $\alpha$, beam density and the beam energy. Here, we briefly explain the results of measured $\phi_s(r)$, $\phi_s(z)$, $I_p(r)$ and $I_p(z)$. In Fig.3, typical data are shown. The details of the result can be found in Ref.5. Here, $\sqrt{\psi}=0$ shows $R_{ax}$ and $\sqrt{\psi}=1$ stands for the Last Closed Flux Surface or LCFS. The white and black marks correspond to data obtained from the $r$-probe and $z$-probe, respectively. Throughout these measurements, the e-gun is placed 1cm outside the LCFS, in the stochastic region. These data show the electron penetration via stochastic magnetic region. Moreover, the penetration depends on $\alpha$, which can be seen in Fig.2 of Ref.5. The profiles of $\phi_s$ formed by the penetrated electrons vary as $\alpha$ is changed.

In order to study the detail of the dependence on $\alpha$, we have systematically measured $\phi_s$ with changing $\alpha$. Figure 4 shows a change in $\phi_s$ measured at $r = 118$ cm when $\alpha$ is varied every 15° from 0° to 360°. The two profiles of $\phi_s$ shown in Fig.4 are obtained with the same experimental parameters except for the direction of the coil current, which means that the direction of **B** is flipped: (1) the normal case (black circle) and (2) the reverse case (white circle). The gun is placed at $r = 117.5$ cm which is 1.5 cm outside the LCFS. As seen in Fig.4, the value of $\phi_s$ does not strongly depend on $\alpha$ except a narrow 'window' of $\alpha \sim 120$ for the normal case and 300 deg for the reverse case, respectively. And, the difference in those two values of $\alpha$ can be recognized to be 180°. Thus, these results reflect symmetry of the measured data against **B**.

We now discuss the weak dependence on $\alpha$ outside the 'window' region. Orbits of the injected electrons in a helical field strongly depend on $\zeta$ in the e-gun. Generally the orbits are classified into three types with respect to $\zeta$ [7].

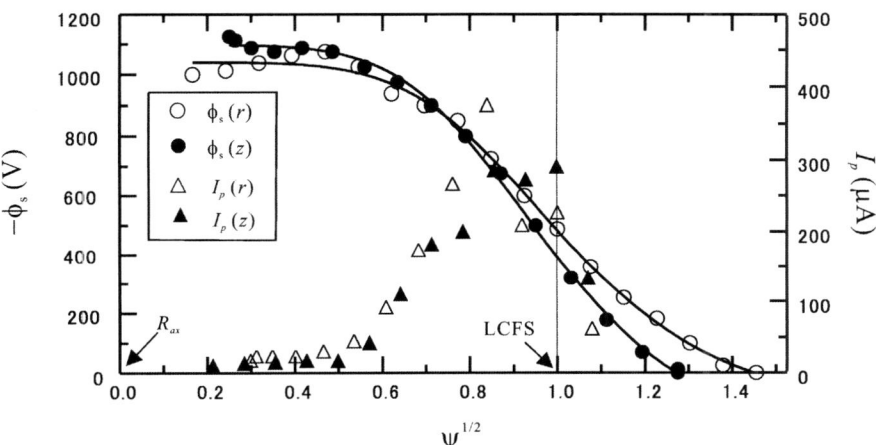

**FIGURE 3.** Profiles of $\phi_s$ and $I_p$ measured along the $r$ and the $z$ axes. The horizontal axis is normalized by the flux function $\psi$. It is remarkable that finite value of $I_p$ is observed even near $\psi^{1/2} \sim 0.2$.

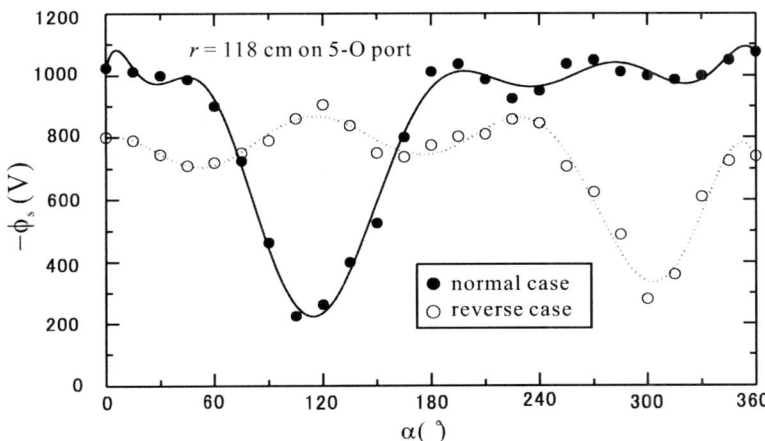

**FIGURE 4.** Dependence of $\phi_s$ on $\alpha$. The black and white circles correspond to data obtained in two cases identified with the direction of **B**:(1) normal case and (2) reverse case in which **B** is flipped. One can find that $\phi_s$ has a weak dependence on $\alpha$ except a narrow 'window' of $\alpha$ where $\phi_s$ significantly drops.

On the other hand, let us consider electron accleration in the e-gun used in the helical magnetic field. The thermal electrons emitted from the cathode of e-gun are accelerated by an electric field $\mathbf{E}_g$ between the pair of electrodes. Then, it is generally considered that the electrons are injected along $\mathbf{E}_g$ from the anode. However, in the presented experiment, a transverse component of the helical magnetic field **B** exists between the electrodes as well. Thus, the emitted electrons never move straight towards the anode but drift across $\mathbf{E}_\perp$ (a normal component of $\mathbf{E}_g$ to **B**) and **B** by $\mathbf{E}_\perp \times \mathbf{B}$ drift. This also means that a parallel component of $\mathbf{E}_g$ along $\mathbf{B}$ exists and it accelerates the electrons along $\mathbf{B}$ between the electrodes. This effect brings a finite parallel velocity of electrons $\mathbf{v}_{\parallel}$, which can be comparable to the $\mathbf{E}_{\perp} \times \mathbf{B}$ perpendicular velocity $\mathbf{v}_{\perp}$.

In fact, $|\mathbf{E}_g|$ is calculated to be $2.4 \times 10^5$ V/m from the gap width of the electrodes and the acceleration voltage. When $\alpha$ is $300°$, $\zeta$ is found to be $80°$ from Fig.2. Thus, $|\mathbf{v}_{\perp}|$ and $|\mathbf{v}_{\parallel}|$ are calculated to be $3.4 \times 10^6$ m/s and $8.6 \times 10^6$ m/s, respectively. So the angle, which is determined by the ratio $|\mathbf{v}_{\perp}|/|\mathbf{v}_{\parallel}|$ at the launching point is calculated to be $\sim 20°$ in the present experiment. As a result, with such a small angle, the injected electron move still as passing particle [7] rather than helically trapped particle [7], which actually can also be recognized by orbit calculations of the electron in Fig.5 where the effect of $\mathbf{v}_{\parallel}$ is included. Thus, in this case, electrons are still injected almost parallel to $\mathbf{B}$ even for the cases of $\alpha = 300°$ (the normal case) and $= 120°$ (the reverse case) where the e-gun is set quasi-perpendicular to $\mathbf{B}$. Therefore, such a weak dependence on $\alpha$ is expected in the measured $\phi_s$ profiles shown in Fig.4.

Let us now return to the 'window' where $\phi_s$ significantly drops for $\alpha \sim 120$ (the normal case) and $300$ (the reverse case) deg. The reason is still unclear but it might be due to an interaction of the injected beam electrons with the confined electron plasmas. In Fig.4, one finds that $\phi_s$ drops when $\alpha$ is almost anti-parallel to the expected poloidal flow. In this case, some instability may occur and it might degrade the confinement properties of helical electron plasmas. And, this might result in the observed profiles of $\phi_s$ for $\alpha \sim 120°$ of the normal case and $\alpha \sim 300°$ of the reverse case. In order to examine the details experimentally, measurements of the instability and electron flow will be performed.

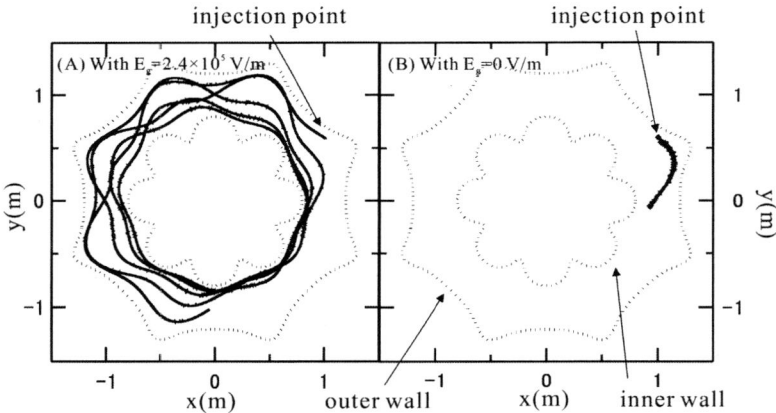

**FIGURE 5.** Orbit diagrams of an electron when $\zeta$ is $80°$ ($\alpha = 300°$). (A) The electron is accelerated along $\mathbf{B}$ by $\mathbf{E}_{\parallel}$ when $\mathbf{E}_g$ is taken account, and thus the electron circulates in the toroidal direction as a 'passing particle'. (B) Without $\mathbf{E}_g$, the electron is trapped, and lost outward by means of $\nabla \mathbf{B}$ poloidal drift [7].

# SUMMARY

Thoroughout the studies on electron injection into the helical magnetic field, it is observed that $\phi_s$ has the weak dependence on $\alpha$. To explain the result, an orbital analysis of the emitted electron is carried out. With $|\mathbf{E}_g|=2.4\times10^5$ V/m, the emitted electron is accelerated between the electrodes along $\mathbf{B}$, and then launched out into the chamber, regardless of $\alpha$. Without $\mathbf{E}_g$, the electron is trapped. These results seem to explain the weak dependence of $\phi_s$ on $\alpha$ (in Fig.4) for the presented settings.

Near $\alpha \sim 120°$ for the normal case, the value of $\phi_s$ significantly drops to $\sim 200$ V which is about 20% for other cases of $\alpha$. This drop of $\phi_s$ appears at $\alpha \sim 300°$ when the coil current of $\mathbf{B}$ is flipped. Thus, the drop of $\phi_s$ depends on $\mathbf{B}$. The explanation of the result is so far unknown. As one of possible reasons, a current-driven instability might be occured at $\alpha \sim 120°$ for the normal case and $\alpha \sim 300°$ for the normal case. Because the direction of expected electron flow by means of $\mathbf{E}\times\mathbf{B}$ drift is anti-parallel to $\alpha$, the instability may occur and the insufficient penetration might be resulted. In order to examine the details experimentally, measurements of the instability and electron flow will be performed.

# ACKNOWLEDGEMENTS

The authors are grateful to CHS group members especially to Drs. M. Isobe, S. Okamura, S. Nishimura, C. Suzuki and T. Akiyama for operating the device in the expetiment, and to M. Hirota, H. Saito and R. Numata for their help and suggestions in numerical calculations. This work is performed under the auspices of the NIFS LHD Project Research Collaboration. HW is also supported by Futaba Electric Corporation foundation.

# REFERENCES

1. Taylor, R., Phys. Rev. Lett. **63**,2365-2368 (1989)
2. Nakashima, C. et al., Phys. Rev. E. **65**,036409 (2002).
3. Clark, W. et al., Phys. Rev. Lett. **37**,592-595 (1976)
4. Khirwadkar, S.S. et al., Phys. Rev. Lett. **71**,4334-4337 (1993)
5. Himura, H., Wakabayashi, H. et al., *"Experimental Investigation of Helical Non-neutral Plasmas"* in *this proceedings.*
6. Nishimura, K. et al., Fus. Tech. **17**,86-100 (1990).
7. Wakatani, M., "Stellarator and Heliotron Devices" (International Series of Monographs on Physics, Vol 95)

# List of Participants

Marco Amoretti
Istituto Nazionale di Fisica Nucleare
INFN-Sezione de Genova Via Dodecaneso, 33
Genova, 16146, ITALY

Francois Anderegg
University of California at San Diego
9500 Gilman Drive
La Jolla, CA 92093 , USA

Eric Bass
University of California at San Diego
9500 Gilman Drive
La Jolla, CA 92093, USA

William Bertsche
UC Berkeley
123 Le Conte Hall
Berkeley, CA 94720, USA

Giovanni Bettega
Universita Degli Studi Di Milano
DIP Fisica Via Celoria 16
Milano, 20133, ITALY

Georg Bollen
Michigan State University
South Shaw Lane
East Lawsing, MI 48824, USA

John Bollinger
NIST
325 Broadway
Boulder, CO 80305, USA

Edwin Chacon
Los Alamos National Laboratory
P.O. Box 1663, MS H838
Los Alamos, NM 87545, USA

Yongbin Chang
University of North Texas
P.O. Box 310499
Denton, TX 76201-1427, USA

Ben Chang
University of Delaware
223 Sharp Lab
Newark, DE 19716, USA

Paul Channell
Los Alamos National Laboratory
P.O. Box 1663, MS H808
Los Alamos, NM 87545, USA

Jonathan Clarke
University of Wales Swansea
Singleton Park
Swansea, SA2 8PP, UK

Gianni Coppa
Politecnico Di Torino
Corso Duca Degli Abruzzi, 24
Torino, 10129, ITALY

Jose Correa
University of North Texas
P.O. Box 310499
Denton, TX 76201-1427, USA

James Danielson
University of California at San Diego
9500 Gilman Dr.
La Jolla, CA 92093, USA

Ronald Davidson
Plasma Physics Laboratory
P.O. Box 451
Princeton, NJ 08510, USA

Gian Luca Delzanno
Politecnico Di Torino
Dipartimento Al Energetica,
Corso Duca Degli Abruzzi 24
Torino, 101029, ITALY

Carl Dietrich
MIT
77 Massachusetts Ave.
Cambridge, MA 02139, USA

Michael Drewsen
University of Aarhus
Ny Munkegade , Bldg. 520
DK-800 Aarhus, , DENMARK

Fred Driscoll
University of California at San Diego
9500 Gilman Drive
La Jolla, CA 92093, USA

Daniel Dubin
University of California at San Diego
9500 Gilman Drive
La Jolla, CA 92093, USA

Dennis Eggleston
Occidental College
Physics Department, 1600 Campus Rd
Los Angeles, CA 90041, USA

Joel Fajans
UC Berkeley
Physics Dept. MS 7300
Berkeley, CA 94720, USA

Makota C. Fujiwara
RIKEN
Division EP, CERN, CH 1211
Geneva, 23, SWITZERLAND

Gerald Gabrielse
Harvard University
17 Oxford Street
Cambridge , MA 02138, USA

John Goree
University of Iowa
Department of Physics and Astronomy
Iowa City , IA 52246, USA

Vladimir Gorgadze
UC Berkeley
366 Le Coute Hall
Berkeley, CA 94720, USA

Roy Gould
Caltech
MS128-95
Pasadena, CA 91125, USA

Rod Greaves
First Point Scientific Inc.
5330 Derry Ave. Suites
Agoura Hills, CA 91301, USA

Jeffrey Hangst
University of Aarhus
Department of Physics
Aarhus, DK-8000, DENMARK

Saitoh Haruhiko
University of Tokyo
#032 9th Bldg. of Engineering UT
Tokyo, 113-0032, JAPAN

Taro Hasegawa
NIST
325 Broadway
Boulder, CO 80305, USA

Haruhiko Himura
University of Tokyo
7-3-1 Hongo, Bunkyo Ward
Tokyo, 113-0033, Japan

Gerald Jackson
HBAR Technologies, LLC
1275 Roosevelt Road, Suite 103
West Chicago, IL 60185, USA

Marie Jensen
NIST
325 Broadway
Boulder, CO 80305, USA

Lars Jorgensen
CERN
Bat 545 R-021, CERN, CH-1211
Geneva, , SWITZERLAND

Andrey Kabantsev
University of California at San Diego
9500 Gilman Drive
La Jolla, CA 92093, US

Jason Kremer
Columbia University
200 Seally Mudd Building
New York, NY 10027, USA

Stanislav Kuzmin
University of California at San Diego
9500 Gilman Drive
La Jolla, CA 92093, USA

Karoly Makonyi
National Institute of Standard and Technology
100 Bureau Drive
Gaithersburg, MD 20899,

Grant Mason
Brigham Young University
N249 ESC
Provo, UT 84602, USA

Marco Matranga
National Institute of Standards and Technology
100 Bureau Drive
Gaithersburg, MD 20899,

Thomas McGuire
Massachusetts Institute of Technology
77 Mass Ave., Bldg. 37350
Cambridge, MA 02139, USA

Kirby Meyer
Positronics Reasearch LLC
4001 Office Court Dr.
Santa Fe, NM 87507, USA

Travis Mitchell
University of Delaware
223 Sharp Lab
Newark, DE 19716, USA

Kyle Morrison
PPPL
P.O. Box 451
Princeton, NJ 08543, USA

Rick Nebel
Los Alamos National Laboratory
P.O. Box 1663, MS K717
Los Alamos, NM 87545, USA

Koichi Noguchi
Los Alamos National Laboratory
P.O. Box 1663, MS P225
Los Alamos, NM 87545,

Thomas O'Neil
University of California at San Diego
9500 Gilman Drive
La Jolla, CA 92093, USA

Carlos Ordonez
University of North Texas
P.O. Box 310499
Denton, TX 76201-1427, USA

Jaeyoung Park
Los Alamos National Laboratory
P.O. Box 1663, MS E526
Los Alamos, NM 87545, USA

Stephen Paul
Princeton University
P.O. Box 451
Princeton, NJ 08543, USA

Fabio Peano
Politecnico Di Torino
Corso Duca Degli Abruzzi, 24
Torino, 10129, ITALY

Thomas Pedersen
Columbia University
S.W. Mudd Rm. 204
New York, NY 10027, USA

Federico Peinetti
Politecnico Di Torino
Corso Duca Degli Abruzzi, 24
Torino, 10129, ITALY


Robert Pollock
Indiana University
Swain Hall West
Bloomington, IN 47405-7105,


Roberto Pozzoli
Universita Di Milano, INFN
Via Celoria 16
Milano, 20133, ITALY


Charles Roberson
1908 Toll Bridge Ct.
Alexandria, VA 22308, USA


Scott Robertson
University of Colorado
Department of Physics
Boulder, CO 80309, USA


Massimiliano Rome
Universita-Degli Studi di Milano
Dipartimento Di Fisica
Milano, I-90133, ITALY


Gary Rouleau
Los Alamos National Laboratory
P.O. Box 1663, MS H838
Los Alamos, NM 87545, USA


Akio Sanpei
Kyoto University
Yoshida, Nihonmatutyou
Kyoto, 606-8501, JAPAN


Martin Schauer
Los Alamos National Laboratory
P.O. Box 1663, MS H803
Los Alamos, NM 87545, USA


Andrea Schmidt
U.C. Berkeley
MS 7300
Berkeley, CA 94720, USA


Lutz Schweikhard
University of Greifswald
Domstr. 10a
17487 Greifswald, , GERMANY


Raymond Sedwick
MIT Space Systems Laboratory
77 mass Ave. Rm. 37-431
Cambridge, MA 02139, USA


Joseph Sherman
Los Alamos National Laboratory
P.O. Box 1663, MS H838
Los Alamos, NM 87545, USA


Nobuyasu Shiga
University of California at San Diego
9500 Gilman Dr.
La Jolla, CA 92093, USA


Clayton Simien
Rice University
6100 Main Street
Houston, TX 77005, USA


Yukihiro Soga
Kyoto University
Yoshida, Nihonmatutyou Sakyo-ku
Kyoto, 606-8501, JAPAN


George Spalek
Spalek Consulting
1311 Lejano Ln
Santa Fe, NM 87501, USA


Ross Spencer
Brigham Young University
N271 ESC, BYU
Provo, UT 84602, USA

Matthew Stoneking
Lawrence University
115 S. Drew St.
Appleton, WI 54911, USA


Ronald Stowell
PPPL
P.O. Box 451
Princeton, NJ 08543, USA


Clifford Surko
University of California, San Diego
Phsyics Dept. 0319, 9500 Gilman Drive
La Jolla, CA 92093, USA


Dirk van der Werf
University of Wales Swansea
Singleton Park
Swansea, SA2 9AY, UK


Hidenori Wakabayashi
The University of Tokyo
+032 9th Building of Engineering UT
Tokyo, 113-0032, JAPAN


Tai-Sen F. Wang
Los Alamos National Laboratory
P.O. Box 1663, MS H817
Los Alamos, NM 87545, USA


Hermann Wolluik
University Giessen
H. Buff-Ring 16
Giessen, 35392, GERMANY


Yuichi Yatsuyanagi
Kyoto University
Yoshida Nihonmatsu-cho, Sakyo-ku
Kyoto, 606-8501, JAPAN


Jonathan Yu
University of California at San Diego
9500 Gilman Dr.
La Jolla, CA 92093, USA

# AUTHOR INDEX

## A

Amoretti, M., 121, 131, 172
Amsler, C., 121
Aoki, J., 106, 112

## B

Beddows, D., 178
Bertsche, W., 22, 235
Bettega, G., 50, 221
Bollinger, J. J., 193
Bonomi, G., 121, 131, 172
Boozer, A. H., 302
Bouchta, A., 121, 131, 172
Bowe, P. D., 121, 131, 172

## C

Carraro, C., 121, 131, 172
Cavaliere, F., 50, 221
Cavenago, M., 50, 221
Cesar, C. L., 121, 131, 172
Chang, B. T., 239
Chang, Y., 55
Charlton, M., 121, 131, 172, 178
Clarke, J., 178
Coppa, G. M., 93, 99, 273
Correa, J. R., 246

## D

Dahlgren, F., 320
Danielson, J. R., 149
Davidson, R. C., 75, 81, 162, 211
De Luca, F., 50, 221
Delzanno, G. L., 93
Dietrich, C. C., 259
Doser, M., 121, 131, 172
Driscoll, C. F., 3, 61

## E

Ebisuzaki, T., 87
Efthimion, P. C., 211
Eggleston, D. L., 40

## F

Fajans, J., 22, 30, 235, 252
Filippini, V., 121, 131, 172
Fontana, A., 121, 131, 172
Friedland, L., 22
Fujiwara, M. C., 121, 131, 172
Fukao, M., 293, 326, 332
Funakoshi, R., 121, 131, 172

## G

Genova, P., 121, 131, 172
Gilson, E. P., 211
Gorgadze, V., 30
Gould, R. W., 15
Greaves, R. G., 140
Griffiths, B., 178
Growdon, M. A., 310

## H

Hangst, J. S., 121, 131, 172
Hasegawa, T., 193
Hayano, R. S., 121, 131, 172
Herlert, A., 203
Hilsabeck, T. J., 3
Himura, H., 293, 326, 332

## I

Illiberi, A., 221

345

## J

Jensen, M. J., 193
Jørgensen, L. V., 121, 131, 172

## K

Kabantsev, A. A., 3, 61
Kiwamoto, Y., 87, 106, 112
Kotelnikov, I., 50, 221
Kremer, J. P., 302, 320

## L

Lagomarsino, V., 121, 131, 172
Landua, R., 121, 131, 172
Lindelöf, D., 121, 131
Lodi Rizzini, E., 121, 131

## M

Macri, M., 121, 131, 172
Madsen, N., 121, 131, 172
Majeski, R., 211
Manuzio, G., 121, 172
Marchesotti, M., 131
Maroli, C., 221
Marx, G., 203
Mason, G. W., 69, 229
Meyer, K. J., 184
Milne, M. L., 310
Mitchell, T. B., 239
Montagna, P., 121, 131, 172
Moon, N., 184
Morikawa, J., 326
Morrison, K. A., 75, 81
Moxom, J., 140

## N

Noguchi, K., 285

## O

O'Neil, T. M., 3
Ordonez, C. A., 55, 246, 267

## P

Pariev, V. I., 285
Pasquini, T. A., 30
Paul, S. F., 75, 81
Peano, F., 99, 273
Pedersen, T. S., 302, 320
Peinetti, F., 93, 99, 273
Peterson, R. T., 310
Petrillo, V., 221
Pomphrey, N., 320
Pozzoli, R., 50, 221, 279
Pruys, H., 121, 131, 172

## R

Rasband, S. N., 229
Regenfus, C., 121, 131, 172
Reiersen, W., 320
Rielder, P., 131
Rizzini, E. L., 172
Romé, M., 50, 221, 279
Rotondi, A., 121, 131, 172

## S

Saitoh, H., 326
Sanpei, A., 106, 112
Schmidt, A., 252
Schmidt, P., 149
Schweikhard, L., 203
Sedwick, R. J., 259
Serafini, L., 221
Smith, G. A., 184
Soga, Y., 106, 112
Spalek, G., 184
Spencer, R. L., 121, 229
Startsev, E. A., 211
Stoneking, M. R., 310
Stowell, R., 162
Sullivan, J. P., 149
Surko, C. M., 149